INTEGRATED URBAN MODELS

INTEGRATED URBAN MODELS

Policy analysis of transportation and land use

S.H. PUTMAN

Routledge
Taylor & Francis Group
LONDON AND NEW YORK

First published in 1983

This edition published in 2007
Routledge
2 Park Square, Milton Park, Abingdon, Oxfordshire OX14 4RN
711 Third Avenue, New York, NY 10017

First issued in paperback 2014

Routledge is an imprint of Taylor & Francis Group, an informa business

British Library Cataloguing in Publication Data
A CIP catalogue record for this book
is available from the British Library

Integrated Urban Models
ISBN10: 0-415-41830-5 (volume)
ISBN10: 0-415-41935-2 (subset)
ISBN10: 0-415-41318-4 (set)

ISBN13: 978-0-415-41830-0 (volume)
ISBN13: 978-0-415-41935-2 (subset)
ISBN13: 978-0-415-41318-3 (set)

ISBN 13: 978-1-138-88205-8 (pbk)

Routledge Library Editions: The City

integrated urban models

S H Putman

policy analysis of transportation and land use

Pion Limited, 207 Brondesbury Park, London NW2 5JN

ISBN 0 85086 098 9

Printed in Great Britain by Page Bros (Norwich) Limited

To

Robert S Putman and Samuel Pearl

Acknowledgement

The research and development work which yielded the results described and discussed in this book was done over a ten-year period beginning in 1971. Various portions of the work were sponsored by different federal agencies. The initial work was performed under US Department of Transportation Contract number DOT-FH-11-7843. The next sponsor was the National Science Foundation, under Grant NSF-GI-38978. This was followed by further sponsorship by the National Science Foundation, under Grant APR73-07840-A02. Additional sponsorship came from the US Urban Mass Transit Administration and from the Office of the Secretary, both of the US Department of Transportation, via interagency funds transfers to the second of the above-mentioned NSF grants. Reports were prepared for the sponsors of these projects, portions of which are used in this text. The opinions, conclusions, and findings presented here, however, should in no way be construed as representing the positions of any of the sponsors.

During the conduct of the sponsored research several individuals from the sponsoring agencies, in their roles as project or contract managers, provided invaluable assistance. These individuals were Dr Robert Goldman, Mr Thomas Hillegass, Mr Terry Moore, Dr David Seidman, and Mr Edward Weiner.

During the years in which this research was being done, many graduate students worked on different portions of the projects. At the last count there had been more than two dozen of these, a number of whom in their professional capacities still concern themselves with aspects of the development and/or use of transportation and land-use models in planning and policy analysis. Any academic researcher should readily acknowledge the importance of bright graduate students, in their roles as research assistants, to projects such as these. Without their efforts nowhere near as much research could have been accomplished. At certain places in the text excerpts from these students' term papers, project memoranda, and Masters or Doctoral theses are used. Credit is given to those students where their work appears in the text. Appreciation is given here to all those students who worked on aspects of the research described in this book.

Thanks are also due to Patricia Paciorek who typed (and in some cases retyped) the manuscript with never more than a low murmur of complaint.

Finally, commendation is due to my daughter and, even more so to my wife, for continuing to endure life with an often preoccupied social scientist and author.

Stephen H Putman
Philadelphia, Pennsylvania
Summer 1982

Contents

1

Introduction

1.1 Purpose

Urban simulation modelling is a broad, multifaceted field of endeavor which progresses, albeit at different rates, in many directions at once. If we ignore for the moment questions of success and failure of application, the development of knowledge in this field seems to have followed a path which somewhat resembles the S-shaped logistic curve. Before 1960, the possibility of urban simulation models per se really did not even exist. During this period there had been only some theoretical work on spatial patterns of activities. Even so, this work, some of which had been done fifty or more years earlier, was of no direct use for developing detailed descriptions of urban spatial processes.

In the late 1950s and early 1960s two sociopolitical developments in the United States resulted in increased interest in the possibilities of urban simulation. The first of these was the increasing demand for scientific means of preparing highway impact statements. The second was the emergence of explicit federal governmental concern with urban problems. This concern, in turn, resulted in a variety of programs intended to solve the problems of the cities. These programs also generated a demand for scientific procedures for urban analysis. The impetus from these two major developments, along with the new availability of electronic data-processing equipment, yielded an enormous increase in urban simulation endeavors. Ultimately these endeavors added considerably to the existing meager stock of knowledge of urban spatial processes. The majority of these urban simulation projects were, in many respects, premature and doomed to failure in application. Yet if viewed as preliminary research endeavors, rather than as failed applications, these were very valuable projects(1). Thus in the late 1950s and early 1960s there was more learned about the empirically observable determinants of urban spatial patterns than had been learned in the previous fifty years.

The late 1960s saw the development of the maximum entropy, statistical-mechanics-cum-information-theory derivations of spatial interaction models. These new results provided a much needed unifying structure for many of the various empirically derived macrodescriptive models which had by then been developed. In the wake of these developments major work in the field shifted from the United States to England. The following several years, through the late 1960s and into the middle 1970s, saw an enormous amount of additional work in the field, both empirical and theoretical. Since then, during the late 1970s and into the 1980s, there has been a period of maturation and consolidation as professionals in the field try to integrate and unify the broad spectrum of recently developed knowledge.

(1) An extensive review of these earlier urban simulation model projects can be found in Putman (1979).

The multifaceted aspect of this field may be seen in the many directions of research endeavors. One study may attempt empirical verification of some aspect of a theory, while another attempts to reconcile aspects of two different theories. Some studies have tested several theories, or models, with a single source of data, whereas others have worked with a single model over several data sets. Another dimension of comparison may involve an examination of the consequences of similar studies being performed at varying levels of spatial, temporal, or sectoral detail. Another group of studies may be imagined where portions of systems, or whole systems, of models are examined. Indeed, much of the work described in this book is oriented along this same dimension from a partial to a holistic view of urban spatial phenomena. More specifically, the principal theme of the work described here is the examination of the consequences of integrating urban transportation and location models. The goal of these integration efforts is to develop a better understanding of the interactions between transportation and location phenomena, and thus greater knowledge of the determinants of urban spatial patterns.

One of the most frequently described strategies of systems analysis is that of separating a complex, multicomponent system into small groups of components, or even individual components, for analysis. In some circumstances it is only by this means that it is possible even to begin to analyze a complex system. The one great danger of this strategy is that, for virtually all complex systems, the whole system does not equal the simple sum of its parts. All types of component-component or component-group relationships exist. These may be direct or indirect, feedthrough or feedback; they may amplify or attenuate the actions of other components of the system. In short, although it is useful and often necessary to describe the separate components of a system, it is virtually never adequate to describe a system only in terms of its components.

The great bulk of the research on urban spatial phenomena alluded to above has been done in the form of analysis of system components. There have been analyses of urban residential location and of urban employment location (usually subdivided into 'basic' and 'nonbasic' employment components). There have been analyses of differences between residential location paradigms and of relative efficiencies of techniques for the solution of equations. More importantly in the context of the work described here, there have been analyses of transportation and analyses of location.

More or less in parallel to the work on urban spatial phenomena, there has been a stream of work on transportation systems. This work seems to have developed a bit earlier and somewhat more rapidly than the urban simulation modelling efforts. By the mid 1960s, the United States Bureau of Public Roads had distributed, to transportation planning agencies across the country, a package of computer programs for transportation network analysis, including numerous supplementary computer programs for auxiliary data processing and analysis. Even though this work did develop at about the same time as the urban simulation work was developing, the connections between the two, though desirable, were almost nonexistent.

At first, despite the empirical evidence of their interrelationships, research on transportation systems and research on urban spatial phenomena were conducted with very little interchange of theory or practice. Later, with the entropy-maximizing derivations of spatial interaction models, the connections, in theory, between location models and transportation system models became more apparent. Yet, in practice the two endeavors were conducted virtually in mutual isolation. While many of the professionals in the field gave lip service to the need for their integration, in practice, little was done. This was a classic example of analysis of portions of a system, with a variety of factors, not the least of which was bureaucratic tradition, militating against analysis of the whole system. There were technical reasons as well, as will be described in later chapters of this book. Nonetheless, it was not until 1971, with the beginning of the first phase of the research work described here, that either an empirical or a theoretical analysis of the interrelationships between urban transportation and location was undertaken.

The philosophy of method implicit in the work discussed in this book is one of balancing theoretical development with empirical verification and, more importantly, of balancing examination of the individual components of the urban spatial phenomena system with the examination of the system as a whole. Thus the work described here deals with some components of the system in considerable detail, whereas others are given only cursory examination. The emphasis, however, is on the functioning of these components as an integrated system which describes urban spatial phenomena. In doing this it is sometimes necessary to pass over some matters in order to develop the system as a whole. This is done with the knowledge that the long-term pattern of activity in research and in applications will involve regular examination and reexamination of each of the components as well as their relationships each to each other.

George Homans (1967) has said,

> "What makes a science are its aims, not its results. If it aims at establishing more or less general relationships between properties of nature, when the test of the truth of a relationship lies finally in the data themselves, and the data are not wholly manufactured—when nature, however stretched out on the rack, still has a chance to say 'No!'—then the subject is a science".

The aim of work in the field of urban simulation is a more complete knowledge of the determinants of urban phenomena, with a particular emphasis, in the work described here, on urban spatial patterns. Homans goes on to say,

> "Any science has two main jobs to do: discovery and explanation. By the first we judge whether it is a science; by the second, how successful a science it is".

The study of urban spatial dynamics is well into the discovery phase, having accomplished extensive stating and testing of relationships amongst the components of the system. The move into the explanation phase will require, of course, a great deal more work in this same vein. In addition, it will become increasingly important to examine more and more comprehensive representations of urban systems, as some phenomena result not from the functioning of one or another of the components of the system; rather they arise solely from the interactions of those components. It is the purpose of this book to describe the results of studies, not only of the components of a representation of urban spatial phenomena, but also of the interactions of these components one with another.

1.2 Overview

After this introductory chapter, the book contains ten additional chapters. The second chapter is perhaps best grouped with this one. It provides simple numerical examples of the three principal types of model used here in the integrated analysis of location and transport. An example is given of (a) a residence-location model, (b) an employment-location model, and (c) a transportation network assignment model. Having understood these, the reader will have a contextual understanding which will form the background for the remainder of the book.

The third chapter describes the design and construction (assembly) of the first prototype version of an integrated transportation and land-use model package. Here the initial problems of specifying the links between the models are discussed, along with the results of the first runs of the model package. Some of the revisions which were immediately necessary and obvious are also described. The fourth chapter follows up this discussion with a presentation of preliminary policy analyses of the San Francisco Bay Area. A set of three analyses is described, involving the consequences of redistribution of basic employment, land-use controls, and major changes in the highway network.

The fifth chapter is the first of several devoted to the description of modifications and improvements to the component models of the integrated package. Thus these chapters chronicle the development of theory and the implementation of subsequent practical improvements to the model package. In the fifth chapter the transportation component is the focus of attention. The practical aspects of converting trip-probability matrices to actual vehicle-trip matrices are discussed first. Next the problems of simultaneous residence location and trip generation are discussed, as this is a major aspect of the functioning of the integrated model system. This is followed by a discussion of alternative traffic-assignment procedures along with results of several comparison tests. A prototype mode-split procedure is also described.

The development of models of intraurban employment location is taken up in the sixth chapter. The results of an extensive series of tests of such models are presented. A description of a new urban employment-location model is then given, along with the equations for partial derivatives necessary for estimation of its parameters.

The seventh and eighth chapters both deal with residence location. The seventh chapter contains the theory and development of the residence-location model used in later versions of the integrated model system. The remainder of the seventh chapter deals with the problems of trip-end calibration (parameter estimation) of such models. The eighth chapter then describes a series of verification experiments for the residence-location models. Some of these experiments take the form of parameter estimation for a number of cities from widely varying points in space and time. The results show a high degree of descriptive validity for the model structure. The second set of verification experiments involves a set of sensitivity tests of the residence-location model which demonstrate the correctness of its response to changes in inputs.

The ninth chapter presents an entire new series of policy analyses with the modified version of the integrated model system. The new version of the system contains the improved models developed and described in chapters 5–8. Several different policies are examined, including changes in the transportation system due to construction as well as to energy-cost changes, toll changes, and user changes. The spatial consequences of regional growth or decline were also examined. Also in this chapter a procedure for estimating the impacts of these policies on air quality is described and those impacts are analyzed and discussed. This chapter, in addition to discussing the likely real impacts of various policies, also demonstrates the usefulness of integrated transportation and land-use analysis.

The tenth chapter is a potpourri of discussions of topics for further theoretical development and application. These topics range from formulation of dynamic models and system behavior to improved calibration techniques.

The eleventh and last chapter summarizes both the substantive and the methodological findings presented throughout the text, and ends with a brief comment on the place of this type of research in modern planning practice and society.

2

A review of transportation and location models

2.1 Introduction

In this chapter, each of the major submodels of an integrated transportation and location model system will be developed and discussed in terms of a simple numerical example. A residence-location model, an employment-location model, and a transportation network assignment model will be presented.

The reader is reminded that this book was prepared some years after the initial work on an integrated model package was begun. The models illustrated in this chapter represent theoretical and operational models which were, in some cases, not developed until after various portions of the work described in subsequent chapters had been completed. Therefore, the simple models illustrated here are more advanced, say, than one or more of the submodels used in the work on the first version of the integrated package described in chapter 3. As will be shown, however, the essential concepts are similar. The reader who follows through the examples in this chapter will easily be able to recognize the essential elements in the various forms of submodel used in the evolving versions of the integrated package which are discussed in subsequent chapters.

In this chapter, three simple models are discussed in terms of worked numerical examples. First, a simple residence-location model is developed. Several illustrative 'policy' manipulations are worked through, as well as several different trip functions and attractiveness measures. Next, a simple employment-location model is worked through in a similar fashion. The two models are then linked together and the numerical example is reworked. The final model to be illustrated is a transportation network assignment model, which is worked through both in unconstrained and in capacity-constrained forms.

The three simple models presented here represent the state of the art for applied models. There have been further theoretical developments, and even some trial applications. But these models are the types in current use. In subsequent chapters, numerous embellishments are developed and discussed. Newer theoretical forms are also presented. All are discussed in the context of the simple examples presented here. Moreover, the crucial idea of integrating these models into a cohesive package can readily be discussed and understood once the workings of the separate submodels are grasped.

2.2 Residence-location model: illustrative examples

2.2.1 Regionally constrained form

All the forms of residence-location model used in the work described in this book are variants of the same general construct. The central hypothesis of this construct is that, ceteris paribus, place of work determines place of residence. It is perfectly obvious that all things are not equal in any realistic analysis of location. In the more sophisticated model formulations actually used in the analyses described in subsequent chapters, many of the other important factors are accounted for explicitly. Nonetheless, it is quite surprising how robust this simple model formulation is.

One of the earliest uses of this model form was by Lowry (1964), who used it as the basis for the residence-location component for his well-known early work on location analysis. In his model, the further assumption is made that tripmakers' propensity to travel is negative. That is, the greater the 'length' (in distance, time, or cost) of a possible trip, the less likely the tripmaker is to make it. Lowry used a very simple declining power function of the form

$$p_{ij} = d_{ij}^{-1\cdot330} \qquad (d_{ij} \geqslant 1\cdot0) , \tag{2.1}$$

where p_{ij} is the probability of a trip being taken between points (representative of zones) i and j, and d_{ij} is the distance between points i and j. The residence-location model used by Lowry was simplicity itself, with the form

$$N_i = g\sum_j E_j p_{ij} , \tag{2.2}$$

where N_i is the number of persons living in zone i, E_j the number of employees working in zone j, and g a scaling factor. The scale factor g serves two purposes. First, it makes any necessary adjustment for labor-force participation rates, unemployment levels, and workers-per-household ratios. Second, it ensures that the number of persons estimated to be living in each zone i does not, when summed over all zones, exceed a previously determined regional total. A numerical example will help to illustrate this, as well as to show a built-in inconsistency of this formulation.

Consider a region divided into five subareas or zones, as shown in figure 2.1. Suppose that the employment in each zone of the region is as shown in figure 2.1:

$$E_1 = 100 , \qquad E_2 = 200 , \qquad E_3 = 300 , \qquad E_4 = 0 , \qquad E_5 = 0 .$$

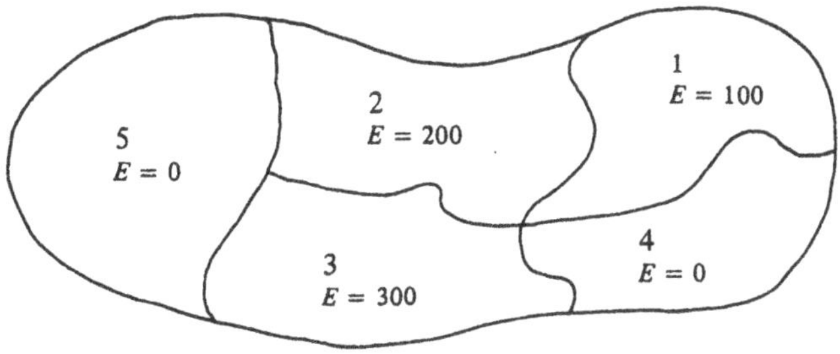

Figure 2.1. Hypothetical five zone region.

Suppose that the 'cost' of travel from each zone to each other zone, c_{ij}, is as given in the matrix in table 2.1.

In order to simplify matters, it will be assumed that the above-mentioned labor-force, unemployment, and household-size ratios are all equal to unity, so that there are equal numbers of employees and residents in the region; thus

$$\sum_i N_i = \sum_j E_j = 600 \; .$$

For ease of computation, a value of $-2{\cdot}000$, rather than Lowry's exponent of $-1{\cdot}330$, will be assumed for this example. It is now possible to estimate the locations of residence of the 600 employees in the region. Equation (2.2), with the declining power function for trip probabilities of equation (2.1), will be used:

$$N_i = g\sum_j E_j c_{ij}^{-2} \; . \tag{2.3}$$

The first step is to calculate the matrix of c_{ij}^{-2} values; these are shown in table 2.2. Then, for zone 1, assuming a trial value of $g = 1{\cdot}0$, we obtain

$$\begin{aligned} N_1 &= 1{\cdot}0[100(1{\cdot}00)+200(0{\cdot}44)+300(0{\cdot}25)+0(0{\cdot}44)+0(0{\cdot}16)] \\ &= 263 \; ; \end{aligned}$$

and for the remaining zones (dropping the E_4 and E_5 terms, as all will be zero),

$$\begin{aligned} N_2 &= 1{\cdot}0[100(0{\cdot}44)+200(1{\cdot}00)+300(0{\cdot}44)] = 376 \; , \\ N_3 &= 1{\cdot}0[100(0{\cdot}25)+200(0{\cdot}44)+300(1{\cdot}00)] = 413 \; , \\ N_4 &= 1{\cdot}0[100(0{\cdot}44)+200(0{\cdot}25)+300(0{\cdot}25)] = 169 \; , \\ N_5 &= 1{\cdot}0[100(0{\cdot}16)+200(0{\cdot}25)+300(0{\cdot}16)] = 114 \; . \end{aligned}$$

Note that if g is assumed to be equal to unity, then

$$\sum_i N_i = 263+376+413+169+114 = 1335 \; .$$

Table 2.1. Zone-to-zone travel costs c_{ij}, for the hypothetical region shown in figure 2.1.

Zone	1	2	3	4	5
1	1·0	1·5	2·0	1·5	2·5
2	1·5	1·0	1·5	2·0	2·0
3	2·0	1·5	1·0	2·0	2·5
4	1·5	2·0	2·0	1·0	1·5
5	2·5	2·0	2·5	1·5	1·0

The proper value of g must then be given by

$$g' = \frac{600}{1335} = 0 \cdot 4494 \ .$$

Multiplication of the initial calculated values by g' yields

$$N_1 = 118 \ , \qquad N_2 = 169 \ , \qquad N_3 = 186 \ , \qquad N_4 = 76 \ , \qquad N_5 = 51.$$

The sum of these gives the correct regional total of 600 persons.

2.2.2 Trip-end-constrained form

The inconsistency alluded to above has to do with the fact that in the form of model given by equation (2.3) there are no 'origin' or 'destination' constraints. Consider equation (2.3) expanded for zone 1:

$$N_1 = g[E_1 c_{11}^{-2} + E_2 c_{12}^{-2} + E_3 c_{13}^{-2} + E_4 c_{14}^{-2} + E_5 c_{15}^{-2}] \ . \tag{2.4}$$

Each term in equation (2.4)—for example, $E_1 c_{11}^{-2}$—represents the number of persons working in zone j, multiplied by the probability of a trip between zone i and zone j. This equation is then evaluated for $N_1, ..., N_5$. The first terms of these equations for all five zones are:

$$E_1 c_{11}^{-2} \ , \qquad E_1 c_{21}^{-2} \ , \qquad E_1 c_{31}^{-2} \ , \qquad E_1 c_{41}^{-2} \ , \qquad E_1 c_{51}^{-2} \ .$$

These terms represent all the possible residence locations for employees who work in zone 1. After application of the scale factor g', the sum of these terms is 'the sum, over all residence zones, of the persons who work in zone 1'. Doing the arithmetic, we obtain

$$E_1' = 0 \cdot 4494 [100(1 \cdot 00) + 100(0 \cdot 44) + 100(0 \cdot 25) + 100(0 \cdot 44) + 100(0 \cdot 16)] = 103 \ .$$

Similarly, for employees in zones 2 and 3, it is possible to calculate that

$$E_2' = 0 \cdot 4494 [200(0 \cdot 44) + 200(1 \cdot 00) + 200(0 \cdot 44) + 200(0 \cdot 25) + 200(0 \cdot 25)] = 214 \ ;$$

$$E_3' = 0 \cdot 4494 [300(0 \cdot 25) + 300(0 \cdot 44) + 300(1 \cdot 00) + 300(0 \cdot 25) + 300(0 \cdot 16)] = 283 \ .$$

The inconsistency is that

$$E_1 = 100 \neq E_1' = 103 \ ,$$
$$E_2 = 200 \neq E_2' = 214 \ ,$$
$$E_3 = 300 \neq E_3' = 283 \ .$$

The number of persons who are resident in each zone of a region thus implies, through the equations of the model, a specific number of persons who are employed in those zones. Although the type of model shown in equation (2.3) preserves consistency at the regional level, it allows for inconsistencies at the zonal level.

This problem existed in all Lowry model variants and derivatives until the development of spatial interaction models based on entropy-maximizing principles. With these developments, stemming in the first instance from transportation (that is, trip-generation) models, this particular problem was easily resolved. This was not a very difficult matter, and with hindsight seems almost trivially simple. The problem is quite simply resolved by the introduction of a trip-end constraint which resulted in the inclusion of what is now known as a balancing factor, B_j, in equation (2.3). This is of the form

$$B_j = \left(\sum_i c_{ij}^{-2}\right)^{-1} . \tag{2.5}$$

With this, the need for g in equation (2.3) is eliminated, as the regional constraint is now replaced by a vector of B_j which serve as zonal constraints. It should be noted that p_{ij} from equation (2.2) may still be considered to be a probability of a trip between zones i and j, but that it is now a normalized probability of the form

$$p_{ij} = B_j c_{ij}^{-2} = \frac{c_{ij}^{-2}}{\sum_j c_{ij}^{-2}} . \tag{2.6}$$

If this result is substituted into equation (2.3), the result is

$$N_i = \sum_j E_j B_j c_{ij}^{-2} = \sum_j E_j \left(\frac{c_{ij}^{-2}}{\sum_i c_{ij}^{-2}}\right) . \tag{2.7}$$

Now the same data which were used before may be used in this new equation to estimate the place of residence of the employees in the region.

In solving this revised equation it will be convenient to precalculate the column totals from the matrix of c_{ij}^{-2} given in table 2.2, which are the sums of the denominators in equation (2.7):

$$\sum_i c_{i1}^{-2} = 2{\cdot}29 , \qquad \sum_i c_{i2}^{-2} = 2{\cdot}38 , \qquad \sum_i c_{i3}^{-2} = 2{\cdot}10 ,$$
$$\sum_i c_{i4}^{-2} = 2{\cdot}38 , \qquad \sum_i c_{i5}^{-2} = 2{\cdot}01 .$$

Table 2.2. Matrix of zone-to-zone travel costs given in table 2.1, transformed to c_{ij}^{-2}.

Zone	1	2	3	4	5
1	1·00	0·44	0·25	0·44	0·16
2	0·44	1·00	0·44	0·25	0·25
3	0·25	0·44	1·00	0·25	0·16
4	0·44	0·25	0·25	1·00	0·44
5	0·16	0·25	0·16	0·44	1·00

Now, to calculate N_1 these values are used to evaluate the expression

$$N_1 = \sum_j E_j B_j c_{1j}^{-2} = \frac{E_1 c_{11}^{-2}}{\sum_i c_{i1}^{-2}} + \frac{E_2 c_{12}^{-2}}{\sum_i c_{i2}^{-2}} + \ldots + \frac{E_5 c_{15}^{-2}}{\sum_i c_{i5}^{-2}} .$$

This yields

$$N_1 = \frac{100(1 \cdot 00)}{2 \cdot 29} + \frac{200(0 \cdot 44)}{2 \cdot 38} + \frac{300(0 \cdot 25)}{2 \cdot 10} + \frac{0(0 \cdot 44)}{2 \cdot 38} + \frac{0(0 \cdot 16)}{2 \cdot 01} = 116;$$

similarly,

$$N_2 = \frac{100(0 \cdot 44)}{2 \cdot 29} + \frac{200(1 \cdot 00)}{2 \cdot 38} + \frac{300(0 \cdot 44)}{2 \cdot 10} = 166 ,$$

$$N_3 = \frac{100(0 \cdot 25)}{2 \cdot 29} + \frac{200(0 \cdot 44)}{2 \cdot 38} + \frac{300(1 \cdot 00)}{2 \cdot 10} = 191 ,$$

$$N_4 = \frac{100(0 \cdot 44)}{2 \cdot 29} + \frac{200(0 \cdot 25)}{2 \cdot 38} + \frac{300(0 \cdot 25)}{2 \cdot 10} = 76 ,$$

$$N_5 = \frac{100(0 \cdot 16)}{2 \cdot 29} + \frac{200(0 \cdot 25)}{2 \cdot 38} + \frac{300(0 \cdot 16)}{2 \cdot 10} = 51 .$$

Now, owing to the inclusion of the B_j values, these estimates of the N_i do sum to 600 persons.

To check on the origin consistency (that is, do the actual numbers of employees equal the implied numbers of employees?), it is necessary to sum the first terms from each of the estimates of N_i. Thus, we find

$$E'_1 = \frac{100(1 \cdot 00)}{2 \cdot 29} + \frac{100(0 \cdot 44)}{2 \cdot 29} + \frac{100(0 \cdot 25)}{2 \cdot 29} + \frac{100(0 \cdot 44)}{2 \cdot 29} + \frac{100(0 \cdot 16)}{2 \cdot 29}$$

$$= 100 ;$$

similarly,

$$E'_2 = \frac{200(0 \cdot 44)}{2 \cdot 38} + \frac{200(1 \cdot 00)}{2 \cdot 38} + \ldots = 200 ,$$

and $E'_3 = 300$. Thus this form of the model preserves both regional and zonal consistency.

2.2.3 Alternative trip functions

To continue with the above example, it is of some interest to observe the effects of alternative forms of trip function for the model. In the calculations above,

$$f(c_{ij}) = c_{ij}^{-2} . \tag{2.8}$$

Suppose now that the original exponent value from Lowry's residence-location model [equation (2.1)] had been used. Then,

$$f(c_{ij}) = c_{ij}^{-1 \cdot 330} . \tag{2.9}$$

As this implies a less steeply declining function of travel propensity with increasing travel cost, the expectation is that the residents in the five zone region would be more evenly distributed amongst the zones. This can be examined numerically. First, the matrix elements $c_{ij}^{-1\cdot330}$ are calculated, along with the column sums, as shown in table 2.3. These values are substituted into equation (2.7), such that

$$N_1 = \frac{100(1\cdot00)}{2\cdot86} + \frac{200(0\cdot58)}{2\cdot96} + \frac{300(0\cdot40)}{2\cdot68} = 119 \ ;$$

and, similarly,

$$N_2 = 153 \ , \qquad N_3 = 165 \ , \qquad N_4 = 92 \ , \qquad N_5 = 71 \ .$$

These results may be compared to the previous results for c_{ij}^{-2}, and it can be seen that there is indeed a somewhat more uniform spread of residences across the zones in the region.

As one last illustration of this sort, let

$$f(c_{ij}) = c_{ij}^{1.5} \exp(-2c_{ij}) \ . \qquad (2.10)$$

This function is that of a modified gamma distribution, which yields a skewed, peaked curve of the form shown in figure 2.2(b), as compared

Table 2.3. Matrix of zone-to-zone travel costs given in table 2.1, transformed to $c_{ij}^{-1\cdot330}$.

Zone	1	2	3	4	5
1	1·00	0·58	0·40	0·58	0·30
2	0·58	1·00	0·58	0·40	0·40
3	0·40	0·58	1·00	0·40	0·30
4	0·58	0·40	0·40	1·00	0·58
5	0·30	0·40	0·30	0·58	1·00
Σ	2·86	2·96	2·68	2·96	2·58

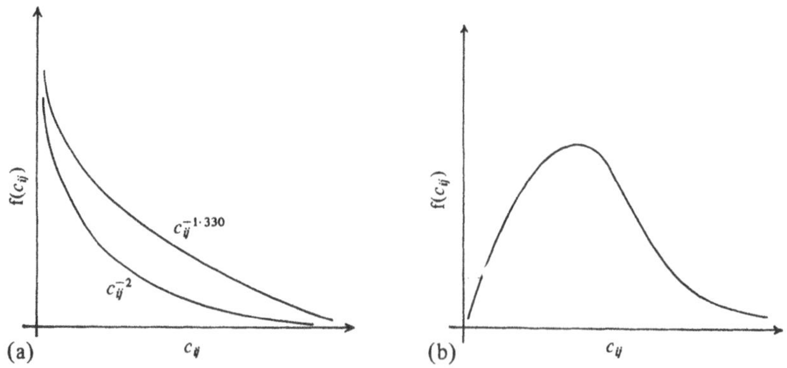

Figure 2.2. Alternative forms of the trip function. (a) The negative power functions, c_{ij}^{-2} and $c_{ij}^{-1\cdot330}$; (b) the modified gamma distribution, $c_{ij}^{1.5}\exp(-2c_{ij})$.

with the two curves for c_{ij}^{-2} and $c_{ij}^{-1\cdot330}$ shown in figure 2.2(a). The transformed cost matrix which results from equation (2.10) is given in table 2.4, along with the column sums. These sums, in turn, are substituted into equation (2.7) as above, to yield

$$N_1 = 120\,, \quad N_2 = 164\,, \quad N_3 = 172\,, \quad N_4 = 88\,, \quad N_5 = 56\,,$$

which produces yet another slightly different distribution of residences throughout the region. The choice of trip function is derived from empirical evidence as well as, in part, being a matter of the preference of the model builder. However, it should be noted that virtually all empirically observed trip distributions have the forms shown in figures 2.2(a) and 2.2(b). Further, the function (2.10) can, if the exponent of c_{ij} is less than or equal to zero, have the same shape as the functions (2.8) and (2.9). This whole matter will be discussed again when the topic of model calibration is presented.

Table 2.4. Matrix of zone-to-zone travel costs given in table 2.1, transformed to $c_{ij}^{1.5}\exp(-2c_{ij})$.

Zone	1	2	3	4	5
1	0·14	0·09	0·05	0·09	0·03
2	0·09	0·14	0·09	0·05	0·05
3	0·05	0·09	0·14	0·05	0·03
4	0·09	0·05	0·05	0·14	0·09
5	0·03	0·05	0·03	0·09	0·14
Σ	0·40	0·42	0·36	0·42	0·34

2.2.4 Addition of an attractiveness measure

It is possible to add additional substance to the simple model by considering a measure of attractiveness of each potential residence zone. This can be done by partitioning the probability term in equation (2.6) into an attractiveness component, W_i, and a trip-function component, $\mathrm{f}(c_{ij})$:

$$p_{ij} = \frac{W_i\,\mathrm{f}(c_{ij})}{\sum\limits_i W_i\,\mathrm{f}(c_{ij})}\,. \tag{2.11}$$

Note that the notion of the balancing factor introduced in equations (2.5) and (2.6), is preserved here by the form of the denominator. The purpose of the attractiveness measure should be quite clear. In the version of the model given by equation (2.7), any two zones which are equidistant from a particular place of employment are equally likely to serve as zones of residence for those employees, regardless of the characteristics of the zones themselves. Thus it is quite clear that a more realistic model would result from the inclusion of some measure (or measures) of each zone's characteristics (that is, some set of zonal attributes). These attributes might, in actual model applications, include such variables as available developable land, residential density, and perhaps the income distributions of the residents currently living in the zone. These will be discussed in subsequent chapters.

To continue the numerical example, let the measure of attractiveness be taken to be the vacant developable land in each of the zones, to give the values:

$$W_1 = 2{\cdot}0\,, \qquad W_2 = 1{\cdot}5\,, \qquad W_3 = 1{\cdot}0\,, \qquad W_4 = 3{\cdot}0\,, \qquad W_5 = 5{\cdot}0\,.$$

It is then possible to substitute these values, along with the transformed cost matrix set out in table 2.4, into the following expanded residence-location model:

$$N_i = \sum_j E_j \left[\frac{W_i c_{ij}^{1.5} \exp(-2c_{ij})}{\sum_i W_i c_{ij}^{1.5} \exp(-2c_{ij})} \right] . \tag{2.12}$$

Thus, evaluating equation (2.12) for N_1, we have

$$N_1 = \left[\frac{E_1 W_1 c_{11}^{1.5} \exp(-2c_{11})}{\sum_i W_i c_{i1}^{1.5} \exp(-2c_{i1})} \right] + \left[\frac{E_2 W_1 c_{12}^{1.5} \exp(-2c_{12})}{\sum_i W_i c_{i2}^{1.5} \exp(-2c_{i2})} \right] + \ldots .$$

It is again helpful to calculate the denominator sums in advance:

$$\sum_i W_i c_{i1}^{1.5} \exp(-2c_{i1}) = 0{\cdot}89\,; \qquad \sum_i W_i c_{i2}^{1.5} \exp(-2c_{i2}) = 0{\cdot}88\,;$$

$$\sum_i W_i c_{i3}^{1.5} \exp(-2c_{i3}) = 0{\cdot}68\,; \qquad \sum_i W_i c_{i4}^{1.5} \exp(-2c_{i4}) = 1{\cdot}18\,;$$

$$\sum_i W_i c_{i5}^{1.5} \exp(-2c_{i5}) = 1{\cdot}14\,.$$

These may then be substituted into equation (2.12), which yields

$$N_1 = \frac{100(2{\cdot}0)(0{\cdot}14)}{0{\cdot}89} + \frac{200(2{\cdot}0)(0{\cdot}09)}{0{\cdot}88} + \frac{300(2{\cdot}0)(0{\cdot}05)}{0{\cdot}68} = 117\,;$$

similarly,

$$N_2 = 122\,, \qquad N_3 = 89\,, \qquad N_4 = 131\,, \qquad N_5 = 141\,.$$

If these results are compared with those from the case where the same travel function, equation (2.10), is used, but with no attractiveness measure, substantial differences will be observed. In particular, this latter example, with much more vacant developable land in zones 4 and 5, shows substantially increased residence estimates for those zones.

Having demonstrated the potential importance of attractiveness measures, it remains to be shown that as many attractiveness variables as are desired may be included in these models, subject, of course, to operational constraints on the availability of data, statistical problems of parameter estimation, and computational costs. Furthermore, an empirically estimated parameter may be associated with any or all of these variables. A general form of such an attractiveness measure might be

$$W_i = \mathrm{f}(X_{i,1}^{\alpha_1}, X_{i,2}^{\alpha_1}, X_{i,3}^{\alpha_3}, ..., X_{i,n}^{\alpha_n}) , \tag{2.13}$$

where the $X_{i,k}$ are the zonal attributes, and the α_k are empirically derived parameters. Extensive testing has been done of one such function, of the form

$$W_i^g = \prod_k X_{i,k}^{\alpha_{g,k}} , \tag{2.14}$$

where W_i^g is a measure of attractiveness of zone i to resident type g, $X_{i,k}$ is the kth attribute of zone i, and $\alpha_{g,k}$ are empirically derived parameters. These results are described in Putman (1977a; 1977b; 1980a), and in Putman and Ducca (1978a; 1978b), and will be summarized in a subsequent chapter.

2.2.5 Prototypical policy tests

Before the end of this discussion and illustration of a simple residence-location model, it will be instructive to demonstrate two possible policy tests.

First, suppose that a floodplain zoning program is enacted in the test region which reduces the vacant developable land in zone 5 from 5·0 to 4·2 units. The new values of W_i may then be used in equation (2.12) to calculate a revised distribution of residents. The first numerical changes will be in the denominator sums. These are now

$$\sum_i W_i c_{i1}^{1.5} \exp(-2c_{i1}) = 0{\cdot}86 ; \qquad \sum_i W_i c_{i2}^{1.5} \exp(-2c_{i2}) = 0{\cdot}84 ;$$

$$\sum_i W_i c_{i3}^{1.5} \exp(-2c_{i3}) = 0{\cdot}65 ; \qquad \sum_i W_i c_{i4}^{1.5} \exp(-2c_{i4}) = 1{\cdot}10 ;$$

$$\sum_i W_i c_{i5}^{1.5} \exp(-2c_{i5}) = 1{\cdot}02 .$$

The values of N_i then can be calculated as above; this yields

$$N_1 = 122 , \quad N_2 = 127 , \quad N_3 = 92 , \quad N_4 = 136 , \quad N_5 = 123 .$$

If these results are compared with the results before floodplain zoning, it can be seen that there are noticeable effects. The reduction of available land in zone 5 produces a substantial reduction in the number of residents locating in that zone (we assume that there is no change in the region's total number of residents). As there are no other variables in the attractiveness measure, the residents who do not now locate in zone 5 are distributed rather evenly to the other four zones.

The effects of a change in transportation facilities may also be simulated by altering the travel-cost data input. Suppose that a transport improvement is to be constructed such that there is a one-third reduction in travel cost between zones 3 and 4. The cost of travel between these zones is therefore reduced from its original value of 2·0 to a new value of 1·3. This yields the new cost matrix shown in table 2.5, together with its corresponding transformed matrix evaluated from equation (2.10). Now the revised denominator sums may be calculated (with the original values for vacant developable land):

$$\sum_i W_i c_{i1}^{1\cdot 5} \exp(-2c_{i1}) = 0\cdot 89 \; ; \qquad \sum_i W_i c_{i2}^{1\cdot 5} \exp(-2c_{i2}) = 0\cdot 88 \; ;$$

$$\sum_i W_i c_{i3}^{1\cdot 5} \exp(-2c_{i3}) = 0\cdot 86 \; ; \qquad \sum_i W_i c_{i4}^{1\cdot 5} \exp(-2c_{i4}) = 1\cdot 24 \; ;$$

$$\sum_i W_i c_{i5}^{1\cdot 5} \exp(-2c_{i5}) = 1\cdot 14 \; .$$

Finally, it becomes possible to calculate the values of N_i which result from the proposed transportation change:

$$N_1 = 108 \,, \quad N_2 = 110 \,, \quad N_3 = 75 \,, \quad N_4 = 181 \,, \quad N_5 = 126 \,.$$

A comparison of these results with those obtained without the proposed transportation improvement shows rather dramatic changes. All the zones containing employment (zones 1, 2, and 3) show relative declines, as does the 'suburban' zone 5 which does not benefit from the proposed improvement. Zone 4, which is the direct beneficiary of the improvement, shows an increase in residents of almost 40%.

These two prototype policy tests demonstrate the general types of responses of this class of model. The more sophisticated models used in the policy analyses described in subsequent chapters of this book operate in very much the same way. An understanding of the simple illustrations given here will enable the reader to understand and interpret better the results presented later. In this same vein, the next section of this chapter presents a similar illustration for a simple employment-location model.

Table 2.5. Revised matrix of zone-to-zone travel costs, together with transformation of this matrix according to equation (2.10)

Zone	Cost matrix					Transformed cost matrix				
	1	2	3	4	5	1	2	3	4	5
1	1·0	1·5	2·0	1·5	2·5	0·14	0·09	0·05	0·09	0·03
2	1·5	1·0	1·5	2·0	2·0	0·09	0·14	0·09	0·05	0·05
3	2·0	1·5	1·0	*1·3*	2·5	0·05	0·09	0·14	*0·11*	0·03
4	1·5	2·0	*1·3*	1·0	1·5	0·09	0·05	*0·11*	0·14	0·09
5	2·5	2·0	2·5	1·5	1·0	0·03	0·05	0·03	0·09	0·14

2.3 Employment-location model: illustrative examples

2.3.1 'Retail', nonbasic, or local-serving employment

In discussing even simple models of intraurban (or intrametropolitan) employment location, it is first necessary to discuss types of employment (Putman, 1972). The main distinction between these employment types is the extent to which their intrametropolitan locational behavior is a function of the spatial distribution of activities in the metropolitan area. The classification schemes for these employment types are chosen, in each study, according to characteristics presumed to be important to location in the particular study region. The first point to be noted in this regard is that the 'types' of employment being discussed here are partitioned according to industry class—for example, per the standard industrial classification schemes. The second point is that, for the purposes of modelling intrametropolitan employment location, the employment classification scheme must be consistent with the determinants of intrametropolitan location. It is often difficult to reconcile the differences between these two classifications.

Intrametropolitan location models assume that the regional (metropolitan) totals for various employment types will be exogenously provided. One implication of the requirement for these exogenously provided totals is that effects of growth or decline of regional employment are not dealt with in these models. This procedure also assumes that questions of firm or facility location in one region versus another have already been dealt with in the regional forecasts. Thus, the question of intrametropolitan location is of the type:

> "Given that there is to be growth (or decline) in type *X* employment levels throughout the region, where in the region will this take place?"

One of the key considerations in determining where the growth or decline of employment will take place is the degree to which the location of a particular type of employment is sensitive to local markets and their spatial distribution. On the basis of this notion it is appropriate to group employment into two classes: (a) employment whose intrametropolitan location is insensitive to the spatial distribution of local markets, and (b) employment whose intrametropolitan location is sensitive to the spatial distribution of local markets. In most cases these two classes will be similar to more traditional distinctions such as basic versus nonbasic, or basic versus service (Massey, 1973). In the remainder of this discussion these distinctions will be used, but it must be kept in mind that applications of these models in different metropolitan areas may result in different differentiations of basic and nonbasic employment classes in different regions.

A number of different approaches have been taken towards modelling location of nonbasic employment. One of the approaches most frequently taken, and in many respects one of the most successful, involves the development of a market-potential term (or measure). The model to be illustrated here depends upon such a measure, but has, in addition, a lagged employment term which adds both reliability and stability to its forecast outputs.

The development of this model stems from a pair of hypotheses. The first is that the amount of local-market-sensitive (LOMS) employment in a zone is directly proportional to the spatial distribution of the markets for that employment type in and around the zone. The second is that, primarily because of agglomeration effects, the quantity of LOMS employment in a zone at time t is directly proportional to the quantity of LOMS employment which was in the zone at time $t-1$. Further hypotheses will be necessary to define the quantitative measures of the spatial distributions of markets. One typical definition—especially, say, for retail trade—is simply the spatial distribution of the population. This is the measure to be used in the numerical example which follows.

The structure of the simple LOMS employment model is as follows

$$E_{j,t}^{k} = \mathrm{f}(M_{j,t}^{k}, E_{j,t-1}^{k}) \,. \tag{2.15}$$

Here $E_{j,t}^{k}$ is the employment of type k in zone j at time t; $M_{j,t}^{k}$ is a measure of the markets for type k goods spatially distributed in and around zone j, at time t; and $E_{j,t-1}^{k}$ is the employment of type k in zone j at some prior time $t-1$.

Each type of employment will vary in its dependence on the markets variable and/or on the lagged variable for its present status in a particular zone; thus it will be useful to separate these two independent variables in the equation structure. There is also a practical problem here, when the parameter estimations are to be done, with regard to the relative magnitude of the markets variable and the lagged variable. For ease of interpretation it will be useful to adopt the following structural form

$$E_{j,t}^{k} = a^{k}M_{j,t}^{k} + b^{k}E_{j,t-1}^{k} \,, \tag{2.16}$$

in which the parameters a^k and b^k will indicate the relative importance of the markets variable or the lagged variable in determining location. Further, it will be helpful to constrain these parameters such that

$$a^{k} + b^{k} = 1 \cdot 0 \,, \tag{2.17}$$

which will allow the relative weights of the M_j^k and E_j^k terms to be seen directly. In order to prevent the parameters a^k and b^k from merely acting as scaling for the variables and thus possibly masking their weighting effect, it will be necessary to scale these variables before their use in equation (2.16) so that they are of equal magnitudes. Finally, in order to avoid the use of an arbitrary scaling parameter, analogous to g in equation (2.2), the scaling of these two variables should result in each of their sums over the region totalling the regionwide employment forecasts for this type of employment.

We recall that an asterisk replacing a superscript or subscript denotes summation over that superscript or subscript; thus let $E_{*,t}^{k}$ be the exogenous regionwide forecast of type k employment. Then it is desired that the scaled values of the markets and lagged variables, $\tilde{M}_{j,t}^{k}$ and $\tilde{E}_{j,t-1}^{k}$, have the property

$$\sum_{j} \tilde{M}_{j,t}^{k} = \sum_{j} \tilde{E}_{j,t-1}^{k} = E_{*,t}^{k} \,. \tag{2.18}$$

Now, in fact, the spatial distribution of $E^k_{j,t-1}$ is already known. Thus the forecast distribution should be of only the regionwide increment or decrement in type k employment. Thus equation (2.18) should be replaced by

$$\sum_j \tilde{M}^k_{j,t} = \sum_j \tilde{E}^k_{j,t-1} = \Delta E^k_{*,t} , \tag{2.19}$$

where

$$\Delta E^k_{*,t} = (E^k_{*,t} - E^k_{*,t-1}) .$$

This scaling may then be easily accomplished for the markets variable by use of the expression

$$\tilde{M}^k_{j,t} = M^k_{j,t}\left(\frac{\Delta E^k_{*,t}}{M^k_{*,t}}\right) . \tag{2.20}$$

The lagged variable will be scaled in the same fashion:

$$\tilde{E}^k_{j,t-1} = E^k_{j,t-1}\left(\frac{\Delta E^k_{*,t}}{E^k_{*,t-1}}\right) . \tag{2.21}$$

Thus the form of the original equation (2.16) has now become

$$\Delta E^k_{j,t} = a^k M^k_{j,t}\left(\frac{\Delta E^k_{*,t}}{M^k_{*,t}}\right) + b^k E^k_{j,t-1}\left(\frac{\Delta E^k_{*,t}}{E^k_{*,t-1}}\right) . \tag{2.22}$$

It now remains to define a functional form for the markets variable.

As mentioned above, the measure of the market in this numerical example will simply be the spatial distribution of the population. It should be noted that other measures of markets are not only possible, but in many cases preferable to this one for actual model applications. One such measure, often used in the nonbasic-employment-location component of Lowry derivative models, involves the use of population as the 'home-to-shop-trip' measure of markets, plus the total employment in zone j as the 'work-to-shop-trip' measure.

If we recall that what is needed is the spatial distribution of the population, it is then possible to suggest the following function

$$M^k_{j,t} = \sum_i P_{i,t} p^s_{ij} , \tag{2.23}$$

where $P_{i,t}$ is the population living in zone i at time t; and p^s_{ij} is the probability of a 'shopping' trip between zone i and zone j, where

$$p^s_{ij} = \mathrm{f}(c_{ij}) .$$

As was the case for the residence-location model illustrated in section 2.2, the simplest form of p^s_{ij} would be a simple declining function, such as c_{ij}^{-2} used in equation (2.3). Again, to avoid the need for arbitrary scaling, and to maintain interval consistency, a balancing factor will be used as an integral part of the p^{s}_{ij} function:

$$p^{s}_{ij} = \frac{c_{ij}^{-2}}{\sum_j c_{ij}^{-2}} = \frac{c_{ij}^{-2}}{c_{i*}^{-2}} . \tag{2.24}$$

The functional form of the markets variable then becomes

$$M^k_{j,t} = \sum_i P_{i,t}\left(\frac{c_{ij}^{-2}}{c_{i*}^{-2}}\right) . \tag{2.25}$$

This may, in turn, be substituted into equation (2.22) to yield

$$\Delta E^k_{j,t} = a^k \sum_i P_{i,t}\left(\frac{c_{ij}^{-2}}{c_{i*}^{-2}}\right)\left(\frac{\Delta E^k_{*,t}}{M^k_{*,t}}\right) + b^k E^k_{j,t-1}\left(\frac{\Delta E^k_{*,t}}{E^k_{*,t-1}}\right) . \tag{2.26}$$

Now, it is also true that

$$M^k_{*,t} = \sum_j M^k_{j,t} = \sum_j \left[\sum_i P_{i,t}\left(\frac{c_{ij}^{-2}}{c_{i*}^{-2}}\right)\right] , \tag{2.27}$$

which, on substitution, yields the final equation for the LOMS employment model:

$$\Delta E^k_{j,t} = a^k \sum_i P_{i,t}\left(\frac{c_{ij}^{-2}}{c_{i*}^{-2}}\right)\left\{\Delta E^k_{*,t} \Bigg/ \sum_j \left[\sum_i P_{i,t}\left(\frac{c_{ij}^{-2}}{c_{j*}^{-2}}\right)\right]\right\} + b^k E^k_{j,t-1}\left(\frac{\Delta E^k_{*,t}}{E^k_{*,t-1}}\right) . \tag{2.28}$$

A numerical example can now be demonstrated, to show the workings of this type of model.

First, let us use the same zone system and cost matrix as that given in the residence-location example here, though the employment from that example will here be used as the $t-1$ employment. Second, of the total employment in each zone used in the residence-location example, a portion will be designated as not local-market-sensitive (NLMS), or local-market-insensitive—that is, basic—employment; the remainder will be designated as LOMS employment (see table 2.6).

The zonal populations will be the populations, N_i, estimated by the version of the simple residential model given in equation (2.12), which uses a gamma function for the trip function and vacant developable land

Table 2.6. Employment data for numerical example of LOMS employment model.

Zone	Employment		
	total	LOMS	NLMS
1	100	50	50
2	200	80	120
3	300	110	190
4	0	0	0
5	0	0	0

for the attractiveness measure. These population estimates were:

$$P_{1,t} = 117, \quad P_{2,t} = 122, \quad P_{3,t} = 89, \quad P_{4,t} = 131, \quad P_{5,t} = 141.$$

Next, it is necessary to assume values for a^k and b^k (note that in a real application of the model, these would be estimated empirically). Reasonable values of a^k and b^k are 0·8 and 0·2 respectively; this implies that LOMS employment is more responsive to markets, but has some sensitivity to the prior existence of LOMS employment in any particular zone. Finally, it is necessary to assume a LOMS employment forecast for time t, which in this case is taken to be 288 employees for the entire region.

To begin, we obtain

$$\Delta E^k_{\bullet,t} = (E^k_{\bullet,t} - E^k_{\bullet,t-1}) = (288 - 240) = 48,$$

as

$$E^k_{\bullet,t-1} = \sum_j E^k_{j,t-1} = 50 + 80 + 110 = 240;$$

thus,

$$\frac{\Delta E^k_{\bullet,t}}{E^k_{\bullet,t-1}} = \frac{48}{240} = 0{\cdot}2$$

Next, we calculate the $c_{i\bullet}^{-2}$ terms, which are the row sums of the c_{ij}^{-2} matrix. From table 2.2, these are calculated to be

$$c_{1\bullet}^{-2} = 2{\cdot}29, \quad c_{2\bullet}^{-2} = 2{\cdot}38, \quad c_{3\bullet}^{-2} = 2{\cdot}10,$$

$$c_{4\bullet}^{-2} = 2{\cdot}38, \quad c_{5\bullet}^{-2} = 2{\cdot}01.$$

The denominator of the first principal term of equation (2.28) is

$$\sum_j \left[\sum_i P_{i,t}\left(\frac{c_{ij}^{-2}}{c_{i\bullet}^{-2}}\right)\right] = \sum_j \tilde{P}_{j,t}.$$

This will be a constant term in subsequent calculations, and can be precalculated as follows. For $j = 1$, we can write

$$\sum_i P_{i,t}\left(\frac{c_{i1}^{-2}}{c_{i\bullet}^{-2}}\right) = P_{1,t}\left(\frac{c_{11}^{-2}}{c_{1\bullet}^{-2}}\right) + P_{2,t}\left(\frac{c_{21}^{-2}}{c_{2\bullet}^{-2}}\right) + \ldots = 119{\cdot}68;$$

similarly, for the other four terms, we find

$$\tilde{P}_{2,t} = 123{\cdot}69, \quad \tilde{P}_{3,t} = 102{\cdot}69,$$

$$\tilde{P}_{4,t} = 131{\cdot}80, \quad \tilde{P}_{5,t} = 122{\cdot}14.$$

Now, the sum over j yields

$$\sum_j \tilde{P}_{j,t} = 600,$$

which is equal to the total population in the region and thus illustrates that this scaling is a rather simple proportioning scheme. The constant scaling factor for the first term of equation (2.28) is thus simply

$$\frac{\Delta E^k_{\bullet,t}}{600} = \frac{48}{600} = 0{\cdot}08 \ .$$

The markets variable, or market-potential term, may then be calculated, for $j = 1$, as

$$\sum_i P_{i,t}\left(\frac{c_{i1}^{-2}}{c_{i\bullet}^{-2}}\right) = P_{1,t}\left(\frac{c_{11}^{-2}}{c_{1\bullet}^{-2}}\right) + P_{2,t}\left(\frac{c_{21}^{-2}}{c_{2\bullet}^{-2}}\right) + \ldots = 119{\cdot}68 \ .$$

Note that this is the same calculation as that performed in the first step of calculating the scaling factor $\sum_j \tilde{P}_{j,t}$ above.

Now, for the incremental LOMS employment forecast for zone 1, we obtain

$$\Delta E^k_{1,t} = 0{\cdot}8(119{\cdot}68)(0{\cdot}08) + 0{\cdot}2(50)(0{\cdot}2) = 7{\cdot}66 + 2{\cdot}00 = 10 \ .$$

The incremental employment forecasts for the remaining zones are calculated in a similar fashion, with the result:

$$\Delta E^k_{2,t} = 7{\cdot}92 + 3{\cdot}20 = 11 \ , \qquad \Delta E^k_{4,t} = 8{\cdot}44 + 0{\cdot}00 = 8 \ ,$$

$$\Delta E^k_{3,t} = 6{\cdot}57 + 4{\cdot}40 = 11 \ , \qquad \Delta E^k_{5,t} = 7{\cdot}82 + 0{\cdot}00 = 8 \ .$$

Note that the sum of these rounded incremental employment forecasts is indeed 48 employees, the regionwide forecast for the LOMS employment increment. It may also be seen that, of the markets variable and the lagged variable, each contributes its effect to the final forecast for each zone.

2.3.2 Extensions of the LOMS employment model

As was the case with the residence-location model discussed in section 2.2, once the elementary form has been shown, many extensions may be explored. In the case of the LOMS employment-location model, one obvious possibility is to replace the c_{ij}^{-2} trip function with a modified gamma function which more closely resembles actual 'shopping-trip' functions. This can be worked through the numerical example to demonstrate its effect on the location forecasts for LOMS employment.

To begin, let

$$\mathrm{f}(c_{ij}) = c_{ij}^2 \exp(-2{\cdot}5c_{ij}) \ , \qquad (2.29)$$

which gives a more sharply skewed curve than the work-trip curve described by equation (2.10) in the residence-location example. The first step is to calculate the transformed cost matrix, shown in table 2.7. The values of two parameters a^k and b^k will be unchanged, as will all the other inputs to the calculations.

Similarly, the two previously calculated scaling factors will remain unchanged even though the p_{ij}^s terms do change, as was shown when the denominator for equation (2.28) was calculated and was shown to be equal to $P_{*,t}$, or 600. Therefore, only the market-potential terms need to be recalculated: for zone 1,

$$\sum_i P_{i,t}\left[\frac{c_{i1}^2 \exp(-2\cdot 5c_{i1})}{c_{i*}^2 \exp(-2\cdot 5c_{i*})}\right] = P_{1,t}\left[\frac{c_{11}^2 \exp(-2\cdot 5c_{11})}{c_{1*}^2 \exp(-2\cdot 5c_{1*})}\right] + P_{2,t}\left[\frac{c_{21}^2 \exp(-2\cdot 5c_{21})}{c_{2*}^2 \exp(-2\cdot 5c_{2*})}\right] + \ldots .$$

The values of these market-potential terms are now:

$$M_{1,t} = 116\cdot 44\,, \quad M_{2,t} = 129\cdot 61\,, \quad M_{3,t} = 101\cdot 18\,,$$
$$M_{4,t} = 138\cdot 02\,, \quad M_{5,t} = 114\cdot 98\,.$$

These, together with the other numerical values, may then be substituted into equation (2.22), which yields

$$\Delta E_{1,t}^k = 0\cdot 8(116\cdot 44)(0\cdot 08)+0\cdot 2(50)(0\cdot 2) = 7\cdot 45+2\cdot 00 = 9\,,$$
$$\Delta E_{2,t}^k = 0\cdot 8(129\cdot 61)(0\cdot 08)+0\cdot 2(80)(0\cdot 2) = 8\cdot 30+3\cdot 20 = 12\,,$$
$$\Delta E_{3,t}^k = 0\cdot 8(101\cdot 18)(0\cdot 08)+0\cdot 2(110)(0\cdot 2) = 6\cdot 48+4\cdot 40 = 11\,,$$
$$\Delta E_{4,t}^k = 0\cdot 8(138\cdot 02)(0\cdot 08)+0\cdot 2(0)(0\cdot 2) = 8\cdot 83+0\cdot 00 = 9\,,$$
$$\Delta E_{5,t}^k = 0\cdot 8(114\cdot 98)(0\cdot 08)+0\cdot 2(0)(0\cdot 2) = 7\cdot 36+0\cdot 00 = 7\,.$$

Thus the change in shape of the shopping-trip function yields slight differences (slight in this example, in practice the differences can be substantial) in the estimated incremental forecasts of LOMS employment in the zones of the test region.

A final illustration of this LOMS employment-location model can be given using the previously calculated results from the residence-location problem with improved transportation network described in section 2.2.5. In that case, the population distribution was

$$P_{1,t} = 108\,, \quad P_{2,t} = 110\,, \quad P_{3,t} = 75\,,$$
$$P_{4,t} = 181\,, \quad P_{5,t} = 126\,.$$

Table 2.7. Matrix of zone-to-zone travel costs given in table 2.1, transformed to $c_{ij}^{2\cdot 0}\exp(-2\cdot 5c_{ij})$, with row sums.

Zone	1	2	3	4	5	Σ
1	0·08	0·05	0·03	0·05	0·01	0·22
2	0·05	0·08	0·05	0·03	0·03	0·24
3	0·03	0·05	0·08	0·03	0·01	0·20
4	0·05	0·03	0·03	0·08	0·05	0·24
5	0·01	0·03	0·01	0·05	0·08	0·18

It will be necessary to calculate the new elements resulting from the revised c_{34} and c_{43} values in the travel-cost matrix. These new values, substituted in the shopping-trip gamma function, yield values of 0·07 to replace the previous values of 0·03. Furthermore, the third and fourth row sums are then each increased by 0·04, to yield 0·24 and 0·28 respectively.

As before, no other values in the calculations change except the market-potential terms, which can now be recalculated:

$$M_{1,t} = 110{\cdot}89\,, \qquad M_{2,t} = 117{\cdot}23\,, \qquad M_{3,t} = 114{\cdot}89\,,$$
$$M_{4,t} = 146{\cdot}88\,, \qquad M_{5,t} = 110{\cdot}11\,.$$

These values may then be substituted into equation (2.22) to yield

$$\Delta E^k_{1,t} = 0{\cdot}8(110{\cdot}89)(0{\cdot}08)+0{\cdot}2(50)(0{\cdot}2) = 7{\cdot}10+2{\cdot}00 = 9\,,$$
$$\Delta E^k_{2,t} = 0{\cdot}8(117{\cdot}23)(0{\cdot}08)+0{\cdot}2(80)(0{\cdot}2) = 7{\cdot}50+3{\cdot}20 = 11\,,$$
$$\Delta E^k_{3,t} = 0{\cdot}8(114{\cdot}89)(0{\cdot}08)+0{\cdot}2(110)(0{\cdot}2) = 7{\cdot}35+4{\cdot}40 = 12\,,$$
$$\Delta E^k_{4,t} = 0{\cdot}8(146{\cdot}88)(0{\cdot}08)+0{\cdot}2(0)(0{\cdot}2) = 9{\cdot}40+0{\cdot}00 = 9\,,$$
$$\Delta E^k_{5,t} = 0{\cdot}8(110{\cdot}11)(0{\cdot}08)+0{\cdot}2(0)(0{\cdot}2) = 7{\cdot}05+0{\cdot}00 = 7\,.$$

Thus it can be seen that there is a slight shift in the forecasted increments of LOMS employment due to the proposed transportation improvement. It is important to note here that owing to the small size of this example, the effects of rounding to the nearest whole number are more significant than would be the case in actual application of the models. In this example in particular, the consequences of rounding show up for zones 4 and 5, where there is no change in the incremental LOMS employment forecast as a result of the proposed improvement in the transportation network. This concludes the numerical illustration of a LOMS employment-location model.

2.3.3 Revision of the employment model for NLMS employment

Now that a simple model for location of LOMS (local-market-sensitive) employment has been developed, modest revisions of the model will make it capable of locating manufacturing and other basic, or local-market-insensitive (NLMS, that is, not local-market-sensitive) employment. The heart of this version of the employment-location model is closely analogous to that of the LOMS version. In this case, the first of the pair of hypotheses is somewhat different. This first hypothesis is that the level (quantity) of NLMS employment in a zone is directly proportional to the spatial distribution in and around the zone of the 'production' input requirements of that employment type. This hypothesis reflects the fact that whereas the intrametropolitan spatial distribution of markets is of little importance to this type of industry, the spatial distribution of inputs becomes an important determinant of each zone's attractiveness.

The second hypothesis for NLMS employment is the same as that for LOMS employment: the quantity of NLMS employment in a zone at time t is directly proportional to the quantity of NLMS employment that was in the zone at time $t-1$. It is again necessary to have subsidiary hypotheses about the definition(s) of the spatial distributions of input requirements. Recent surveys have indicated that one of the most important requirements for location or relocation of industrial establishments is the availability of a work force within commuting distance (PIDC-PCPC, 1976). As population is a reasonably good measure of potentially available work force, it can be used as the measure in this numerical example. Thus, at the level of these simple illustrative models, the same general formulation can be used for NLMS employment location as was used for LOMS employment location.

There will, however, be differences in the parameters used for NLMS employment as compared to those used for LOMS employment. In the first instance, the travel function in the term for potential labor force should be a work-trip function, rather than the shopping-trip function which was used for the market-potential function in the LOMS employment-location model. Thus, in this numerical example the work-trip function (2.10) can be used for calculations of the potential labor force. Further, it is likely that the lagged variable term will be of greater significance to NLMS employment location than it was for LOMS employment location[(2)]. Consequently, the values of a^k and b^k for this NLMS example will be taken to be 0·3 and 0·7 respectively. Now, by using the same numerical input data as were used in section 2.3.2, it is possible to demonstrate the calculation of a forecast of the increment in NLMS employment.

The forecast of regionwide NLMS employment at time t will be taken to be 468 employees. As before, we first obtain $\Delta E^k_{\bullet,t}$:

$$\Delta E^k_{\bullet,t} = (E^k_{\bullet,t} - E^k_{\bullet,t-1}) = (468-360) = 108 \ ;$$

thus

$$\frac{\Delta E^k_{\bullet,t}}{E^k_{\bullet,t-1}} = \frac{108}{360} = 0{\cdot}3 \ .$$

The trip-function transform of the travel-cost matrix was calculated in the residence-location model example, and is given in table 2.4. As was shown in section 2.3.2, the scaling factor for the first term of equation (2.28) is simply

$$\frac{\Delta E^k_{\bullet,t}}{P_{\bullet,t}} = \frac{108}{600} = 0{\cdot}18 \ .$$

Next, the terms representing potential labor force, analogous to $M^k_{j,t}$, may be calculated, by substituting these values into equation (2.25), as before.

(2) All of these assumptions are examined in some detail in later chapters.

This yields

$$M_{1,t} = 119 \cdot 97\ , \qquad M_{2,t} = 125 \cdot 57\ , \qquad M_{3,t} = 103 \cdot 42\ ,$$
$$M_{4,t} = 130 \cdot 55\ , \qquad M_{5,t} = 120 \cdot 55\ .$$

These values, together with the other necessary quantities, may now be substituted into equation (2.22), which yields

$$\Delta E^k_{1,t} = 0 \cdot 3(119 \cdot 97)(0 \cdot 18) + 0 \cdot 7(50)(0 \cdot 3) = 6 \cdot 48 + 10 \cdot 5 = 17\ ,$$
$$\Delta E^k_{2,t} = 0 \cdot 3(125 \cdot 57)(0 \cdot 18) + 0 \cdot 7(120)(0 \cdot 3) = 6 \cdot 78 + 25 \cdot 2 = 32\ ,$$
$$\Delta E^k_{3,t} = 0 \cdot 3(103 \cdot 42)(0 \cdot 18) + 0 \cdot 7(190)(0 \cdot 3) = 5 \cdot 58 + 39 \cdot 9 = 45\ ,$$
$$\Delta E^k_{4,t} = 0 \cdot 3(130 \cdot 55)(0 \cdot 18) + 0 \cdot 7(0)(0 \cdot 3) = 7 \cdot 05 + 0 \cdot 00 = 7\ ,$$
$$\Delta E^k_{5,t} = 0 \cdot 3(120 \cdot 55)(0 \cdot 18) + 0 \cdot 7(0)(0 \cdot 3) = 6 \cdot 51 + 0 \cdot 00 = 7\ .$$

Thus the same employment model with somewhat different parameters has located forecast increments of NLMS employment principally in the zones where NLMS employment has located in the past, but with some 'suburbanization' of basic employment in response to the spatial distribution of the potential labor force.

Although the residence-location and employment-location (both for LOMS and for NLMS employment) models will be expanded upon and discussed in much more detail in subsequent chapters, their presentation here does serve to describe their essential characteristics. They have been introduced in the form of these numerical examples in order to allow readers who may have been unfamiliar with these types of models to trace through their workings easily. Having done this, the reader will then find that the further descriptions of the more sophisticated forms of these models and their applications are much more readily understood. In the next section of this chapter, a similar simplified numerical example will be described for a transportation network assignment model.

2.4 Network assignment model: illustrative example

2.4.1 The assignment problem: a simple example

In the numerical examples of residence-location and employment-location models presented in sections 2.2 and 2.3, one of the principal inputs was a zone-to-zone travel-cost matrix. In actual application, the data in this matrix result from a number of rather complex processes. In a typical situation these processes might begin with a procedure for generating trips. The result of this procedure would be two (or several sets of two each) vectors; one of trips originating from each zone, and one of trips terminating in each zone. According to the nature of the application, these trip types may or may not include work trips, shopping trips, and any of a number of other possible categories. The next procedure would be the use of an algorithm to generate the origin–destination (OD) trip matrix (or matrices). The OD trip matrix contains information of the destinations, by zone, of all the trips originating in each zone. A representation of a transportation network would then be introduced, and the trips in the OD trip matrix would be *assigned* to the network. This assignment would involve the use of information about the characteristics of the network, along with a set of decision rules, and an algorithm for doing the assignment in an efficient fashion. On the completion of the assignment procedure, times and costs of travel through the network will have changed as a result of congestion. It will then be possible to calculate the zone-to-zone least-time or least-cost paths. The results of this calculation yield the travel-cost matrix which is later input to the employment-location and residence-location models.

In the following discussion, the problems of trip generation and distribution will be temporarily put aside. The discussion and its included numerical example will deal exclusively with the assignment problem. To begin with, assume the existence of the OD trip matrix, given in table 2.8. First, note that the row sums in this trip matrix equal the population forecasts which were calculated in the residence-location model example using vacant developable land as an attractiveness measure and the gamma function as a trip function. Second, note that the column sums in this trip matrix equal the zonal employment figures which were used as input to the residence-location calculations. This trip matrix could therefore have originated from the results of that residence-location model example. Such a trip matrix would be a home-to-work trip matrix, with each row sum equal to the number of trips originating in the zone represented by that row (for example in table 2.8, 117 trips originate from zone 1). The column sums are equal to the numbers of trips terminating in their respective zones (for example, in table 2.8, 100 trips terminate in zone 1, and 0 trips terminate in zone 4). Before continuing it must be noted that this home-to-work trip matrix could have resulted from the input data for the residence model, but actually it did not. A simple proportioning scheme was used to produce the matrix elements so they would be consistent with the row and column totals. A more realistic procedure for trip generation and distribution, incorporating the travel costs, among

Table 2.8. Origin-destination trip matrix for example of network assignment.

Origin	Destination					Σ
	1	2	3	4	5	
1	19	38	60	0	0	117
2	20	41	61	0	0	122
3	15	30	44	0	0	89
4	22	44	65	0	0	131
5	24	47	70	0	0	141
Σ	100	200	300	0	0	600

other factors, will be discussed later in this book. However, for the purposes of this example this rather arbitrarily produced trip matrix will be quite adequate.

We now turn to the description of the transportation network. Figure 2.3 shows the assumed transportation network for the five zone region used in the preceding numerical examples. It will be assumed in this discussion that the transportation system is unimodal, and for the sake of convenience it will be assumed to be a highway system. In this network there are five points, or load nodes, where traffic is presumed to enter or leave the network. One load node is used for each zone. It is possible to have other nodes, where links of the network intersect but where no trips enter or leave the network. For the sake of simplicity, no 'nonload' nodes of this type are included in this problem. Each connection between any pair of nodes is a link. Traffic can flow in either direction over the link, and the travel time and/or cost is the same for each link, regardless of the direction of travel. Each link is referred to by its origin and destination nodes; thus link 5-2 is the connection from node 5 to node 2. In these terms, then, there are eighteen links in this network. The links are listed in table 2.9 by increasing origin node number, and for each origin node, by increasing destination node number. The travel costs for each link are also shown in this table. These travel costs are of the same order of magnitude as those used in the employment-location and residence-location examples, but no attempt has been made to have them correspond exactly to those examples.

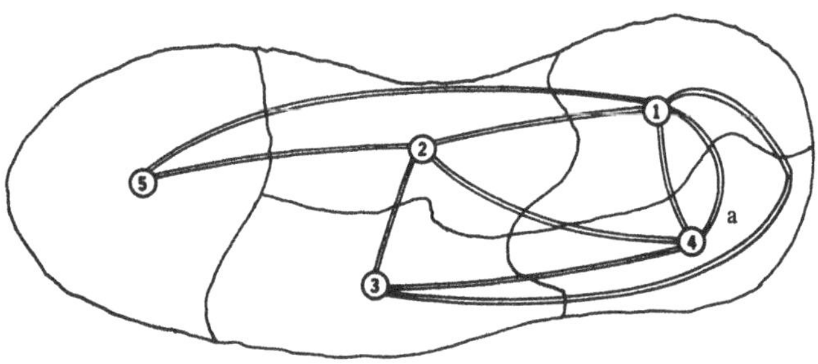

Figure 2.3. Five zone, eighteen link network example.

Table 2.9. Travel costs for the links shown in figure 2.3.

Link	Cost	Link	Cost	Link	Cost
1-2	1·5	2-3	1·5	4-1	2·0
1-3	3·0	2-4	2·0	4-1a	1·5
1-4	2·0	2-5	2·0	4-2	2·0
1-4a	1·5	3-1	3·0	4-3	2·0
1-5	2·5	3-2	1·5	5-1	2·5
2-1	1·5	3-4	2·0	5-2	2·0

The assignment problem is to take the trips from the matrix shown in table 2.8 and to assign them to (that is, distribute them among, or load them onto) the links of the network. In this simplest illustration of the problem, the only important assumption is that each tripmaker will choose the 'shortest' path through the network from origin to destination. The shortest path is defined in this example as that of least travel cost. More complex assumptions and/or constraints will be introduced in subsequent discussion. Also, to keep the example simple no intrazonal links are defined. Intrazonal trips, such as the nineteen originating in zone 1 and terminating in zone 1, will therefore not be assigned, and need not be considered further in this example.

There were thirty-eight trips originating in zone 1 and terminating in zone 2. An examination of the network in figure 2.3 and the link table in table 2.9 indicates that the 'shortest' path from zone 1 to zone 2 is via the direct link 1-2. Thus all thirty-eight trips from zone 1 to zone 2 will travel on link 1-2. Sixty trips go from zone 1 to zone 3. Note there are several alternative paths between these two zones: (i) link 1-2, link 2-3; (ii) link 1-4, link 4-3; (iii) link 1-4a, link 4-3; (iv) link 1-3. There are other paths as well, link 1-5, link 5-2, link 2-3, for example, but these are obviously too long for consideration here. There are two paths of equal length from zone 1 to zone 3. Path (i) and path (iv) both have travel costs of 3·0, which raises the general question of how to deal with more than one path with equal lengths. In this case, it will be assumed that each path of equal length will carry an equal share of the trips. Thus link 1-2 and link 2-3 [path (i)] each get thirty trips, and link 1-3 [path (iv)] will also get thirty trips.

Now that the trips from origin (zone) 1 have been assigned, those from origin 2 can be analyzed. There are twenty trips from zone 2 to zone 1; these will all go on link 2-1. The forty-one trips from zone 2 to zone 2 are not considered here. Finally, there are sixty-one trips from zone 2 to zone 3. Examination of the network makes it quite clear that all of these trips will go on link 2-3. If this analysis is continued in this fashion, the trips will be loaded on the network links as shown in table 2.10.

In table 2.10 the trips (all except the intrazonal trips) in the system are shown in terms of volumes of trips flowing on each of the eighteen network links. Notice the considerable variation in trips per link, and the obvious inefficiencies implied by this 'all-or-nothing' least-cost (minimum-path) loading procedure. These inefficiencies lead one to consider adding a bit more sophistication to this assignment procedure. One very desirable improvement involves adding a capacity-restraint or capacity-constraint calculation.

Before moving to a discussion of capacity-constrained network assignment, a final set of calculations based on the first simple network example should be done. This last step involves the transformation of the information on flows on links and link travel costs into a zone-to-zone travel-cost matrix for use as input to employment-location and/or residence-location calculations. In this particular example, the zone-to-zone travel-cost matrix could have been calculated before the trip-assignment procedure, as the flows on the network which resulted from the assignment procedure had no effect on travel time or travel cost on the network. In practice this will not be the case. After the capacity-constraint mechanism is introduced in the next section of this discussion, travel flows (trips) loaded on (assigned to) the network will affect the travel times and/or costs of travel over the network. This fact is the basis of the connection of land use (location) back to the transport system, which is one of the essential ingredients of the integrated model packages discussed in later chapters.

The transformation mentioned above involves the use of a technique which in transportation planning parlance is referred to as 'tree tracing'. This involves a systematic examination of the paths from each origin node to each destination node in an effort to find the shortest path through the network between each possible origin–destination pair. The set of all such minimum paths from a particular origin to all possible destinations is referred to as a 'tree'. The name refers to the image of these paths: one link common to all or most paths from the origin, with an ever-increasing number of links as the tree reaches the various destinations.

Now, for the network illustration in figure 2.3, the tree from origin 1 can be traced. Zone 1 to zone 1 is not defined in this network, though it should be noted that in transport planning practice there are often intrazonal trips and links to be considered. From zone 1 to zone 2, the shortest path is via link 1–2. From zone 1 to zone 3, the shortest path is via link 1–3 or links 1–2 plus 2–3. The shortest path from zone 1 to zone 4 is via link 1–4a, and link 1–5 is the shortest path from zone 1 to zone 5. Thus the tree from zone 1, albeit a simple one, is shown in figure 2.4(a). The trees from the remaining origin zones may also be

Table 2.10. Flows on links shown in figure 2.3 (unconstrained example).

Link	Number of trips [a]	Link	Number of trips [a]
1-2	68 (38_{12}, 30_{13})	3-1	8 (8_{31})
1-3	30 (30_{13})	3-2	37 (7_{31}, 30_{32})
1-4	0	3-4	0
1-4a	0	4-1	0
1-5	0	4-1a	22 (22_{41})
2-1	27 (20_{21}, 7_{31})	4-2	44 (44_{42})
2-3	161 (30_{13}, 61_{23}, 70_{53})	4-3	65 (65_{43})
2-4	0	5-1	24 (24_{51})
2-5	0	5-2	117 (47_{52}, 70_{53})

[a] Numbers in brackets are portions of the total numbers of trips arising from the OD matrix elements given in table 2.8; subscripts are origin zone and destination zone, respectively.

traced by inspection of this simple network. Figures 2.4(a)–2.4(e) show all the 'trees' from these calculations.

Having traced the trees, it is a simple matter to construct the zone-to-zone travel-cost matrix for this network. For origin zone 1, the c_{11} element of the matrix is, for the present, undefined. From the tree for origin zone 1, we can see that the c_{12} element is simply the travel cost of link 1–2 (from table 2.9, 1·5 units). The c_{13} element is the cost over either link 1–3, or links 1–2 plus 2–3, which equals 3·0 units in both cases. This process is continued for all the off-diagonal elements in the travel-cost matrix, to yield the matrix given in table 2.11. For the sake of completeness here, the diagonal elements of this matrix might arbitrarily be defined to be 1·0, but this is not essential for the next part of the example. This matrix is now in a proper form to be used as the zone-to-zone travel-cost input to an employment-location and/or residence-location model. We may now turn to the capacity-constraint modifications.

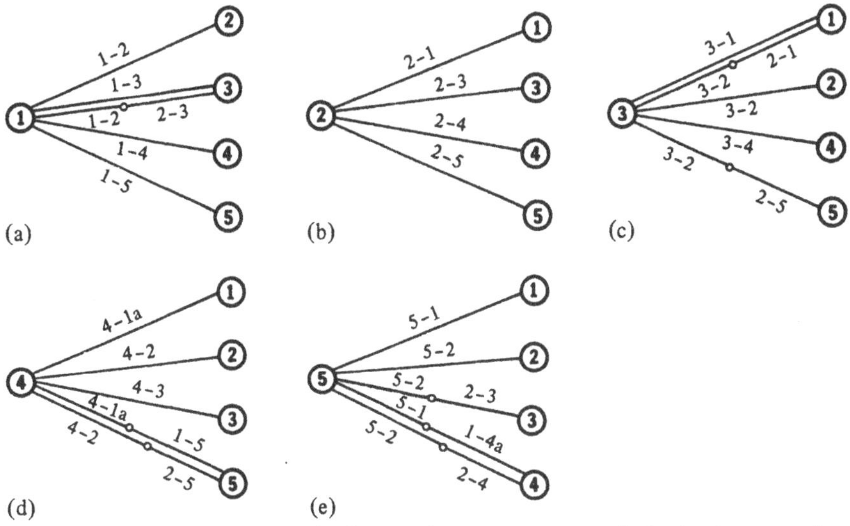

Figure 2.4. (a)–(e) Shortest paths from each origin zone to each destination zone, for the network shown in figure 2.3 (unconstrained example).

Table 2.11. Zone-to-zone travel-cost matrix resulting from network assignment (unconstrained example).

Zone	1	2	3	4	5
1	–	1·5	3·0	1·5	2·5
2	1·5	–	1·5	2·0	2·0
3	3·0	1·5	–	2·0	3·5
4	2·0	2·0	2·0	–	4·0
5	2·5	2·0	3·5	4·0	–

2.4.2 The assignment problem: a capacity-constraint modification

The idea of capacity constraint results from the simple observable fact of traffic congestion. As the volume of traffic attempting to use a particular link increases, the time (or cost) to traverse the link increases as well. There has been a fair amount of research into the nature of the capacity-constraint, or volume-delay, function (Branston, 1976). A typical capacity-constraint function will be a nonlinear, increasing function of link flows and link capacity. Frequently these functions will have some arbitrary upper limit. A function which has been used in transportation planning in the USA for some years is

$$t_1 = \begin{Bmatrix} t_0\left[1\cdot 0+0\cdot 15\left(\dfrac{v}{v_0}\right)^4\right] & (v > v_0) \\ t_0 & (v \leqslant v_0) \end{Bmatrix}, \qquad (2.30)$$

where t_1 is the modified (congested) link travel time, t_0 the uncongested (free flow) link travel time, v the volume of trips on the link, and v_0 the design capacity (volume of trips) of the link.

Thus, for a fifteen-minute (free flow) link with a design capacity of 500 trips per hour, at any traffic volume less than 500 trips per hour the link travel time would be equal to fifteen minutes. At 600 trips per hour, the link travel time will have increased to just under twenty minutes; and at 900 trips per hour, the link travel time will have increased to almost forty minutes. A simple plot of this function is shown in figure 2.5.

The next question is how such a function can be added to the simple assignment procedure described above. In fact there are quite a number of ways in which a capacity-constrained network assignment procedure can be organized. The simplest of these would involve merely recomputing the travel times or costs on the network links after all trips had been assigned. The difficulty with such a procedure is that the resulting network times or costs would show some network links to be grossly overloaded (congested) while other, perhaps even parallel, links would be

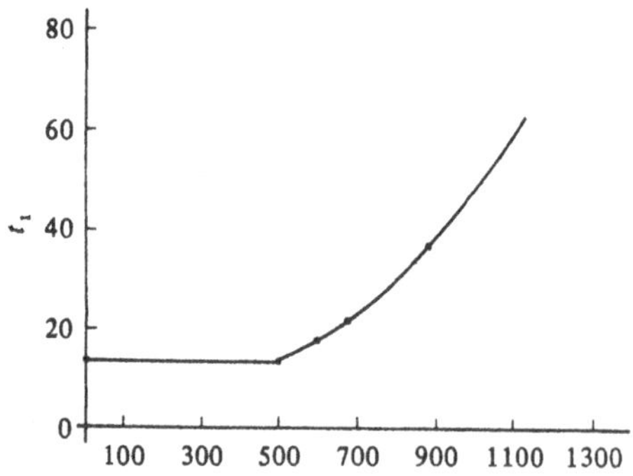

Figure 2.5. Plot of link time, t_1, versus link volume, v, for equation (2.30).

completely unused. To remedy this an iterative procedure might be used in which after all the trips had been assigned and the network congestion levels calculated all the trips could be reassigned to the congested network links. The resulting different pattern of network flows could then be used to recalculate the congestion levels. In practice this procedure often seems to lead to unstable, oscillatory solutions which leaves the transport analyst in the position of having to select from several different, but all inadequate, network flow representations.

Another possibility involves calculating the congestion level of each link immediately after any new trips are loaded onto it. By this technique, each succeeding tree trace (or minimum-path search) would face a successively more congested network. This approach would significantly improve the realism of the network loadings, but suffers from two further defects. First, the overall results could be biased by always beginning the tree tracing with the same origin node. This problem may easily be remedied by random origin selection. The second problem is that for any chosen initial origin, the last tree to be traced will be over an almost fully loaded (congested) network, whereas the first tree is traced over an empty (uncongested) network. This problem can be remedied by an incremental loading process. All of these notions will be discussed more extensively in chapter 5. For the moment, just one of these possible modifications to the simple assignment procedure will be illustrated with the continuation of the numerical example.

In this example, capacity constraint will be added to the network assignment procedure. The other possibilities of incremental assignment and random selection of origin nodes will be put aside for the moment. The capacity-constraint function shown in equation (2.30) will be used in this example. Capacity constraint (congestion) will be calculated immediately after any new trips are loaded onto a link. Thus any subsequent trips will 'see' a congested link during the tree tracing. Finally, the zone-to-zone travel-cost matrix will be calculated from the characteristics of the loaded, capacity-constrained network (travel time and travel cost have been used interchangeably in this discussion and will continue to be so used).

It is first necessary to establish capacities for each of the network links. For this example, all but six of the links will be assumed to have a capacity of thirty vehicles per unit time. The following links will be assumed to have a capacity of fifty vehicles per unit time:

link 1-5,	link 5-1,
link 1-4a,	link 4-1a,
link 1-3,	link 3-1.

Again, we begin with zone 1. The intrazonal trips (zone 1 to zone 1) are not considered. The thirty-eight trips from zone 1 to zone 2 will travel on link 1-2. As this exceeds the link's free flow capacity of thirty, the travel cost will be increased via equation (2.30), as follows

$$t_1 = 1\cdot 5\left[1\cdot 0 + 0\cdot 15\left(\frac{38}{30}\right)^4\right]$$
$$= 2\cdot 08 \approx 2\cdot 1\ .$$

Thus the next trips to examine link 1–2 as a possible link in a path will discover that its currently congested travel cost is 2·1 units.

There are sixty trips from zone 1 to zone 3. The shortest path is via link 1–3. Note here that in the unconstrained (no capacity constraint) example, there were two paths of equal length: path (i) over link 1–3, and path (ii) over links 1–2 and 2–3. Now, however, as link 1–2 has been congested by the trips from zone 1 to zone 2, there is only one shortest path from zone 1 to zone 3. Sixty trips then are assigned to link 1–3. The capacity of link 1–3 is fifty, so again the effects of congestion must be calculated, as

$$t_1 = 3\cdot 0\left[1\cdot 0 + 0\cdot 15\left(\frac{60}{50}\right)^4\right]$$
$$= 3\cdot 93 \approx 4\cdot 0\ .$$

This is the congested travel cost on link 1–3, and is the cost that will be seen by succeeding trip-assignment calculations. This process may then be continued for all trips, and will yield the flows listed by link in table 2.12. These results are quite different from the unconstrained results given in table 2.10. For example, owing to the existing congestion, trips from

Table 2.12. Flows on links shown in figure 2.3 (capacity-constrained example).

Link	v_0	t_0	Number of trips [a]	t_1
1-2	30	1·5	38 (38_{12})	2·1
1-3	50	3·0	130 (60_{13}, 70_{53})	23·6
1-4	30	2·0	0	
1-4a	50	1·5	0	
1-5	50	2·5	0	
2-1	30	1·5	27 (20_{21}, 7_{31})	1·5
2-3	30	1·5	61 (61_{23})	5·3
2-4	30	2·0	0	
2-5	30	2·0	0	
3-1	50	3·0	8 (8_{31})	3·0
3-2	30	1·5	37 (7_{31}, 30_{32})	2·0
3-4	30	2·0	0	
4-1	30	2·0	0	
4-1a	50	1·5	22 (22_{41})	1·5
4-2	30	2·0	44 (44_{42})	3·4
4-3	30	2·0	65 (65_{43})	8·6
5-1	50	2·5	94 (24_{51}, 70_{53})	7·2
5-2	30	2·0	47 (47_{52})	3·8

[a] See table 2.10.

zone 5 to zone 3 which formerly went via link 5–2 and then link 2–3 are now forced to detour all the way around the region via links 5–1 and 1–3. Even so, substantial congestion builds up on link 1–3. Even this procedure leads to problems, as the congested travel cost of 23·6 on link 1–3 is obviously excessive. Before dealing with this, the minimum-path trees for this loading of the network, with capacity constraint, can be calculated. These trees are shown in figure 2.6. The resulting zone-to-zone travel-cost matrix for the capacity-constrained network is rather different from that for the unconstrained network, and is given in table 2.13. Note that this travel-cost matrix is no longer symmetric. The asymmetry results from the differences in congestion encountered, depending upon the direction of travel over specific links. For example, the travel cost on link 2–3 is 5·1, whereas on link 3–2 it is 2.0. These differences are due to variations of flow with direction along each particular network link.

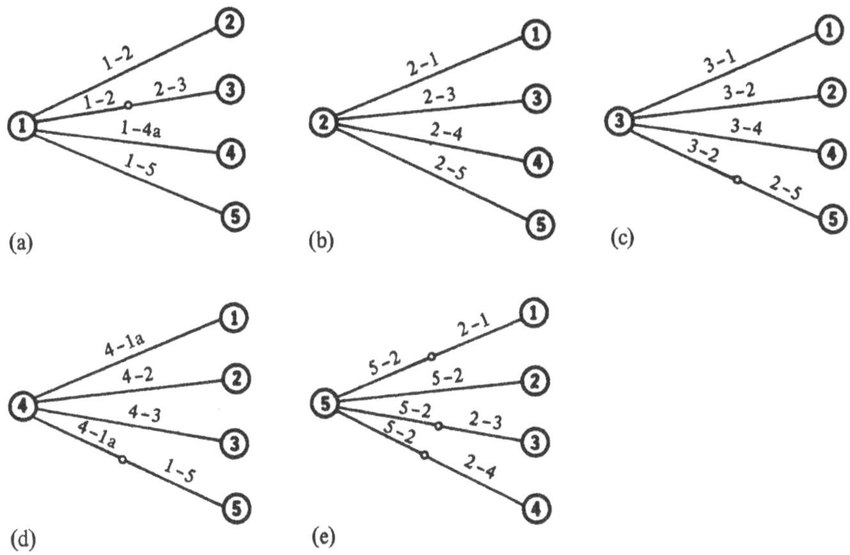

Figure 2.6. (a)–(e) Shortest paths from each origin zone to each destination zone, for the networks shown in figure 2.3 (capacity-constrained example).

Table 2.13. Zone-to-zone travel-cost matrix resulting from network assignment (capacity-constrained example).

Zone	1	2	3	4	5
1	–	2·1	7·4	1·5	2·5
2	1·5	–	5·3	2·0	2·0
3	3·0	1·7	–	2·0	4·0
4	1·5	3·4	8·6	–	4·0
5	5·3	3·8	9·1	5·8	–

Although the introduction of capacity constraint in this fashion does result in improved realism of network loadings, there are still obvious difficulties. Several additional modifications and improvements were made in the assignment procedure before it was actually utilized for the policy and sensitivity analyses described in later chapters. These improvements and modification will be discussed in those chapters in the context of the development and application of the model package. The principles, however, remain unchanged. As was true for the residence-location and employment-location examples given in sections 2.2 and 2.3, if the reader takes the time to understand the examples of network assignment given in this section, then the more sophisticated procedures to be described in the following chapters will be readily understood.

As has been mentioned before, one of the main goals of the work described here has been the integration of transportation and location analyses. In this chapter, simple numerical examples of the three major components of such integrated analyses have been presented; the next chapter will describe methods for their integration and the development and testing of the first practical computer program package designed to accomplish such integrated analyses.

3

Development of the first integrated transportation and land-use package

3.1 Introduction

In March 1971, the United States Department of Transportation issued a "Request for proposal" for a research study which was to be titled "Interrelationships of transportation development and land development" (USDoT, 1971). The standard forms of competitive bidding (proposal submission) and evaluation procedures were followed. The contract to undertake the study was awarded to this author, and work began on the project in November 1971.

One of the principal reasons for the USDoT's desire for this study was an increasing concern with what was viewed as 'premature obsolescence' of urban highways. In particular, this concern was focused on the all too common problem of an urban freeway or expressway which may have been designed in 1960 and constructed in 1965 with the aim of adequately serving a particular part of a metropolitan region until, say, 1990, but which was overcongested and a cause of public annoyance by 1970.

It was evident at the outset of the research that the question of premature obsolescence was but a subproblem of the more general question of urban spatial dynamics and transportation–location interactions. Thus the proposed method of analysis was oriented around the development of an integrated set (program package) of 'land-use' (location) and transportation models. The emphasis of the project was on the integration of these models and their resultant capability for analyzing location–transportation interactions, rather than on the development of any particular new model or models. Consequently it was decided to make use of existing models as far as possible, rather than to attempt the construction of any new ones. The research project was completed in June 1973, and successfully demonstrated the first integrated package of transportation and land-use models, incorporating the effects of transportation on land use, and the feedback effects of land use on transportation. This chapter describes the development, substance, and testing of this first version of the integrated package. Subsequent chapters will deal with later improvements and modifications, and with further testing and results of policy analysis from the later versions of the package.

3.2 Background

One of the most clearly visible of public investments in recent decades has been the enormous expenditure on new highway construction. This same period has also witnessed large-scale shifts in population, both from rural areas to urban areas, and within the urban areas from city center to suburb. Although it has proven difficult to establish clear, complete, and explicit quantitative definitions of the causal relationships between these developments, it is clear that these relationships exist. One visible example of the results of these relationships is observed when the construction of a new section of roadway is followed, and all too soon, by very heavy usage and subsequent congestion.

The development of the first version of the integrated transportation and land-use model package (ITLUP) which is described in this chapter is based, generally, on the following hypothetical description of this process. Specifically, the sequence that this process seems to follow is: (1) Owing to the perceived inadequacy (that is, congestion or inaccessibility) of existing facilities, a decision is taken in a metropolitan region to improve transportation facilities (usually by road building) in a particular area of the region. Then, if the decision is approved, (2) in anticipation of the new roadway, land developers and/or speculators become involved with real estate properties near the proposed route and/or at its termination. If the funding is available, and construction of the facility begins, homeowners and businesses as well may consider and act upon location or relocation decisions, and this results in further land development along the proposed and developing route. Upon completion of the facility, numerous additional similar location decisions are often made. Finally, a relatively short time after completion of the facility, the demand for its use greatly exceeds the forecasts of demand that were made before the decision to construct it. Consequently, (3) the design capacities are soon reached and often exceeded, and congestion and premature obsolescence of the facility is the result.

This is, of course, a rather general and somewhat loose description of a very complex process; nonetheless it is reasonably accurate overall. The principal question that the USDoT wished to have addressed in the first phases of this research was whether it was feasible, through integrated transportation planning and land-use planning, to avoid or ameliorate the results of the processes described above, or of similar sequences of phenomena in the future. Furthermore, it was the intention of those initial research efforts to analyze these processes and to determine (a) whether balanced development of transportation facilities and land use might be feasible; and (b) if it is feasible, what means are available to accomplish it or, at least, to make it more likely to take place.

Considerable evidence is available indicating that the demand for highway travel is rather sensitive to changes in highway capacity. This sensitivity, if considered in terms of the sequence of phenomena described above, frequently results in heavy utilization and congestion of new facilities soon after their construction. One argument holds that the solution to this problem in the future, as in the past, is to construct more transportation facilities. It is possible that if this policy were followed, at some point, an equilibrium situation would indeed be reached. This conclusion can be supported by the assertion that the elasticity of demand for highway travel is finite. If, however, the 'population' generating that demand is continually increasing at the same time, it is not clear that the total demand can continue to be satisfied in this manner. It is obvious that this unbounded situation is not the desired solution. It is to be expected that finite limits on public expenditure, coupled with some vestige of societal sanity (aided perhaps by worldwide increases in energy costs), would preclude ever-increasing road construction in any case. Yet, rather than rely solely on these exogenous constraints, it is clearly necessary to analyze possible intermediate solutions.

A balance between transportation facility development and land development implies a market equilibrium between the demand for use of a transportation facility and its characteristics (cost, speed, and capacity) as a source of supply. There are two basic strategies available to the planner who wishes to modify an existing situation of transportation and land use in an attempt to achieve such a balance. The first strategic possibility is to *allow demand for transport to fluctuate freely*, with no interference, and to attempt to cope with it. This would be accomplished, as in the past, by new highway construction or (hopefully with more success than has been achieved previously) by implementation of mass transportation systems. The second strategy alternative is to *attempt to restrict the demand for transport* so that facilities do not become overloaded. This might be accomplished in several ways, for example:

1. existing transportation facilities could be made more costly to use, perhaps by the imposition of tolls or by allowing congestion to develop and thus impose a time penalty on users;
2. land development controls could be imposed, thus reducing (or slowing the growth of) trip-generating activities.

Of course, a mixture of these two actions could be implemented. A mixture of the two basic alternatives might also be considered, where an attempt would be made to cope with a certain amount of transportation demand (by improving transportation facilities) at the same time that an attempt was being made to control its further increase. In general, the policies or programs with the greatest chance of success will be mixtures of these strategies and, within strategies, will also involve mixtures of tactics.

The view of the USDoT (1971), when requesting the research proposals, was that:

> "the following issues have been of concern to transportation planning agencies:
>
> 1. In urban areas, it is necessary to achieve a balance between the need for mobility and the preservation of environmental quality and social stability. How can this process be incorporated into the transportation/land use planning program beginning with plan development?
> 2. To what extent can transportation system planning and facility capacity planning be used to influence land development? What are the available controls for accomplishing this, and how can they be applied effectively?

3. Providing too high a level of service of the auto-highway system in urban areas might result in travel demand exceeding the supply which can feasibly be provided. What should determine the minimum or maximum level of service (speed) which should be used as a criterion for planning transportation facilities in specific parts of the urban area?
4. What would be the consequence of controlling transportation facility capacities as a means of directing land-use development even though such controls could mean congestion with more pollution and higher travel costs?

This study should determine the feasibility of the balanced development of land use and transportation facilities. It should also determine whether or not land development and travel can be brought into a satisfactory balance such that any transportation improvements made can provide a continuing high level of service over time".

To summarize, the heart of the problem being addressed was a search to determine ways of controlling (that is, of altering, directing, and modifying) the spatial organization of the metropolis. In particular, the concern was with its spatial expansion and with the feasibility of balanced development of metropolitan transportation facilities and land. Finally, were it to appear that such balanced development was feasible, methods for testing the means (policies) which may have been relevant to the achievement of that balance were to be identified and examined.

Two important aspects of the problem were the principal determinants of the method of approach. First was the inherent complexity of comprehensive analyses of both transportation and land use. Second was the requirement for a self-consistent procedure for testing the sensitivity and response of transportation and land-use phenomena to policies for their manipulation. The only analysis technique that stood any chance of meeting these requirements was computer simulation modelling. The remainder of this chapter is devoted to describing the package of models which was ultimately developed and tested as the central focus of this initial phase of the research.

3.3 Design of the preliminary model package

The state of the art in simulation modelling of urban and regional phenomena at the time of the inception of the study posed a problem for projects of this sort. In general, those location models which had had operational success were rather too highly aggregated in areal and/or sectoral detail as well as in concept. On the other hand, those models which were more conceptually complete, and which might have been capable of providing more highly detailed forecasts, were not operational—many such models are still not operational in any practical sense. After some consideration of the project's goals, it was decided to begin with an existing *operational* land-use model and to attempt to modify and improve it during efforts to integrate it with a suitable transportation network model.

Although it was realized that this decision implied some sacrifice in the completeness of the model's theoretical underpinnings, as well as in the level of detail of its forecasts, this decision allowed the research on the project to begin rather promptly. As it turned out, the subsequent developments and successes of the overall research efforts did justify this decision.

The requirements of the project had suggested, from the outset, that an appropriate model package would need to be far more complete (and therefore more complex) than any then extant. What was ultimately required was a complete (residence and employment sectors) location or 'land-use' model and a complete (public and private) transportation network model. Furthermore, the overall system of models would need to be highly integrated with respect to the submodels' inputs, outputs, and feedback loops.

At the time that this research was begun, transportation planning and land-use planning had both evolved their own internally consistent, but not well-connected, forecasting techniques. In very general terms, a transportation study would take as given a future land-use pattern consisting, in part, of spatial distributions of residence and employment. The transportation planning task was to design a system of facilities that would adequately serve these spatially distributed activities. The design of these facilities would require that the spatial distributions of activities be used as input to procedures for estimating: (1) total numbers of trips (by several trip purposes) originating and terminating in each zone or subarea of the region, and (2) the origin–destination (OD) matrices of trips between each zone and each other zone. In some cases a mode-split analysis would be done as well. Finally, given the various sets of OD trip matrices, the trips would, via a traffic-assignment algorithm, be 'loaded' onto the transportation network. The loads on the network, congestion levels, and other results would then be analyzed to determine if the particular transportation system configuration being tested was adequate to serve the assumed (input) spatial patterns of activities. A block diagram of these processes is given in figure 3.1.

In very general terms, the forecasting procedure used in land-use planning may be said to begin with the predetermined specification of a transportation network. In addition to this, description of the spatial distributions of activities (residence and employment, for example) for a base year (starting point), and a set of regionwide population and economic forecasts are inputs to the forecasting procedure. Typically there will also be a set of policy inputs, such as land-use constraints. Given all these inputs a set of forecasts is generated, sometimes by means of some type of simulation model for urban land use. The outputs of this forecasting process are then evaluated both in terms of consistency and in comparison with existing regional goals and policies. A block diagram of these processes is given in figure 3.2.

The principal difficulty with these two traditional procedures is that they ignore certain rather critical feedback loops. From each of these phenomena (transportation and location) to the other there are important connections which these unintegrated planning processes overlook. The transportation planning process ignores the redistributive effects on the spatial distributions of activities which are induced by the transport system itself. The construction of any significant addition or modification to a region's transportation facilities inevitably results in some rearrangements of the population and economic activities of the region. To complicate matters further, major changes in transportation facilities (additions or modifications) usually become available for use in incremental stages over time. This implies a time stream of accumulating changes, both in the transportation system and in the spatial patterns of activities. Obviously the assumption of a fixed spatial pattern of activities as input to the transportation planning process ignores what may be very important changes in activity patterns which may be induced by the proposed changes in the transportation system themselves.

On a similar note, the land-use planning process ignores the congestive effects that altered spatial patterns of activities (and the consequent changes in OD trip patterns) have on the transportation system. Any significant change(s) in the spatial distribution of activities will result in changes in the origins and destinations of trips in the region. These changed

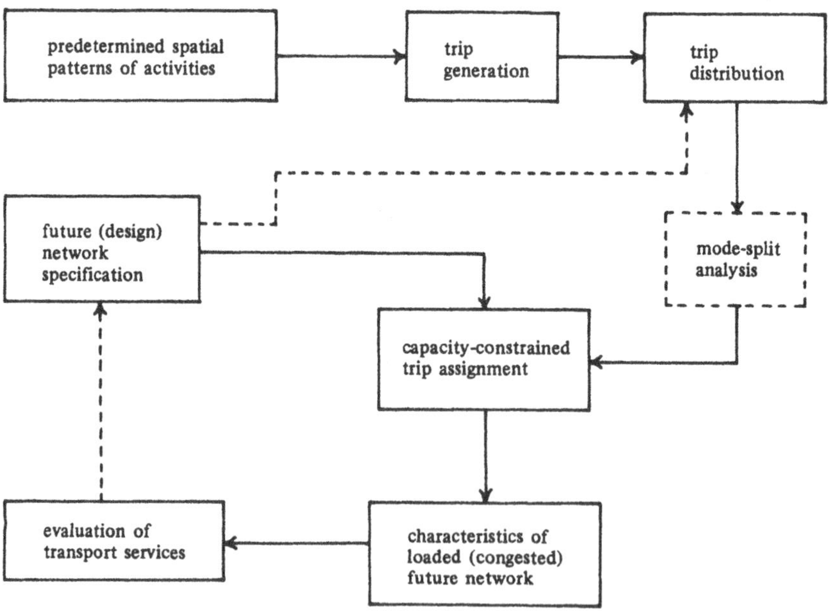

Figure 3.1. Conventional transportation planning procedures.

trip patterns will clearly result in changes in the volumes (loads) on the various links of the transportation network(s). The changed link volumes will, in turn, result in different congestion levels and consequently in differences in link travel times and costs. Thus the assumption of a description of the transportation system for a future (or forecast) year, with fixed zone-to-zone travel times and costs, ignores what may be very important changes in the characteristics of the transportation system which are induced by changes in the spatial distributions of activities themselves.

An integrated transportation and land-use planning method or procedure would attempt to ameliorate the omissions of the two separate planning processes, and to represent explicitly the connections and feedbacks from each of these processes to the other. The connections needed between figures 3.1 and 3.2 are quite obvious, and they, in turn suggest the means by which the forecasting techniques may be joined. The output of the transportation planning procedure—the characteristics of the loaded (congested) future network—should be used as the input to the land-use planning procedure. In point of fact, this connection does sometimes exist in the procedures of the more sophisticated planning agencies. The second necessary connection is the feedback connection, where the forecasts of spatial distributions of activities produced by the land-use planning procedure become, in turn, the input to a second iteration of the transportation planning procedures. This feedback connection is virtually never attempted in any of the conventional transportation or land-use planning processes. An integrated transportation and land-use planning procedure is shown, in the form of a block diagram, in figure 3.3.

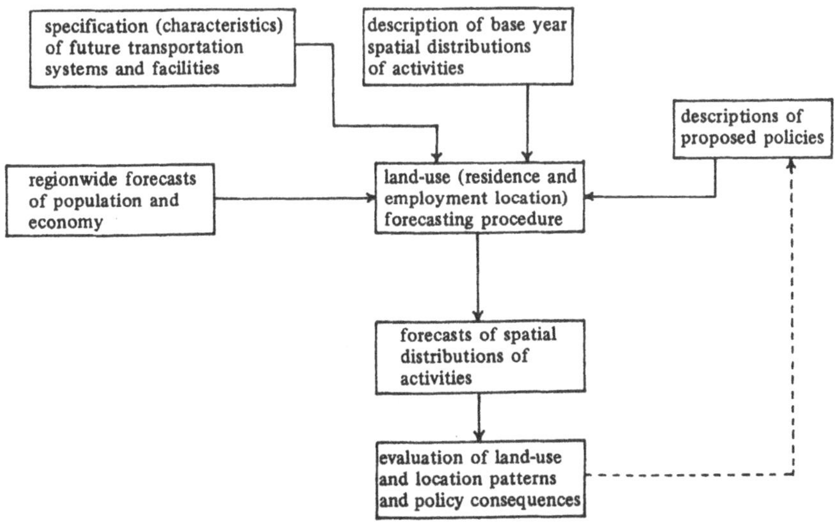

Figure 3.2. Conventional land-use planning procedure.

A set of integrated forecasting models might then be developed for use in the combined land-use and transportation planning process given in figure 3.3. A block diagram of this simulation model package is shown in figure 3.4; a single pass through this package can be described as follows. First, the various base year descriptions of the existing spatial distributions of activity, along with data on the characteristics of the unloaded base year transportation network, are input to the models. These data are then used to generate a preliminary, and probably inflated (owing to unrealistically low congestion levels) estimate of trips taking place in the metropolitan area. Given this preliminary estimate of metropolitan trips, it becomes possible (by use of a capacity-constrained assignment procedure) to load the future (projection) year network so that its characteristics (that is, travel time and cost) reflect the traffic volumes which would be on the network if there had been no change in the spatial distribution of activities from the base year. These characteristics of the partially congested network, along with the base year spatial distribution data and the forecast regionwide control totals, are then used to generate a trial estimate of the spatial distributions of activities for the projection year. From these spatial distributions a new estimate of metropolitan trips will be produced. These would, in turn, be loaded (again by use of a capacity-constrained assignment procedure) on the projection year transportation network. The modified characteristics of the more

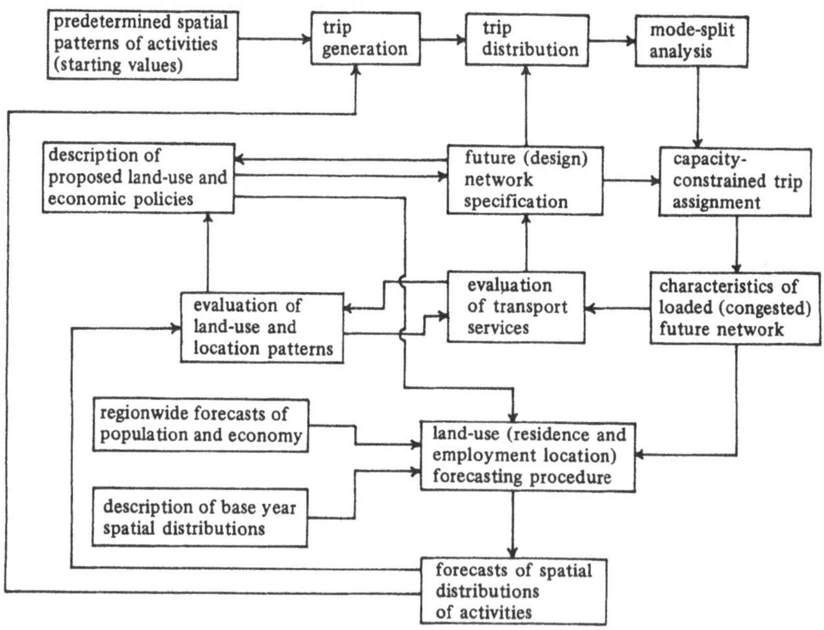

Figure 3.3. Integrated land-use and transportation planning procedure.

realistically congested transportation networks can then be used to reallocate the projected spatial distribution of activities. This distribution of activities is then compared with the first trial estimate thereof. If there are no significant differences, an equilibrium has been reached and the model run is ended. If there are significant differences, new trips are generated and loaded on the networks, and further iterations are made. This integrated package of transportation and land-use (location) models thus attempts to eliminate the principal failing of contemporary land-use or transportation studies, by explicitly including feedback loops.

It is very often the case in complex simulation studies that the operational version of the model system is somewhat less comprehensive than the original conceptual form of the system. Although some simplifications

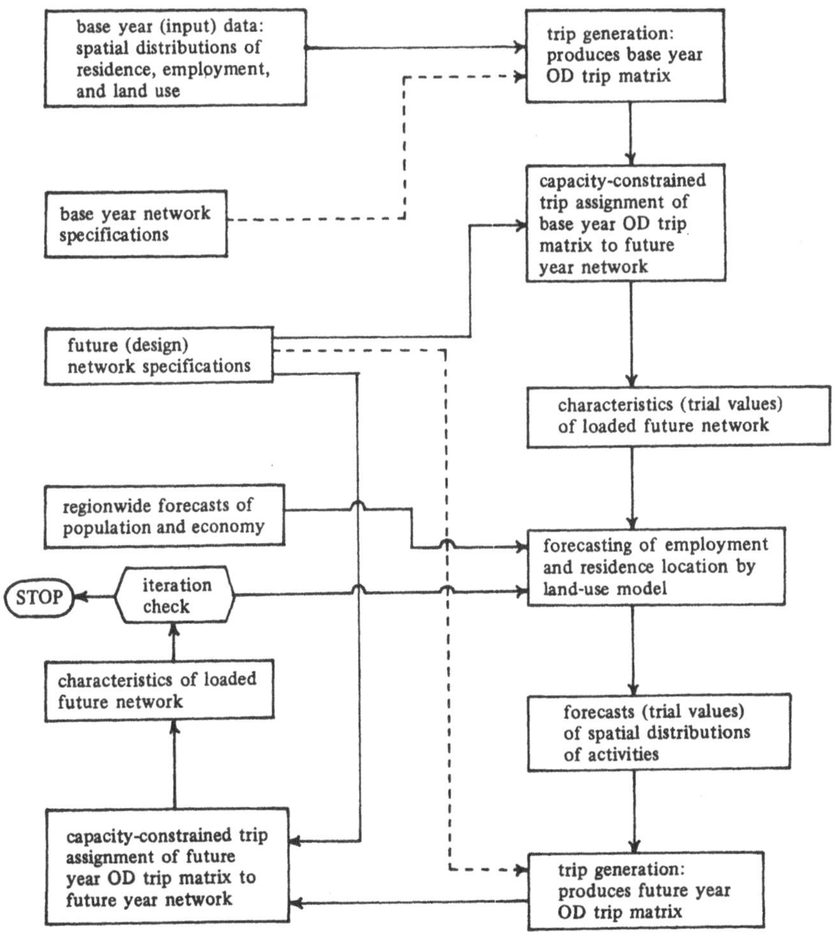

Figure 3.4. Integrated transportation and land-use model package.

were necessary in the first study, the principal goal of developing an operational prototype of an integrated transportation and land-use model was attained. Despite several shortcomings, which are discussed below, the land-use model that was selected for the operational integrated model package was the projective land-use model (PLUM). The PLUM model (and its improved variant, IPLUM) is a derivative of the Lowry model described in chapter 2[(3)]; it exists in several versions, two of which were used, with modifications, in this prototype model package. The PLUM models accept as given the spatial distribution of basic employment and then solve for the interrelated spatial distributions of population and nonbasic, or local-serving, employment. The principal determinants of these spatial distributions are access to workplaces for population and access to customers for nonbasic employment. The access measures are nonlinear functions of travel times weighted by attractiveness measures of the trip destinations, as illustrated in the simple examples of residence-location models in chapter 2.

There were several important implications of the decision to use a form of the PLUM model, which should be mentioned here. First, regional levels of activity would have to be exogenously determined. This is not a particularly unusual situation, as virtually all urban land-use models have this property. This requirement results from the fact that spatial distribution or spatial interaction models are not capable of estimating regionwide growth or decline in population and employment. Second, the selection of PLUM or, for that matter, any Lowry derivative model implied that basic employment location would have to be forecast exogenously too. In fact, in the sensitivity tests and policy analyses of the prototype integrated package, these forecasts did come from a separate model. An important implicit assumption of this procedure was that the spatial distributions of the exogenously forecasted basic employment activities were considered to be independent of the remainder of the activities in the model package (population, nonbasic employment, and the transportation networks). This assumption would clearly be invalid over any substantial projection period and was an aspect of the prototype model package which was remedied as soon as possible in subsequent improvement efforts. Ultimately this was done by integrating the modelling of all intra-metropolitan employment location, linking this to the residential model, and linking the outputs of these residence and employment models to the network model.

A further implication of the choice of the PLUM model was that the level of sectoral aggregation in the PLUM model (no population disaggregation) was rather more gross than was desirable, although this was tolerable in the preliminary development of the prototype integrated package.

(3) The PLUM model was originally described by Goldner et al (1972); see Putman (1979) for a review of the Lowry and PLUM models and their relationships one to the other.

Finally, the actual allocation function used in the various PLUM models had never been subjected to thorough statistical examination, and early experiences with model runs indicated that there was too heavy a reliance on somewhat arbitrary 'correction factors'. The algorithms used in the models were investigated thoroughly; certain anomalies were found in the procedures used for activity allocation, which were subsequently modified; this is described in greater detail later in this chapter. The modifications substantially improved the allocations made by the model, as will be described shortly.

On the positive side, available reports on the PLUM model did indicate that it seemed to have produced reasonable estimates of the general distribution of activities in a large urban region (the San Francisco Bay Area, California). Of course, as was the case with all other operational urban land-use (location) models, its reliability for specific small-area forecasts was considerably less than that for general regional patterns. This was actually not very different from the current state of affairs, nine years later. The model was, at the time of its selection, still being worked on and refined by its originators and it was expected that its performance would continue to improve as a result both of their efforts and of the development of the integrated transportation and land-use prototype model. Finally, a not unimportant part of the decision was the fact that computer programs and documentation for the PLUM model were available and operational at the Urban Simulation Laboratory in the Department of City and Regional Planning of the University of Pennsylvania before the initiation of the research on the integrated package which was to be performed there.

The transportation model package selected for use in the project had been developed several years earlier by the Planning Sciences Group at the University of Pennsylvania. This package contained programs for network coding, tree tracing, and several alternative capacity-constrained traffic-assignment procedures. The only deficiency of this network package as compared with others which were available was a lack of ability to handle highly detailed network specifications. This deficiency was not relevant to the development of the prototype integrated package, as there were no intentions of conducting any investigations at this level of detail. Moreover, the computer operation of the Planning Sciences Group network package was (and, for that matter, still is) significantly more efficient than that of the other network packages which might have been used instead.

The principal connection from the PLUM models to the network model was via trip generation. The main phenomenon, in reality as well as in models, that relates the spatial distribution of activities to the transportation facilities is specified in terms of the trips between activities. It should be noted that in many transportation studies where land-use models are being utilized, the work trips are generated by separate procedures (for example, multiple regressions). These trip matrices, because they are calculated independently of the land-use model, may not be consistent with the work trips implicit in the patterns of residence and workplace generated by the land-use model. This problem was eliminated by the procedure for integrating the transportation and land-use models as described below. The entire matter of the consistency of these trip estimates and the means by which they are extracted from the land-use models is of great importance and will be discussed at length later in this chapter. In brief summary, in the PLUM and IPLUM (a modification of PLUM) models the location of residences is generated by work-trip-allocation functions based on transportation impedances (derived from the characteristics of the transportation network) and the locations of workplaces. In so doing, the models implicitly generate a matrix of work-to-home trips. The work trips used in the prototype version of the integrated model package were extracted directly from these implicit work-trip matrices. It was, of course, then necessary to make certain adjustments to this set of work trips in order to assure that their total was equal to the exogenously estimated totals indicated in the data for the San Francisco Bay Area that were used for testing the model package. Those adjustments included mode-split trip ratios, auto occupancy rates, percentage of all work trips which occur during peak hours, and so on.

The location of nonbasic or local-serving employment was generated in the PLUM and IPLUM models by home-to-shop and work-to-shop allocation functions based on transportation impedances and the locations of the activities. As was the case for work trips, this procedure also generated an implicit matrix of shopping trips. These trips too were to be extracted from the models, although by a somewhat different procedure involving the use of a gravity-model-type algorithm for balancing the origins and destinations of the trips. Again, adjustments were necessary to ensure that the projected trip totals equalled the totals indicated in the San Francisco data.

In the prototype version of the integrated transportation and land-use model package, peak hour work trips and shopping trips were added together and loaded on (assigned to) the networks. Peak hour trips were used in order to obtain the closest approximation to work-trip travel. Work trips were presumed to be a key influence on residential location in these models, and most work trips take place during peak hours. Accordingly, the shopping trips used were only the percentage of all shopping trips which take place during the peak hour. A problem which was left to be resolved for later versions of the model package was that of other nonwork trips besides the shopping trips. In fact, very little further work was ever done on this problem, as it was never felt to be of prime importance. The problem of mass transportation facilities was also largely ignored during the development of the prototype package. There was, however, no reason to preclude its inclusion in the model package in future efforts, and thus adequate provision for such a later modification was made. In later work with the package, these modifications involved construction of a mode-split procedure and a procedure for dealing with multimodal networks in the spatial allocation functions for residence and employment.

Finally, several alternative procedures for loading the trips on the network were tested and are described later in this chapter and, in greater detail, in chapter 5. The procedure used, called "incremental tree-by-tree loading method", was a modification of one developed at the Chicago Area Transportation Study. This procedure is still followed in current versions of the integrated package. In brief, a single tree is first traced on a free flow (uncongested) network (that is, all the minimum paths from a given origin to all destinations). Onto these paths, a portion of the trips from that origin to all of the various destinations is then loaded. By use of a volume–capacity relationship, the new volumes on the links of the tree are used to calculate new travel times. Then a new origin is selected, a new tree is traced, and the process is repeated. When a tree has been traced for each origin, the network is loaded with a portion of all the trips. This complete procedure is then repeated. Currently, as in the case of the prototype, the networks are loaded in three passes of 40%, 40%, and 20% of the trips, respectively. After these network assignment procedures were completed, the matrix of travel times from each zone to each other zone via the minimum length paths through the network is calculated and prepared for subsequent input to the location (land-use) models.

Thus the iterative nature of the integrated package can be seen to consist of the passing of trial estimates back and forth between the land-use and transportation programs. Before the results of the various sensitivity tests and policy analyses which were performed with the prototype package are described, some of the points raised above will be discussed in greater detail, and the workings of portions of the prototype package will be illustrated with simple numerical examples like those presented in chapter 2.

3.4 Details and numerical examples

In order to describe more explicitly the functioning of the prototype of the integrated package, the main elements are described here in terms of small numerical examples. First, however, it is appropriate to describe the original versions of the PLUM model, before their modification for use in the prototype integrated package.

The basic structure of the PLUM model is that of the regionally constrained residence-location model described in chapter 2. Thus the general form of the residence portion of the PLUM model is

$$N_i = g\sum_j E_j p_{ij} \ , \tag{3.1}$$

where N_i is the number of persons living in zone i, E_j the number of employees working in zone j, p_{ij} the probability that a person who works in zone j will live in zone i, and g a scaling adjustment factor.

The probability p_{ij} is, of course, the heart of this equation, and may conveniently be considered as having two components, a trip function and an attractiveness measure. The trip function used in all the variants of PLUM was

$$p_c = \frac{\beta}{c^2} \exp\left(\alpha - \frac{\beta}{c}\right) , \tag{3.2}$$

where p_c is the probability of a trip of length (that is, travel time or travel cost) equal to c, and α, β are empirically derived parameters.

In the earliest versions of PLUM there was no attractiveness term. The attractiveness terms used in the later versions will be described after the discussion of the trip function is completed.

Leaving aside, for the moment, the question of how the values of α and β are obtained, the manner in which the trip function of equation (3.2) was applied is of some interest. First, the probability function was broken into intervals of three minutes of travel time each. Next, all zones were grouped into three-minute intervals of time (for example, 0–3 minutes, 3–6 minutes, 6–9 minutes, and so on) separation from each other zone. Then all zones were ranked into isochronal annuli around each other zone. Thus for each zone, taken as an origin zone, there was a vector which contained sets of the zone numbers of zones situated in each annulus around the origin zone.

If, for example, $\alpha = 0 \cdot 5$ and $\beta = 4 \cdot 0$, then the probabilities may be calculated for various travel times, c, by substitution into equation (3.2);

Table 3.1. Tabulated trip function as given by equation (3.2) with $\alpha = 0 \cdot 5$, $\beta = 4 \cdot 0$, in terms of travel 'time', c.

c	p_c	c	p_c	c	p_c
0·0	0·0000	2·5	0·2130	4·5	0·1339
0·5	0·0088	3·0	0·1932	5·0	0·1185
1·0	0·1208	3·5	0·1717	5·5	0·1053
1·5	0·2037	4·0	0·1516	6·0	0·0941
2·0	0·2231				

Table 3.2. Sample matrix of zone-to-zone travel times.

Zone	1	2	3	4	5
1	1·0	1·5	2·0	1·5	2·5
2	1·5	1·0	1·5	2·0	2·0
3	2·0	1·5	1·0	2·0	2·5
4	1·5	2·0	2·0	1·0	1·5
5	2·5	2·0	2·5	1·5	1·0

table 3.1 lists these probabilities in time steps of 0·5. Suppose also that the data from the residence-location numerical example of chapter 2 are used here, particularly the travel matrix (table 2.1). (This was presented as a 'cost' matrix in chapter 2, but can equally well be called 'time' here.) These data are shown in table 3.2. Now, suppose that the zones are to be considered in 0·5-unit intervals:

interval 1 $0 \cdot 0 < c \leqslant 0 \cdot 5$,
interval 2 $0 \cdot 5 < c \leqslant 1 \cdot 0$,
interval 3 $1 \cdot 0 < c \leqslant 1 \cdot 5$,
interval 4 $1 \cdot 5 < c \leqslant 2 \cdot 0$,
⋮ ⋮

With zone 1 as an origin, there are no zones in interval 1, zone 1 is in interval 2, zones 2 and 4 are in interval 3, zone 3 is in interval 4, and zone 5 is in interval 5. In this manner, a set of ranks of each zone from each other zone can be constructed, as shown in table 3.3. It is also possible to tabulate the numbers of zones in each isochronal annulus around each zone. This is done in table 3.4.

In the original PLUM allocation function, the travel probabilities were associated with travel to the annular rings. Specifically, the probability of a trip from any given origin zone to any given destination zone was set equal to the probability of travelling from the origin zone to the annulus in which the destination zone was located, divided by the number of zones in that annulus. Thus the probability of trips from zone 1 to a zone in annulus (interval) 3 is, from table 3.1, equal to 0·2037 (that is, the probability of a trip of length 1·5). Since, however, the probability of a

Table 3.3. Ranking of zones in time intervals for sample problem.

Origin zone	Interval					
	1	2	3	4	5	6
1	–	1	2,4	3	5	–
2	–	2	1,3	4,5	–	–
3	–	3	2	1,4	5	–
4	–	4	1,5	2,3	–	–
5	–	5	4	2	1,3	–

Table 3.4. Numbers of zones in isochronal annuli for sample problem.

Origin zone	Isochronal annulus					
	1	2	3	4	5	6
1	0	1	2	1	1	0
2	0	1	2	2	0	0
3	0	1	1	2	1	0
4	0	1	2	2	0	0
5	0	1	1	1	2	0

trip from origin i to destination j is also inversely proportional to the number of zones, K_j, in the annulus which contains zone j, we can write

$$p_{ij} = \frac{\beta}{K_j c_{ij}^2} \exp\left(\alpha - \frac{\beta}{c_{ij}}\right) . \tag{3.3}$$

There remains, however, the problem of annuli with no zones in them. This was handled in the original version of PLUM by accumulating the probability of travelling to any empty annulus into the next annulus containing one or more zones. In the numerical example, we note that all the interval 1 annuli are empty. Thus the probability of travelling to the interval 2 annuli is (0·0088 + 0·1208), or 0·1296. As there is only one zone in the interval 2 annulus of each origin zone, the probability of travelling to the zones in these annuli is 0·1296. For zone 1 considered as an origin zone, the interval 3 annulus contains two zones. Thus the probability of travel to either of these zones is 0·2037/2, or 0·10185. Thus a table of probabilities may be calculated, from each origin zone to zones in each annulus, as shown in table 3.5.

Finally, from table 3.5, a matrix of zone-to-zone travel probabilities may be constructed, as shown in table 3.6. There are some interesting peculiarities about this matrix. First, it is not symmetrical. The probability of a trip from zone 1 to zone 3 is 0·2231, but the probability of a trip from zone 3 to zone 1 is 0·1116. Yet there is no direction-related difference in the transportation network or the loads on that network between these zones (at least, none which entered these calculations).

Table 3.5. Probabilities of travel from each origin zone to zones in each isochronal annulus, for sample problem.

Origin zone	Isochronal annulus					
	1	2	3	4	5	6
1	0·0000	0·1296	0·1019	0·2231	0·2130	0·0000
2	0·0000	0·1296	0·1019	0·1116	0·0000	0·0000
3	0·0000	0·1296	0·2037	0·1116	0·2130	0·0000
4	0·0000	0·1296	0·1019	0·1116	0·0000	0·0000
5	0·0000	0·1296	0·2037	0·2231	0·1065	0·0000

Table 3.6. Zone-to-zone travel probabilities p_{ij}, for sample problem.

Zone	1	2	3	4	5	Σ
1	0·1296	0·1019	0·2231	0·1019	0·2130	0·7695
2	0·1019	0·1296	0·1019	0·1116	0·1116	0·5566
3	0·1116	0·2037	0·1296	0·1116	0·2130	0·7695
4	0·1019	0·1116	0·1116	0·1296	0·1019	0·5566
5	0·1065	0·2231	0·1065	0·2037	0·1296	0·7694

The difference between the two probabilities is strictly an artifact of the method by which they were calculated. Furthermore, note that table 3.1 shows that the travel time between zone 1 and zone 3 is equal to that between zone 4 and zone 3, yet the probabilities p_{13} and p_{43} are unequal. What has happened is that an aspect of the zone system (that is, the numbers of zones per isochronal annulus) has entered the trip-probability calculation. Although this in itself is not necessarily a bad notion, the way in which it has been done here is inconsistent with a proper model formulation. In particular, the effect of the zone system on these probabilities reflects nothing but the numbers of zones per isochronal annulus. These numbers may or may not have any relationship to any substantive attributes of the zones themselves. Consequently this seemed to be an unsatisfactory allocation procedure for use in the residence-location submodel of the prototype integrated model package.

Before moving to a description of a later and slightly revised form of PLUM, it should be noted that the last step in the manipulation of the zone-to-zone travel probabilities should be to normalize them across the rows. Thus each element in the first row of the zone-to-zone travel-probability matrix shown in table 3.6 should be divided by 0·7695, and so on. The resulting normalized matrix is given in table 3.7. There is little more to be said of this new matrix that has not already been said about its unnormalized form. The normalization procedure does, however, provide further distortion of the original probability functions.

In a later version of PLUM (actually the version that was first obtained for eventual use in the integrated transportation and land-use model package, and which itself was modified shortly after work on the package began), two modifications were introduced into the calculation of the travel probabilities. These modifications were intended to alleviate one of the problems mentioned above. In the form described above, the probability of travelling to an isochronal annulus was divided by the number of zones in the annulus, K, to get the probability of travel to a particular zone. This, as was shown above, produced some unfortunate distortions of the travel probabilities. Consequently this procedure was modified so that the probability of travel to a zone was equal to the probability of travelling to the annulus in which the zone is located,

Table 3.7. Normalized zone-to-zone travel probabilities p_{ij}, for sample problem.

Zone	1	2	3	4	5	Σ
1	0·1684	0·1324	0·2899	0·1324	0·2769	1·0000
2	0·1831	0·2328	0·1831	0·2005	0·2005	1·0000
3	0·1450	0·2647	0·1684	0·1450	0·2769	1·0000
4	0·1831	0·2005	0·2005	0·2328	0·1831	1·0000
5	0·1384	0·2900	0·1384	0·2648	0·1684	1·0000

divided by the travel cost to the annulus:

$$p_{ij} = \frac{1}{c_{ij}} \left[\frac{\beta}{c_{ij}^2} \exp\left(\alpha - \frac{\beta}{c_{ij}}\right) \right] \quad . \tag{3.4}$$

Although this procedure does not have quite the problems that were found for equation (3.3) it is, nonetheless, peculiar. For one thing, this is equivalent to writing

$$p_c = \frac{\beta}{c^3} \exp\left(\alpha - \frac{\beta}{c}\right) \; ; \tag{3.5}$$

this, however, is not the probability function that was initially intended. Moreover, the same practice of accumulating probabilities from empty (no zones included) annuli was continued, thus further distorting the results of the calculations. As before, the resulting matrix of probabilities was normalized across rows before being used in the next step of the calculations.

The second modification to the probability functions in this later version of PLUM was the inclusion of a measure of attractiveness at the residence-zone end of the work trip. This measure was essentially a residential holding capacity, and will be discussed further below.

Quite apart from the attractiveness measure, it was not long after the development of the integrated package prototype began that the inconsistencies in the trip function were discovered. One clue to this problem emerged in the investigation of the PLUM 'correction' factors. There were several correction factors in various subroutines of the PLUM computer program. These were devised to serve a variety of purposes, but essentially were to adjust the results of the model for inconsistencies which developed during the calculations. As such, it was generally assumed that these factors would have magnitudes approximately equal to 1·00, implying minor adjustments to prior calculations. It did not take long to discover that, for example, one correction factor named C_1 had values approaching 20 when the model was tested on a data set for the San Francisco Bay Area (see later). This implied drastic alteration of earlier calculations with the model by the 'brute force' application of a correction factor. Thus the first order of business was to revise the travel-probability function.

The theoretical definition of a travel-probability function is simply a statement of the probability of a trip as a function of the trip length (in distance, time, or cost, or some combination thereof). If it is desired that these probabilities vary as a function of attributes of the origin or destination, then this should be explicitly stated in the equations. More generally, the consideration here is of the aggregate behavior of a group of tripmakers. For a particular origin zone, then, ceteris paribus, the probability of a trip to a particular destination zone will be a function of the difficulty of reaching the destination and the extent to which that particular destination is capable of satisfying the trip purpose.

The difficulty of reaching the destination is typically expressed in terms of travel time or travel cost. The degree or extent to which the destination is capable of satisfying the trip purpose is typically expressed in terms of some measure of attractiveness, or zonal attributes, of the destination zone.

These travel-probability functions (also sometimes referred to as access or access–attractiveness functions) may be conveniently broken into their two main components for further analysis. This is precisely what was done in chapter 2 [see equation (2.11), for example] when an attractiveness measure was added to the simple residence-location model. The PLUM probability function discussed above may properly be considered as the first, or trip-function, component of such an equation. As such, it has a built-in inconsistency. Considered in the fashion being presented here, the probability of a ten-minute (or seventy-five-cent, or two-mile, depending upon what units are being used) trip should be the same, regardless of the zone from which the trip originates or the zone in which it terminates. In the numerical example above, the 'lengths' of trips from zone 1 to zone 2, zone 2 to zone 1, zone 3 to zone 2, and zone 5 to zone 4 are all equal to 1·5 units. The probabilities of these trips are 0·1324, 0·1831, 0·2647, and 0·2648 respectively. This is clearly inconsistent with the theory behind such a trip function.

The first change made in the allocation function was to eliminate completely the notion of the isochronal annulus of zones. The second change, which goes hand in hand with the first, was to eliminate the notion of dividing the probability function into intervals of any sort. Thus the probability of a trip between any two zones was calculated simply by substituting the zone-to-zone travel time directly into equation (3.2). With $\alpha = 0{\cdot}5$, $\beta = 4{\cdot}0$, this resulted in the matrix of zone-to-zone travel probabilities shown in table 3.8. Note how the sums of these rows are rather close to unity without any further calculations. In practice, with larger data sets, this is still true. In order to guarantee that the model is regionally constrained—that is, that all workers have a work trip which ends at a place of residence—these probabilities were normalized across rows of the matrix to yield the matrix given in table 3.9. Here there were still small inconsistencies, but the situation was much improved over the earlier versions of PLUM. This procedure for calculation of the trip-function

Table 3.8. Travel-probability matrix from revised formulation of PLUM (sample problem).

Zone	1	2	3	4	5	Σ
1	0·1208	0·2037	0·2231	0·2037	0·2130	0·9643
2	0·2037	0·1208	0·2037	0·2231	0·2231	0·9744
3	0·2231	0·2037	0·1208	0·2231	0·2130	0·9837
4	0·2037	0·2231	0·2231	0·1208	0·2037	0·9744
5	0·2130	0·2231	0·2130	0·2037	0·1208	0·9736

component was used in the modified PLUM model called IPLUM, which was used as the residence-location submodel of the prototype integrated package.

To test the effects of these changes on the performance of IPLUM, a series of runs was undertaken, using the entire San Francisco Bay Area database. It had been discovered that in the PLUM computer program the normalization across rows of the travel-probability matrix, although called for in the program description, had not been properly implemented (programmed). When this was corrected the value of C_1 decreased by about ten percent, but was still nearly 20. A second run of the model was prepared; this used the correct normalization process and, more importantly, the new probability calculation procedure described above. In this case the value of C_1 fell from 18·9 to 0·91, a value which was much more consistent with the notion of correction factor, rather than its former role as more of a brutalization factor.

While on this general topic, it should be noted that though the discussion thus far has described modifications and tests of the allocation function for households, the IPLUM model used a similar procedure for the allocation of local-serving employment (referred to as LOMS employment in chapter 2). In that case, instead of work-to-home trips, the model was concerned with home-to-shop and work-to-shop trips. The same modifications discussed above with reference to the travel-probability function for residence location were introduced into the procedure for employment allocation. These modifications to the travel-probability function for employment location of course resulted in significant changes in the employment estimates and subsequent changes in the population estimates.

In order to evaluate the effects of these several alterations to the travel-probability functions, several test runs of IPLUM were performed. The results of four runs were compared with reference to summaries of their allocations, and the magnitude of the correction factors C_1 and C_2. The four runs were as follows:

run 1 a benchmark run of the unaltered version of PLUM;

run 2 row sums of travel probabilities for residence location constrained to equal 1·00, but other aspects of the probability calculation left unchanged;

Table 3.9. Normalized travel-probability matrix from revised formulation of PLUM (sample problem).

Zone	1	2	3	4	5	Σ
1	0·1253	0·2112	0·2314	0·2112	0·2209	1·0000
2	0·2090	0·1240	0·2090	0·2290	0·2290	1·0000
3	0·2268	0·2071	0·1228	0·2268	0·2165	1·0000
4	0·2090	0·2290	0·1240	0·1240	0·2090	1·0000
5	0·2188	0·2291	0·2188	0·2092	0·1241	1·0000

run 3 fully revised travel-probability function for residence location as described above;

run 4 as in run 3, but fully revised travel probability function both for residence location and for employment location.

The residence-location output from IPLUM (before the application of several other constraints which need not be considered here) was termed 'unconstrained dwelling unit allocations'. For the nine counties in the San Francisco Bay Area (see figure 3.7), the incremental unconstrained dwelling unit allocations from the above runs are tabulated, along with values of the correction factor C_1, in table 3.10.

It was clear that the application of these new probability calculations to the residence-location portion of the model resulted in an extraordinary reduction in the role of the constraint or correction factor for residence location. It was then decided that enough time had been spent on the modifications to the land-use (location) portion of the prototype package, and that further improvements should be postponed until after a trial set of connections and program executions had been completed and evaluated. Before describing these first runs, the connections between the land-use model (IPLUM) and the transportation model will be described.

In section 3.3, mention was made of the problems of connecting the location models to the transportation models. At the end of this discussion, figure 3.4 showed that the output of the land-use model was the spatial distribution of activities. In the case of IPLUM this included distributions of both residence and LOMS employment, and from these a set of trips (the OD trip matrix) was produced for input to the network model. It was this direct connection from the outputs of spatial distributions of activities, via the trip matrix, to the network model that was the first major step towards the integration of transportation and location analyses.

Table 3.10. Residential unit forecasts (1980) for the San Francisco Bay Area from four test runs of IPLUM (see text).

County	Run 1	Run 2	Run 3	Run 4
San Francisco	65531	66097	64085	48989
Marin	40261	40350	31490	32448
San Mateo	76361	76566	102725	110302
Sonoma	102954	103370	48652	68910
Napa	26000	26057	12709	14607
Solano	59493	59521	32564	34996
Contra Costa	47736	47630	37412	39153
Santa Clara	123035	122391	223091	191712
Alameda	97129	96750	70318	81100
Total	638493	638727	623042	622213
C_1	19·204	18·980	0·912	1·000

In concept, this connection is quite simple and stems from the central hypotheses of the location models. This will now be described in terms of the trip-probability matrices calculated above.

At the heart of the location models used for these analyses is the notion of persons at a place of work being distributed or allocated to places of residence. Places of residence are determined according to the difficulty of reaching the residence zone from the place of work, and the attractiveness of that zone. Of course, one might also think in terms of a given set of workplaces which must be reached from a probabilistic set of residence places. In this circumstance the probabilistic set of residence places may still be considered in the same way as above. In either case, it is clear that a residence-location model based on these concepts has implicit in its output of projected spatial location a set of work-to-home or home-to-work trips. It is this implicit trip matrix which is to be made explicit by extracting it from the outputs of the location model.

If one considers the outputs of the IPLUM model and the way in which they are calculated, then it becomes quite clear as to how the trip matrices are generated. First, recall equation (3.1), which is rewritten here (and renumbered for ease of further reference):

$$N_i = g\sum_j E_j p_{ij} \ . \tag{3.6}$$

In IPLUM, the probability term p_{ij} has two components: one is the trip function of equation (3.2), the other is an as yet unspecified measure of the attractiveness of zone i. In the numerical example given above, the matrix of zone-to-zone travel probabilities was calculated without the use of an attractiveness measure. After its calculation, the matrix was then normalized across the rows. In the actual applications of IPLUM, the matrix of zone-to-zone travel probabilities was first multiplied by a vector of zonal attractiveness values. Then the resultant attractiveness-weighted matrix of zone-to-zone travel probabilities was normalized across its rows. To continue the numerical example it is not necessary to deal with the further complication of an attractiveness measure. The illustration may just as easily be made by using the normalized matrix of zone-to-zone travel probabilities as calculated above and presented in table 3.9.

In equation (3.6), then, the product $E_j p_{ij}$ is the number of persons working in zone j multiplied by the probability of a work-to-home trip between zones i and j. This product is equal to the number of work trips from zone j to zone i. Thus the desired work-trip matrix is immediately available as part of the normal set of IPLUM calculations. To continue the numerical example, the vector of employments by place of work is multiplied by the normalized matrix of zone-to-zone travel probabilities. Suppose, for example, that one hundred persons were employed in zone 1, that two hundred were employed in zone 2, that three hundred were employed in zone 3, and that none were employed in zones 4 or 5; then simple multiplication of the rows of table 3.9 by these numbers gives the work-trip matrix shown in table 3.11.

There are several interesting points to be made concerning table 3.11. First, it should be clear that the third element in the first row, for example, represents twenty-three work trips by persons who work in zone 1 and reside in zone 3. Second, the sum across the first row represents the one hundred persons who work in zone 1. Third, the sum down the first column represents the one hundred and twenty-three persons who work in any of zones 1, 2, or 3, but who all reside in zone 1. From this, it may be seen that the IPLUM procedure generates an OD trip matrix, as well as a spatial distribution of residents. We note here both for further reference, and to relate this example to the example of the trip-end-constrained residence-location model in chapter 2, that the effect of normalizing the rows of the travel-probability matrix is exactly equivalent to the operation of the balancing factor described in chapter 2 [equation (2.5)]. It should be pointed out that even though this may all seem perfectly obvious now, the idea of extracting these trips from a residence-location model had not been tried in planning analysis practice in the USA until the time of this research effort. It was this rather straightforward extension of land-use modelling which led directly to the integrated transportation and land-use package.

In theory, the next step in the analysis would involve using this trip matrix in a capacity-constrained network assignment procedure. In practice, several additional steps were necessary. First, in the numerical example it was assumed that employees equalled employed residents, equalled households, equalled population, and that the sum of any one of these equalled total regional trips. In practice there are conversion ratios relating each of these quantities to the others. Second, in this example only work trips are considered, yet there are many other trip purposes to be dealt with. Third, questions of peak hour versus daily trips and vehicle occupancy rates must also be resolved. Fourth, the question of employment location or relocation has not been considered in the example. Despite these and other considerations, the principles on which the system operates should be apparent.

Table 3.11. Work-trip matrix resulting from IPLUM residence allocation (sample problem).

Employment zone	Residence zone					Σ
	1	2	3	4	5	
1	13	21	23	21	22	100
2	42	24	42	46	46	200
3	68	62	37	68	65	300
4	0	0	0	0	0	0
5	0	0	0	0	0	0
Σ	123	107	102	135	133	600

Now that the essence of the prototype integrated package has been described, the next section presents a discussion of the first results of the operation of the land-use and transportation models functioning together as a linked system.

3.5 Preliminary tests of linked models

The first tests of the model package containing the linked models were done for a small prototype urban region of eight areas. After ironing out several minor problems, acceptable results were obtained, the system seemed to have functioned properly, and preparation began for full-scale runs using data for the San Francisco Bay Area. The precise form of the model system used in these runs is described diagrammatically in figure 3.5. It must again be noted that there were two forms of the PLUM model, both of which produced forecasts of residence location and LOMS employment location. The first of these was essentially a 'state-variables' model, where the forecasts were made in terms of the values of variables at time *t*. The second form of the model (referred to here as IPLUM) was

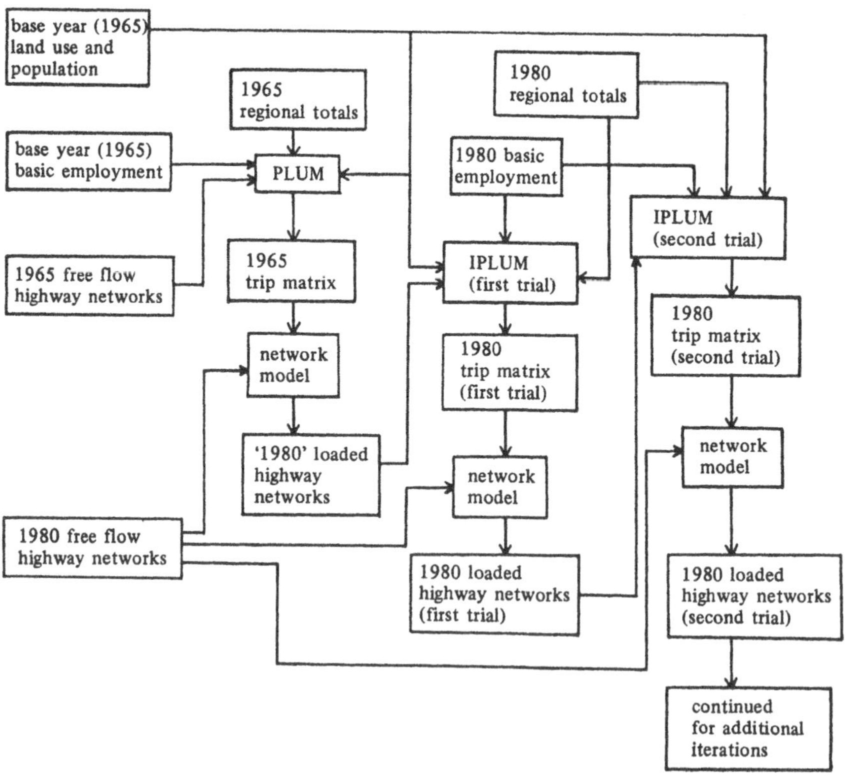

Figure 3.5. Configuration of system for experimental model runs.

a version using incremental variables, where the forecasts were made in terms of changes in values from time $t-1$ to time t. These changes were then added to the base values (that is, those at $t-1$) to yield the forecasts of values at time t. One may trace through the assemblage of models used for these runs as shown in figure 3.5. Basically the procedure involved a run of PLUM to generate base year trip matrices. After loading these trips onto a network, the zone-to-zone travel times on the network were used as input to the IPLUM model. After this, successive iterations of IPLUM and the network model were run.

Before presenting the results of these runs, it is appropriate to consider what expectations were held for the results of such test runs. If, as suggested above, important interrelationships were ignored by the typical transportation study or typical land-use study, either of which assumes fixed values for the other, then these runs should indicate significant and systematic differences in output from a simple run of PLUM or IPLUM not integrated with a network model. These differences should be in accord with the following hypotheses. First, if land-use forecasts are made on the basis of a free flow network, then congestion, and therefore travel times, would be significantly less than would be present in reality. Consequently the overall spatial distribution of activities forecast would be excessively dispersed. An excessively dispersed pattern of activities would, in turn, produce overestimates of tripmaking and network congestion. Second, if land-use forecasts are made on the basis of overcongested networks, the overall spatial distribution of activities would be excessively concentrated. And, in turn, an excessively concentrated pattern of activities would produce underestimates of tripmaking and network congestion. Given these expectations, it was to be anticipated that a linked land-use and transportation model would show convergent oscillation between decentralization and centralization, and consequently between overcongestion and undercongestion.

The first complete set of full runs of the prototype integrated package was done with a version of IPLUM which did not incorporate the modifications described above which were subsequently made to the allocation function. These first test runs suffered also from the use of only work trips for network loading. However, these runs did provide some initial insights into the functioning of the models as an integrated system. In each case, the complete runs consisted of several iterations through the land-use and network assignment components of the integrated package. The basic employment increments were exogenously produced by a model named BEMOD, which is described in chapter 6. These projections were, in fact, the same as those used by the Bay Area Transportation Study Commission in their analysis of the San Francisco Bay Area (BATSC, 1969). The regional control totals of population, employment, and numbers of trips were also produced exogenously, and were obtained from the Metropolitan Transportation Commission (MTC) of

San Francisco. The base year was 1965, and the base year distributions of employment, population, and land use were also obtained from MTC. A full pass through the model package involved a trial estimate of the 1980 distributions of activities and trips, followed by several iterations of these 1980 estimates—each being used as input to the next. The hope was that after not too many of these iterations a stable transportation–location equilibrium would be achieved. The discussion of results focuses mainly on the spatial distributions of population, as these were expected to be the most reliable of the models' outputs.

The first pass through the model package produced the population distributions which, summarized to the county level from the 291 zone system at which the package operated, are given in table 3.12, and displayed graphically in figure 3.6. Figure 3.7 shows the nine counties that make up the region. An immediate conclusion which could be drawn from these results was that there was a substantial difference in the spatial distribution of population according to whether or not the impedance inputs were derived from loaded (congested) or empty networks. Further, it was clear that there were oscillations in the population allocations,

Table 3.12. Population distributions for the nine counties of the San Francisco Bay Area: base year (1965) data, and forecasts (with percentage changes 1965–1980) from four successive iterations of the first run of the integrated package.

County	1965	1980 iteration [a]			
		(1)	(2)	(3)	(4)
San Francisco	754754	764445	763414	765068	765067
		(1·3)	(1·1)	(1·4)	(1·4)
Marin	192523	469669	319634	301379	259124
		(243·9)	(66·0)	(56·5)	(36·6)
San Mateo	523271	801723	756041	835885	938460
		(53·2)	(44·5)	(59·7)	(79·3)
Sonoma	167497	368057	591781	444212	337406
		(119·7)	(253·3)	(165·2)	(125·3)
Napa	69378	135910	183943	156245	130538
		(95·9)	(165·1)	(125·2)	(88·2)
Solano	149386	363338	437717	316304	276112
		(143·2)	(193·0)	(111·7)	(84·8)
Contra Costa	534056	666716	720394	778010	672925
		(24·8)	(34·9)	(45·7)	(26·0)
Santa Clara	922357	1308470	1238821	1341138	1499215
		(41·9)	(34·3)	(45·4)	(62·5)
Alameda	1095236	1528722	1395805	1469301	1488690
		(39·6)	(27·4)	(34·2)	(35·9)
Total	4438458	6407473	6407471	6407470	6407474
		(44·4)	(44·4)	(44·4)	(44·4)

[a] Percentage changes 1965–1980 given in brackets beneath each value.

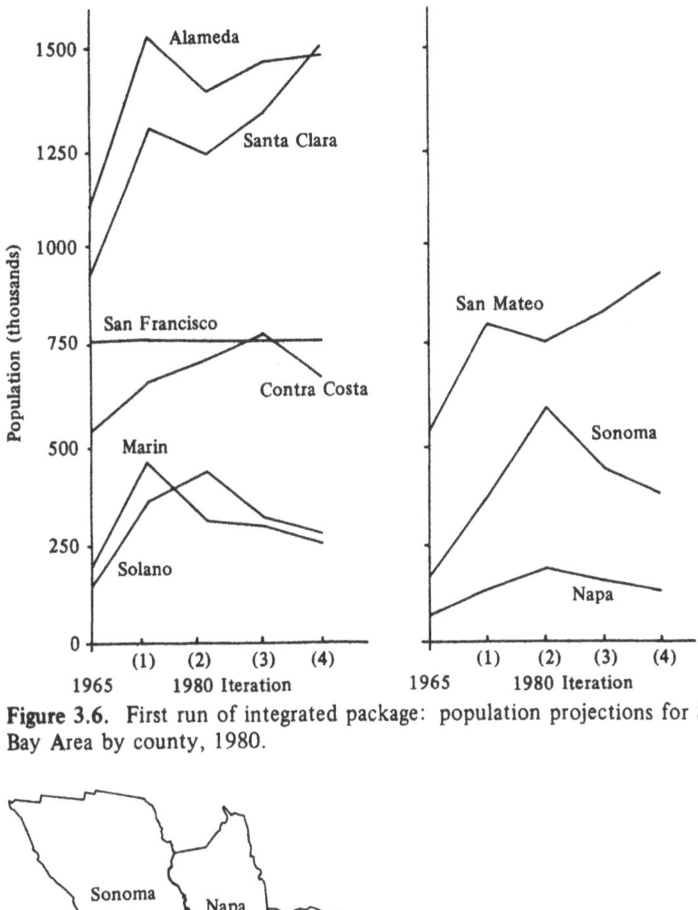

Figure 3.6. First run of integrated package: population projections for San Francisco Bay Area by county, 1980.

Figure 3.7. San Francisco Bay Area, nine zone system: definition of zones (counties).

depending on the levels of congestion on the loaded networks. These conclusions reinforced the central hypothesis of the inadequacy of either a transportation or a land-use forecasting method which ignores the relationships between these two components.

At the zonal level of detail, the general expected pattern was found: uncongested (underutilized?) networks producing dispersed population allocation, in turn produces congested networks, which then produces more centralized population allocation, which in turn produces less congested networks, and so on. At the county level of detail much of this phenomenon was obscured. The outputs of this first complete run demonstrated that the general performance of the model package was in accordance with the original expectations. Even this rather crude assembly of the components in the system had demonstrated the importance of the interactions between the models. However, in view of obvious deficiencies in the first trial version of the prototype package, additional more detailed analysis of these first outputs was curtailed in favor of work on the necessary improvements in the package.

There were three principal areas of further investigation which ultimately led to the second round of model runs. First, preliminary investigations of the operation of the original version IPLUM had revealed evidence of serious inconsistencies in the allocation functions. The conclusion of these investigations was that the IPLUM allocation procedure required important modifications to the travel-probability calculations as described above. These were implemented and resulted in an apparent improvement in the forecasting performance of IPLUM. Second, it was also clear that an important portion of total trips was being ignored by dealing only with work trips. A procedure for making explicit the home-to-shop and work-to-shop trips that were implicit in the earlier versions of IPLUM was developed, along with a gravity model for distributing these trips. The number of shopping trips was then scaled up to equal the total nonwork trips in the region. These, plus the work trips, it was felt, would be a reasonably good estimate of the total trips on the network. Third, and finally, a new incremental procedure to effect substantial savings in computer time for runs of the network assignment procedure was developed. In general this procedure relied on the assumption that after the first two iterations of IPLUM (including the trip-generation procedures) in a given pass through the model package, a large proportion of the trips then on the network would be relatively stable. Consequently a marginal loading of trips was considered to be a possible substitute for the full network loading runs used in this first trial run.

It was hypothesized that, in cases where successive iterations of IPLUM did not produce extremely different configurations of land use, the consequent OD matrices would not show extreme differences either. If this were indeed the case, then it was felt that it would be possible to effect a 30% reduction in computer usage for the network loadings by using such a marginal loading procedure. This procedure, it was hoped, could operate as follows.

1. When loading the network with trips generated by iteration (n) of IPLUM, the information both for the 80%-loaded and for the 100% loaded (fully loaded) networks would be saved (on a computer storage device, for example).
2. The information from the fully loaded network would be used as input to the ($n+1$)th iteration of IPLUM, thus generating a new land-use allocation and new trip matrices.
3. The difference, OD pair by OD pair, between the trip matrices generated in the ($n+1$)th IPLUM iteration and 80% of the trips in the OD matrices generated in the (n)th iteration of IPLUM could then be calculated.
4. This new matrix of trip differences could then be added to the 80% loaded network (which had been saved in step 1) from the (n)th iteration of IPLUM. The 80% loaded network would continue to be saved and the characteristics of the new fully loaded network would be used as input to the ($n+2$)th iteration of IPLUM.
5. This process would be repeated in successive iterations within the forecast period.

Table 3.13. Population distributions for the nine counties of the San Francisco Bay Area: second run of the integrated package (round 2).

County	1965	1980 iteration [a]			
		2.1	2.2	2.3	2.4
San Francisco	754754	759821 (0·7)	762455 (1·0)	758553 (0·5)	760438 (0·8)
Marin	192523	313982 (63·1)	251919 (30·9)	341856 (77·6)	223377 (16·0)
San Mateo	553271	840265 (51·9)	719364 (30·0)	787382 (42·3)	700500 (26·6)
Sonoma	167497	440337 (162·9)	255273 (52·4)	708799 (323·2)	244780 (46·1)
Napa	69378	118582 (70·9)	162268 (133·9)	190562 (174·7)	93982 (35·5)
Solano	149386	223074 (49·3)	351312 (135·2)	288630 (93·2)	229551 (53·7)
Contra Costa	534056	655638 (22·8)	771037 (44·4)	635465 (19·0)	783855 (46·8)
Santa Clara	922357	1412839 (53·2)	1348370 (46·2)	1239020 (34·3)	1429129 (54·9)
Alameda	1095236	1643003 (50·0)	1364748 (24·6)	1457279 (33·1)	1941722 (77·3)
Total	4438458	6407472 (44·4)	6407476 (44·4)	6407468 (44·4)	6407474 (44·4)

[a] Percentage changes 1965–1980 given in brackets beneath each value.

After these investigations and modifications were completed, it was possible to proceed with a second full pass through the model package.

In round 2 of the integrated package runs, the use of this marginal network-loading procedure was initiated after the second iteration of IPLUM, for input to the third iteration. In the fourth iteration of IPLUM in this round, certain zones displayed considerable instability in their population allocations. Because of the extreme complexity of the complete model package, it was not possible to determine by inspection which element(s) of the system were responsible for these instabilities. The results of four iterations of the modified PLUM, the modified trip-generation procedure, and the modified network package incorporating the marginal loading procedure are shown in table 3.13 and figure 3.8. It will simplify further discussion to adopt here the convention of referring to each full pass through the integrated model package as a 'round'.

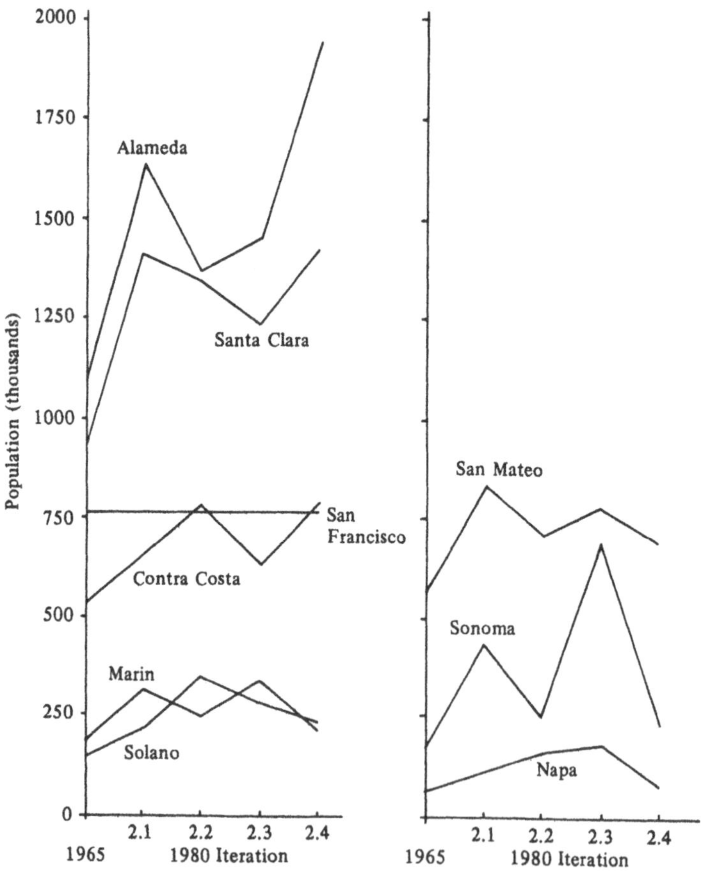

Figure 3.8. Population projections for San Francisco Bay Area by county, 1980, from second run of the integrated package (round 2).

Thus the first pass through, discussed above, will now be called round 1. The pass through now being discussed will then be called round 2. Variations within the general context of a particular round may then be designated by suffix letters—for example, round 2A—and iterations within a round may be designated by a further suffix period and iteration number. Thus the second iteration of round 2D would be designated iteration 2D.2, and so on. It was immediately obvious from the results which are shown in figure 3.8 that there still remained problems of oscillation from one iteration of IPLUM to the next. Although this was, of course, not a particularly unlikely outcome, and certainly was not unexpected, it was definitely not a welcome result at the time.

A series of attempts were made then to isolate the source of these oscillations. The first such attempt began by reverting back to the full network-loading procedure. Since the marginal procedure had been instituted only after iteration 2.2 (to prepare input for iteration 2.3), iterations 2.1 and 2.2 were retained as the initial results of this new round, 2A. Three further iterations of the network package, IPLUM, and trip generation were run: iterations 2A.3, 2A.4, and 2A.5. These results are shown in table 3.14 and figure 3.9. It was clear, as can be seen in

Table 3.14. Population distributions for the nine counties of the San Francisco Bay Area: modified second run of the integrated package (round 2A).

County	1965	1980 iteration [a]				
		2A.1	2A.2	2A.3	2A.4	2A.5
San Francisco	754754	759821 (0·7)	762455 (1·0)	759217 (0·6)	760607 (0·8)	758369 (0·5)
Marin	192523	313982 (63·1)	251919 (30·9)	360995 (87·5)	220628 (14·6)	362523 (88·3)
San Mateo	553271	840265 (51·9)	719364 (30·0)	858236 (55·1)	678678 (22·7)	781160 (41·2)
Sonoma	167497	440337 (162·9)	255273 (52·4)	670765 (300·5)	240426 (43·5)	782643 (367·3)
Napa	69378	118582 (70·9)	162268 (133·9)	155177 (123·7)	95049 (37·0)	237311 (242·1)
Solano	149386	223074 (49·3)	351312 (135·2)	223454 (49·6)	231381 (54·9)	378531 (153·4)
Contra Costa	534056	655638 (22·8)	771037 (44·4)	616419 (15·4)	817730 (53·2)	611018 (14·4)
Santa Clara	922357	1412834 (53·2)	1348370 (46·2)	1408025 (52·7)	1275817 (38·3)	1260181 (36·6)
Alameda	1095236	1643003 (50·0)	1364748 (24·6)	1355254 (23·7)	2087224 (90·6)	1235809 (28·3)
Total	4438358	6407472 (44·4)	6407476 (44·4)	6407465 (44·4)	6407477 (44·4)	6407545 (44·4)

[a] Percentage changes 1965-1980 given in brackets beneath each value.

figure 3.9, that the oscillations were still present in the system; and the overall results of round 2A were nearly identical to those obtained in round 2. It was therefore concluded that the marginal network-loading procedure had not been at fault.

Subsequently, a more detailed investigation of these outputs from the model revealed that, though obscured at the county level of detail, the oscillations were zonal oscillations rather than county-level oscillations. In each county the bulk of the zones remained rather stable (after the first two iterations), whereas a few of the zones underwent rather large oscillations. For example, in Alameda county, of sixty-nine zones only ten to fifteen showed significant oscillations, and of these only two or three showed really large oscillations. In addition, it was discovered

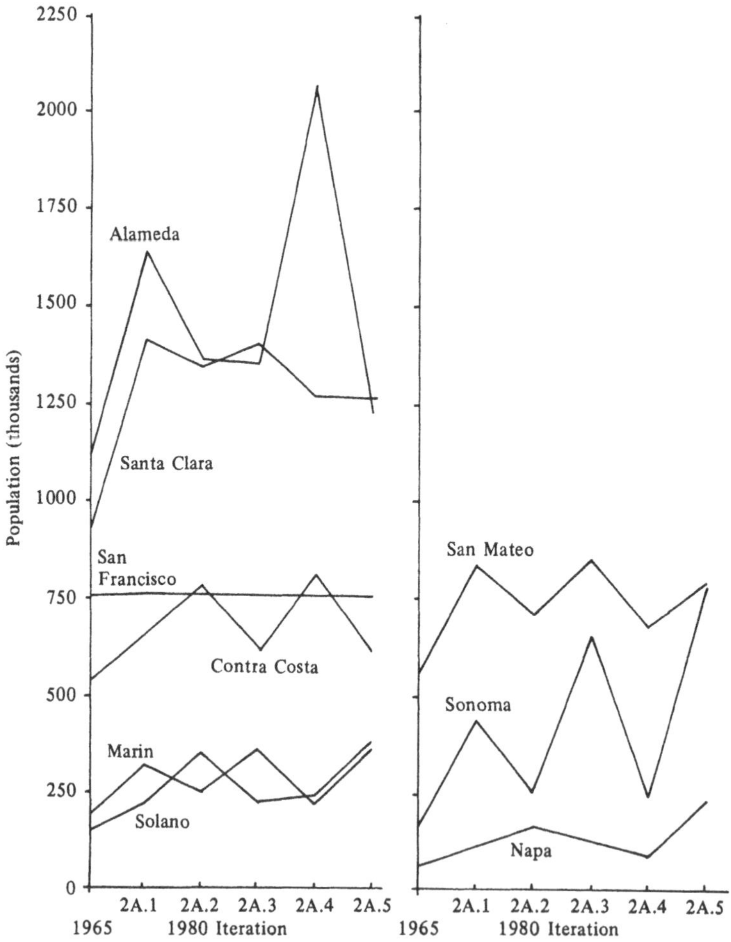

Figure 3.9. Population projections for the San Francisco Bay Area by county, 1980, from the modified second run of the integrated package.

that virtually all of the oscillating zones were at the fringes of the San Francisco Bay Area and/or had substantial quantities of vacant land. Further detailed examination of output data also indicated that though there were significant changes in some zone-to-zone impedances, these alone would not have accounted for the large changes in population in these zones. To confirm this observation, a special run of IPLUM was made. The impedances which had been the inputs to iterations 2A.3 and 2A.4 were averaged together. The new impedances, the arithmetic average of the two separate matrices, were used as input to what was called iteration 2A.AVG of IPLUM. The expectation was that if the oscillations had been due principally to the impedance changes, despite the fact that the averaging procedure ignored the nonlinearity of the network volume-delay function, the population allocations in iteration 2A.AVG should be somewhere near the midpoint between the population allocations in iterations 2A.3 and 2A.4. The results of that test are shown in table 3.15, and were a rather emphatic denial of the hypothesis that in those areas where there were problems of oscillation, the impedances may have been largely responsible. Thus the fact that virtually all of the problem (oscillating) zones tended to be on the fringes of the Bay Area and have large total areas as well as high percentages of vacant land remained their only distinguishing feature. This bit of information, however, eventually led to further investigation of the travel-probability function in IPLUM.

Table 3.15. San Francisco Bay Area: population allocations for 1980, from three runs of IPLUM.

County	Iteration		
	2A.3	2A.AVG	2A.4
San Francisco	759217	762101	760607
Marin	360993	244032	220628
San Mateo	858236	753394	678678
Sonoma	670765	247984	240426
Napa	155177	102127	95049
Solano	233454	227754	231381
Contra Costa	616419	745428	817730
Santa Clara	1408025	1400946	1275817
Alameda	1355254	1923773	2087224

3.6 Revisions and further tests of the model

As discussed several times before, the residential allocation function in IPLUM may be considered as having two components. The first deals with the probability of a trip for a given purpose being of a certain length (in terms of travel time and/or cost). The second deals with measures of attractiveness of possible trip destinations. The first component had already been thoroughly investigated and modified as described above.

In this discussion the focus of concern is the second, attractiveness-measure component of the residential allocation function in IPLUM. In the version of PLUM which was ultimately modified to become the IPLUM submodel in the prototype integrated package, a rather simple formulation was used for the measure of a zone's attractiveness. Called a measure of 'opportunities', the variable was constructed as follows:

$$U_i = L_i^{\mathrm{v}} \frac{H_i}{L_i^{\mathrm{r}}} , \tag{3.7}$$

where U_i is the measure of opportunities for new residential development in zone i; L_i^{r}, L_i^{v} are, respectively, residential land acreage and vacant (non-industrial) acreage in zone i; and H_i is the number of housing units in zone i.

Although a number of objections might be raised to this measure of zonal attractiveness, the principal concern was with only one detrimental characteristic. In a zone containing, say, a small town and a large quantity of vacant land, this measure of opportunities or attractiveness is likely to be quite substantially overstated. That this was indeed the case became abundantly clear on inspection of the troublesome oscillating zones in the model runs described above. It is also clear from statements in his PLUM report (Goldner et al, 1972) as well as in several of his project's working memoranda that Goldner was aware of this problem during early development work with PLUM. In fact, a modification to this attractor formula [equation (3.7)] was proposed which was to compensate for the problem. For reasons not specified, this compensating factor was moved to a part of the PLUM program where it was rarely ever called into operation. It appears that, owing to the previously discussed errors in the probability portion of the original PLUM allocation function, this attractor correction was found to be ineffective or perhaps detrimental to the model results and was therefore shifted elsewhere in the program.

The theoretical basis for this attractor correction was based on the notion that the zones in which these problems appeared tended to have large quantities of vacant land, but lacked the 'infrastructure' to serve the rapid development of this land. The proposed correction factor of Goldner et al (1972, pages 57–58) seemed reasonable and is described below in their words.

> "The land development constraint is an approximation to the area in the zone served by adequate infrastructure, if it were known. Zones may have large quantities of vacant land but are still not developable due to the time-lag in supplying infrastructure. Utilities, water, sewers, schools, local government service, and other elements of the infrastructure cannot be supplied instantly; it takes years, sometimes decades for all these elements to be emplaced. For zones with relatively little development, the constraint (expressed as a proportion between zero and one) should be predominant, but for zones almost totally developed, the constraint should have virtually no effect.

A function with this property is:

$$G(x) = (1-e^{-3x})/(1-e^{-3})$$

where x is the fraction of usable land developed. This function is invoked when zones have little development, but quickly increases so that it has a negligible effect with moderate increases in development. For example, at zero percent development the constraint is fully binding whereas at ten percent development the constraint is almost half gone."

This correction factor was then moved back into the allocation section of the version of IPLUM in the prototype integrated package as a modification to the attractiveness component of the residential allocation function.

It will be recalled that the product of the trip probabilities and the attractiveness was normalized before being used for the residential allocation in IPLUM. This being the case, the introduction of the attractor correction factor involved simply multiplying it by the opportunities variable before normalization. Thus the modified measure of opportunities, or 'attractors', is:

$$U_i' = U_i G_i = \left(\frac{H_i}{L_i^r}\right) L_i^v G_i \; ; \tag{3.8}$$

here

$$G_i = \left[\frac{1-\exp(-3x_i)}{1-\exp(-3)}\right] , \tag{3.9}$$

where

$$x_i = \frac{L_i^t - L_i^u - L_i^s - L_i^{v,\,\mathrm{ind}} - L_i^v}{L_i^t - L_i^u - L_i^s - L_i^{v,\,\mathrm{ind}}} , \tag{3.10}$$

and where L_i^t is the total land acreage, L_i^u the unusable acreage, L_i^s the acreage occupied by streets or highways, $L_i^{v,\,\mathrm{ind}}$ the vacant industrial acreage, and L_i^v the vacant other acreage in zone i.

Two test runs of IPLUM with the modified attractiveness measures U_i' were then undertaken. As the purpose of this investigation had been to discover the causes for oscillation in the population allocations, the inputs to iterations 2A.4 and 2A.5 were used in two further test runs of the model.

Before describing the results of these runs, it is appropriate to mention the role of population control totals in the model. In the IPLUM model, as is the case with virtually all currently operational urban land-use models, there is an input requirement for a regional control total for population. This value, exogenously supplied, acts as a constraint on the summation of the zonal allocations made by the model. An implication of this procedure which is of particular concern here is that if from one iteration to another a particular zone shows, for example, a considerable decrease in population while the control total remains unchanged, then it is inevitable that other zones will have population increases to 'absorb' the population which no longer locates in the decreased zone. This potentially confounding influence should be kept in mind when examining the results of the two test runs with the modified attractiveness measure.

These are shown in table 3.16, along with results of other relevant runs of IPLUM shown for purposes of comparison. Note that iteration 2A.4 and iteration 2A.4G (which contained the new 'developability' correction factor G) both used the same impedance inputs. Similarly the inputs to iterations 2A.5 and 2A.5G were identical. From these results it appears that the oscillations were somewhat damped by the addition of the correction factor.

Further reflection on the problem of development in large vacant areas later suggested an alternative correction factor. Goldner's original developability correction function as described above is shown graphically in figure 3.10a. The implication of this form is that for largely undeveloped zones small differences in x result in substantial differences in developability. Once, however, substantial development has taken place in a zone, further development results in substantially smaller increases in developability. In reality, however, it is quite probable that developability increases rather slowly with increases in x, until some threshold level is reached. At that point small additional increases in x may result in large increases in

Table 3.16. Projected total population of San Francisco Bay Area counties for 1980: comparison of round 2A iterations 4 and 5 with and without land 'developability' adjustment, G.

County	Iteration				
	2A.3	2A.4	2A.4G	2A.5	2A.5G
San Francisco	759217	760607	766434	758369	762909
Marin	360995	220628	225962	362523	406388
San Mateo	858236	678678	739142	781160	751859
Sonoma	670765	240426	243444	782643	536659
Napa	155177	95049	89339	237311	183622
Solano	223454	231381	240993	378531	457115
Contra Costa	616416	817730	836125	611018	691476
Santa Clara	1408025	1275817	1398746	1260181	1347906
Alameda	1355254	2087224	1867354	1235809	1273610

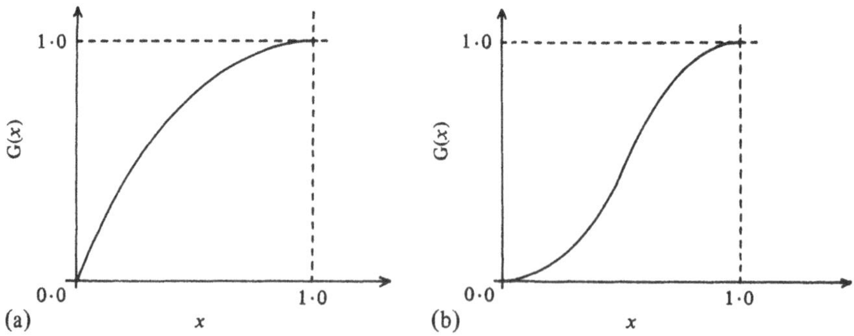

Figure 3.10. Functional forms of the developability correction factor G(x). (a) Original form (Goldner et al, 1972); (b) logistic form.

developability, until a substantially higher level of development is reached at which point the curve might again flatten out. A curve (actually a traditional logistic curve) conforming to this hypothesis is shown graphically in figure 3.10b.

This hypothesis was investigated as work on the project progressed and will be discussed further in a later chapter of this book. In summary, after further investigation of the IPLUM residential allocation procedure and the results thereof, it was decided that further modification was necessary. A correction factor based on earlier and subsequently deleted work of Goldner, presumably intended as a surrogate measure of infrastructure, was introduced. This factor had the effect of damping the attractiveness of fringe areas (zones) of the region which contained large quantities of vacant acreage. Frequently, substantial portions of this vacant land were not really ready for the development which the model would have forecasted in the absence of the correction factor. The introduction of this correction factor appeared to produce the desired effects in the model runs.

After these tests of correction factors in IPLUM, a new set of runs of the prototype integrated package were undertaken. The results from these runs are presented in table 3.17 and shown graphically in figure 3.11.

Table 3.17. Population distributions for the nine counties of the San Francisco Bay Area: third run of the integrated package (round 3).

County	1965	1980 iteration [a]				
		3.1	3.2	3.3	3.4	3.5
San Francisco	754754	759821 (0·6)	767173 (1·6)	763390 (1·1)	766364 (1·5)	763276 (1·1)
Marin	192523	313982 (63·0)	275484 (43·0)	363621 (88·8)	238932 (24·1)	380640 (97·7)
San Mateo	553271	840265 (51·8)	762003 (37·7)	746089 (34·8)	788324 (42·4)	735260 (32·8)
Sonoma	167497	440337 (162·8)	242874 (45·0)	474468 (183·2)	230152 (37·4)	520828 (210·9)
Napa	69378	118582 (70·9)	118737 (71·1)	163118 (135·1)	95836 (38·1)	179825 (159·1)
Solano	149386	223074 (49·3)	363296 (143·1)	328929 (120·1)	262109 (75·4)	450116 (201·3)
Contra Costa	534056	655638 (22·7)	829304 (55·2)	738900 (38·3)	787239 (47·4)	703485 (31·7)
Santa Clara	922357	1412834 (53·1)	1412478 (53·1)	1325489 (43·7)	1520842 (64·8)	1287851 (39·6)
Alameda	1095236	1634003 (49·1)	1636193 (49·3)	1503543 (37·2)	1717740 (56·8)	1386261 (26·5)
Total	4438358	6407472 (44·4)	6407542 (44·4)	6407547 (44·4)	6407538 (44·4)	6407542 (44·4)

[a] Percentage changes 1965–1980 given in brackets beneath each value.

It was, unfortunately, quite clear from these results that the troublesome oscillations were still present and still significant. These oscillations were also visible in the network loadings, with network links to zones of population oscillation showing correspondingly large volume oscillations. From the functions on which the population allocation in IPLUM is based, it had been suggested that the mode of the travel-probability distribution was expected to be approximately equal to one-third of the mean travel time of work trips (Goldner et al, 1972). A look at the travel times on the loaded networks revealed that this was not the case. In the hope that changing the value of β in the travel function [equation (3.2)] would improve the stability of the system, two additional test runs were made with larger values of β. These runs produced results very similar to those of round 3; this indicates first that the model package was not very sensitive to the values of β, and second that these values were not responsible for the oscillation problem.

Further detailed analysis of the linkages and outputs of the system was undertaken. Finally, after extensive examination attention again focused

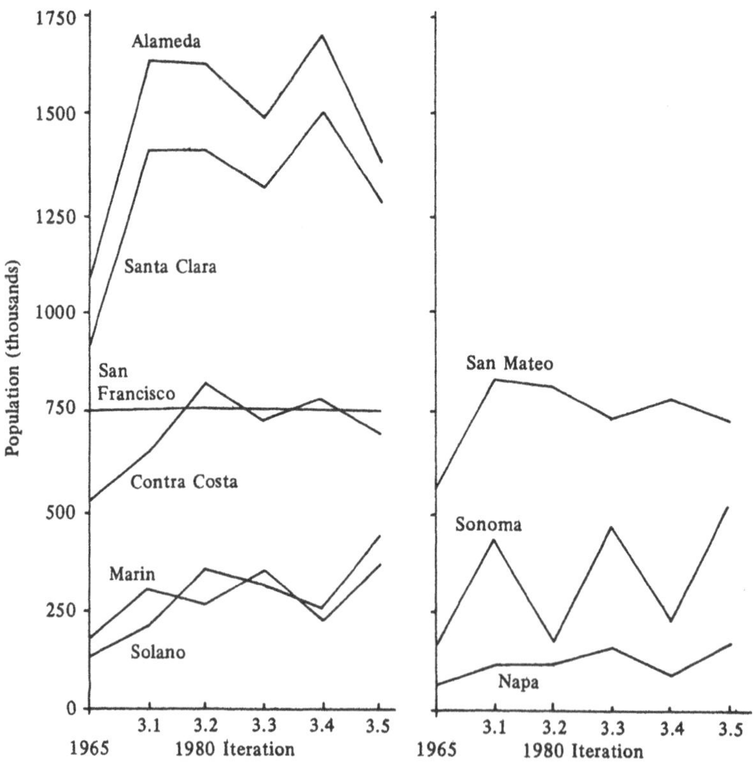

Figure 3.11. Population projections for the San Francisco Bay Area by county, 1980, from the third run of the integrated package (round 3).

on figure 3.11, where it appeared that in going from iteration 3.2 to iteration 3.3 the system might have 'overshot' a possible equilibrium point. Having done so, the system may subsequently attempt, in going from iteration 3.3 to iteration 3.4, to return to the equilibrium point, but then overshoots this possible equilibrium in the opposite direction. Once this process had started, it appeared as if it might well continue indefinitely. It seemed that once having entered an 'unstable' situation, the model system was unable to reach an intermediate and hopefully stable output. The next experiments, taking a cue from the analysis of physical systems, were an attempt to help the model system to return to, or to achieve, such a stable situation. The trips generated in iteration 3.3 and those generated in iteration 3.4 were averaged together, OD pair by OD pair. The resulting averaged trip matrix was loaded on the 1980 network and used as input to a new iteration, denoted 3.5AVG. The trips from iteration 3.5AVG were then averaged with those which had been used as its inputs (that is, the average of iterations 3.3 and 3.4) and again loaded on the 1980 network. The resultant impedances were used as input to IPLUM iteration 3.6AVG. This procedure, as had been hoped, virtually eliminated the oscillations in the output. The results of these runs are

Table 3.18. Population distributions for the nine counties of the San Francisco Bay Area: round 3.AVG.

County	1965	1980 iteration [a]				
		3.2	3.3	3.4	3.5AVG	3.6AVG
San Francisco	754754	767173	763390	766364	765234	764950
		(1·6)	(1·1)	(1·5)	(1·3)	(1·4)
Marin	192523	275484	363621	238932	334808	320451
		(43·0)	(88·8)	(24·1)	(73·9)	(66·4)
San Mateo	553271	762003	746089	788324	788168	778837
		(37·7)	(34·8)	(42·4)	(42·4)	(40·8)
Sonoma	167497	242874	474468	230152	294499	288333
		(45·0)	(183·2)	(34·4)	(75·8)	(72·1)
Napa	69378	118737	163118	95836	126219	130566
		(71·1)	(135·1)	(38·1)	(81·9)	(88·2)
Solano	149386	363296	328929	262109	319934	307813
		(143·1)	(120·1)	(75·4)	(114·1)	(106·1)
Contra Costa	534056	829304	738900	787239	780736	788455
		(55·2)	(38·3)	(47·4)	(46·1)	(47·6)
Santa Clara	922357	1412478	1325489	1520842	1382363	1401803
		(53·1)	(43·7)	(64·8)	(49·8)	(52·0)
Alameda	1095236	1636193	1503543	1717740	1621579	1626328
		(49·3)	(37·2)	(56·8)	(48·0)	(48·5)
Total	4438358	6407542	6407544	6407539	6407551	6407556
		(44·4)	(44·4)	(44·4)	(44·4)	(44·4)

[a] Percentage changes 1965–1980 given in brackets beneath each value.

shown in table 3.18 and figure 3.12. It appeared that if this averaging procedure were imposed on the model system from iteration 3.3, that is, with average of the trips from iterations 3.1 and 3.2, the oscillations would be brought to within tolerable ranges. Consequently, this averaging procedure was adopted as standard procedure in the model package.

As was discussed at the start of this chapter, two of the principal aims of this preliminary research on the integrated package were first to attempt to eliminate forecast errors which resulted from the differences between free flow and loaded (congested) network characteristics, and second to eliminate (partially) the errors due to the incremental land development resulting from changes in the transportation networks. The first aim is achieved by using loaded networks as input to the land-use model. The second aim is partly achieved by repeated iterations of the land-use model and the network model. It should be noted that although the various intermediate steps in this iterative process do not purport to have any one-to-one correspondence to simulated points in time, the overall stream of iterations is probably similar to a simulated stream in time. This general notion, along with considerations of model package equilibria, is considered in a later chapter of this book.

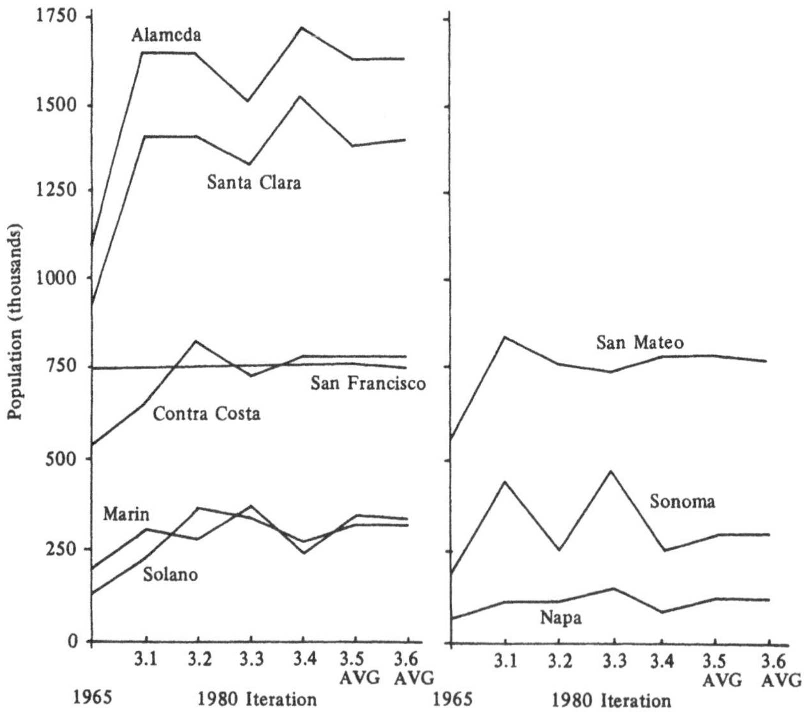

Figure 3.12. Population projections for the San Francisco Bay Area by county, 1980, from round 3.AVG of the integrated package.

In the next chapter, the second discussing work with the prototype model package, a comparison is shown between integrated package forecasts of population and a simple IPLUM forecast. This is then followed by discussion of several policy analyses done to demonstrate the capabilities—and deficiencies—of the prototype integrated package of transportation and land-use models.

4

Policy analysis with the prototype integrated package

4.1 Introduction

The principal focus of this chapter is on a presentation and discussion of several preliminary policy analyses undertaken with a modified form of the prototype integrated transportation and land-use package (ITLUP). Before this there are, however, two other items to be described. The first is a comparison of population forecasts resulting from IPLUM run as a component of ITLUP with the population resulting from IPLUM run alone (that is, 'unintegrated'). Two measures of the performance of the highway networks are then described, as these are used in the subsequent discussion of the policy analyses.

4.2 Comparison of IPLUM with ITLUP-IPLUM

An obvious point of comparison for this prototype integrated forecasting technique was with similar forecasts made in the traditional manner. The modified version of IPLUM was run with the impedances for the uncongested or empty (lacking any trip loadings) 1980 network as input.

Table 4.1. Comparison of population forecasts: integrated package (iteration 3.6AVG) versus IPLUM (iteration 1980-FF).

County	1965	1980 iteration [a]		Difference	
		3.6AVG	1980-FF	absolute	percentage
San Francisco	754754	764950 (1·4)	757892 (0·4)	−7058	−0·9
Marin	192523	320451 (66·4)	319089 (65·7)	−1362	−0·4
San Mateo	553271	778837 (40·8)	716536 (29·5)	−62301	−8·0
Sonoma	167497	288333 (72·1)	589129 (251·7)	300796	104·3
Napa	69378	130566 (88·2)	213256 (207·4)	82690	63·3
Solano	149386	307813 (106·1)	374101 (150·4)	66288	21·5
Contra Costa	534056	788455 (47·6)	725874 (35·9)	−62581	−7·9
Santa Clara	922357	1401803 (52·0)	1163194 (26·1)	−238609	−17·0
Alameda	1095236	1626328 (48·5)	1548474 (41·4)	−77854	−4·8
Total	4438358	6407556 (44·4)	6407545 (44·4)	~0	0

[a] Percentage changes 1965-1980 given in brackets beneath each value.

This, of course, represents the standard procedure for unintegrated use of a land-use model. This yielded land-use forecasts for 1980 based on the projected 1980 network without any consideration of the traffic which would have developed as a result of the changed land-use distribution. These forecasts were then aggregated to the county level. The county population forecasts from this run (called 1980-FF) are listed together with a set of ITLUP forecasts in table 4.1. The 1980-FF forecasts did not reflect any of the feedback from land use to transportation. If the actual links of the 1980 network had had unlimited capacity, then these forecasts would be acceptable, although still not quite correct. The remaining error would be due to the incremental nature of the actual implementation of changes in transportation facilities: some portions of the network come into public use and affect land use before the completion of others. (This does not take into account other, 'unknown' sources of error in the structure and input data of the model.) Any model which makes its forecasts by discrete jumps through time, rather than by some continuous process, will inevitably miss some of these phenomena. It remains for future researchers to determine whether the potential increases in accuracy of forecasts which might result from a 'continuous' model would justify the likely increases in

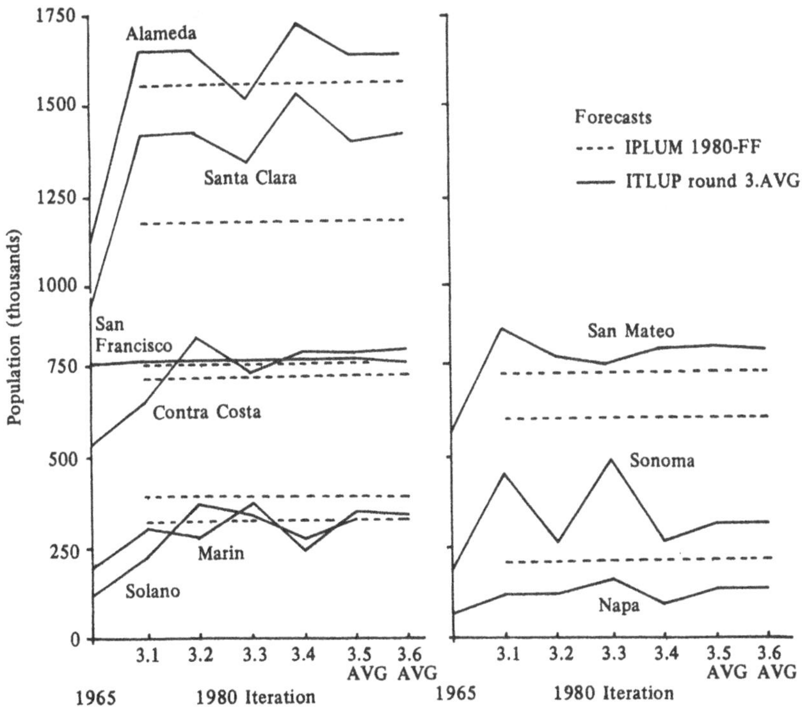

Figure 4.1. Population forecasts for the San Francisco Bay Area by county, 1980: comparison of the integrated package with single IPLUM forecast.

computation and data-collection costs. In figure 4.1 the 1980-FF forecast is plotted with the round 3.AVG iterations. It should be noted that in those areas, shown in figure 4.2, where there has been substantial real growth in population (the southeastern counties), the ITLUP results are consistently greater than those of 1980-FF. This was (and still is) an encouraging result, and was suggestive of the potential advantages in forecasting accuracy to be had from the integrated model system.

Before concluding this section of the chapter, some further discussion of the network assignment modelling is in order. It is clear that it is of importance to have ways of assessing the 'situation' on the transport networks. Since the OD trip matrix will change as the locations of population and employment change, the volume loadings on the network links will provide an indication of the traffic conditions on the network. One possible measure of network conditions is the total time taken by all trips. Unfortunately this measure does not accurately reflect the true traffic situation on the network as it provides no distinction between the various combinations of larger numbers of short trips or fewer numbers of long trips, all of which sum to yield the same total trip time. After further consideration it was decided that the means and standard deviations (SDs) of (a) volume/capacity ratio v/v_0, (b) origin-destination impedance, and (c) trip length, would provide some useful information about the traffic situation on the network. These measures are shown calculated for round 3.AVG in table 4.2.

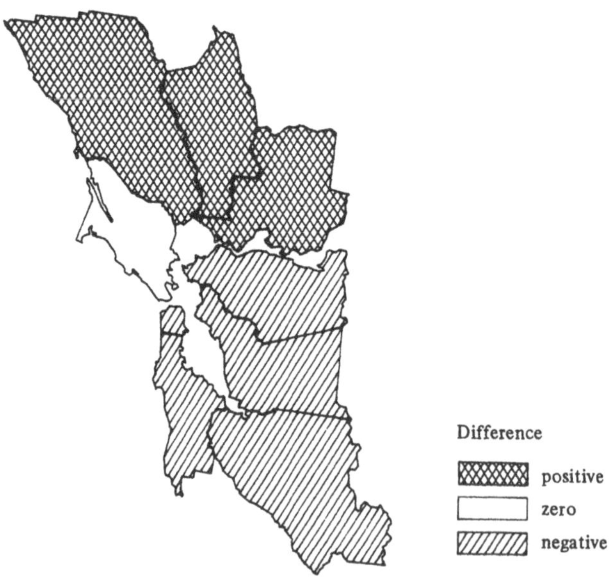

Figure 4.2. San Francisco Bay Area, nine county system: differences in population forecasts—IPLUM run 1980-FF minus ITLUP iteration 3.6AVG.

Although it can be seen that most of the previously described oscillations in the population forecasts follow the oscillations of the network measures, these measures are not zone-specific or county-specific. Consequently it was not possible to compare zonal or county forecasts with the conditions on the relevant portions of the network. An attempt was therefore made to develop a crude zone-specific measure of congestion. First it must be noted that in each zone there is a corresponding load node in the network. All traffic which either originates or terminates in the zone must pass through this load node. Now, for all of the network links which originate at the load node, consider the traffic on these links and the consequent value of volume/capacity ratio for the loaded links. It is thus possible to compute an average volume/capacity ratio, weighted by the traffic experiencing the level of congestion implied by the volume/capacity ratio, for the links originating at each load node. Although this measure does give an indication of zonal congestion, it has the fault of not distinguishing 'through traffic' on the links—traffic passing through the load node (that is, traffic which neither originates nor terminates in that zone)—from trips originating at the load node. Consequently, the specific spatial location of the load node in the zone affects the amount of through traffic and the numerical value of the congestion measure. Nevertheless, this zonal measure could be aggregated to the county level, thereby effectively averaging out the effect of differences in the position of the load node for each zone and thus giving a county-level measure of network congestion. The values of this congestion were calculated for the runs of round 3.AVG and are shown in table 4.3.

When the congestion measures were compared with the population projections, some interesting relationships became apparent. There were two general classes of behavior observed. The first was where the population projections and congestion measures were one iteration 'out of phase' with each other: for example, iteration 3.3 congestion is 'high' and iteration 3.3 population is 'low', whereas iteration 3.4 congestion is 'low' and iteration 3.4 population is 'high'. This situation occurred for five of the nine counties: Santa Clara, San Mateo, Sonoma, Napa, and Marin. This behavior was what was expected of the model system, as, in general, high population results in high congestion which results in lowered

Table 4.2. Network characteristics on input to iterations of round 3.AVG.

Iteration	Volume/capacity ratio, v/v_0		Origin-destination impedance (min)		Trip 'length' (min)	
	mean	SD	mean	SD	mean	SD
3.3	0·798	0·839	74·2	38·3	32·0	23·4
3.4	0·883	0·960	89·0	52·0	43·5	39·2
3.5AVG	0·827	0·822	84·0	44·3	36·2	27·2
3.6AVG	0·811	0·804	80·9	42·5	34·1	24·7

population which, in turn, results in lowered congestion which results in higher population, and so on.

The remaining four counties—Alameda, Contra Costa, San Francisco, and Solano—exhibited a different form of behavior. In these counties the population projection and congestion measures were 'in phase' with each other: for example, iteration 3.3 congestion is 'high' and iteration 3.3 population is 'high', whereas iteration 3.4 congestion is 'low' and iteration 3.4 population is 'low'. This represents a situation where, in general, high population results in low congestion which results in lowered population which, in turn, results in higher congestion, which results in higher population, and so on. This was, at the time, a rather unexpected mode of behavior. Several possible explanations were considered and investigated. First, there was the possibility that the through traffic ignored by the congestion measure was sufficiently important to bias the measure and make it an inaccurate indication of the true area-specific levels of congestion. A second possibility was that congestion could not be properly measured without considering numbers of originating trips and average trip lengths: for example, the actual congestion in an area could increase if number of trips declined slightly and trip lengths decreased substantially. The fact that there appeared to be a systematic spatial distribution of the counties displaying these modes of behavior, shown in figure 4.3, suggested that this problem could be resolved in a systematic fashion. It was not until some time later that the determinants of this behavior were identified. The problem and its resolution will be explained and discussed in chapter 9.

Two important positive points will conclude the discussion of network measures. First, the overall network measures given in table 4.2 did converge in a similar manner to the population forecasts and were probably indicative of the desired equilibrium being reached. Second, the congestion measures shown in table 4.3 also seemed to converge. Finally, it must be noted that in the development (especially of the prototype) of a system of

Table 4.3. County-level measures of congestion v/v_0 on input to iterations of round 3.AVG.

County	Iteration			
	3.3	3.4	3.4AVG	3.6AVG
San Francisco	1·891	2·005	1·939	1·928
Marin	1·425	2·025	1·229	1·090
San Mateo	2·246	2·176	2·144	2·102
Sonoma	2·446	2·711	2·453	2·407
Napa	1·048	2·218	1·341	1·262
Solano	1·837	1·467	1·526	1·551
Contra Costa	1·558	2·311	1·567	1·566
Santa Clara	1·719	1·663	1·692	1·677
Alameda	1·810	1·922	1·777	1·766

models and interrelationships as complex as this one, the existence of simplifying assumptions and certain gaps, both theoretical and operational, was inevitable. In this developmental context, especially in view of the preliminary nature of these results, the question was not one of absolute accuracy of forecasts. The nature of the functioning and response of the whole system was the key result. The demonstration of the relative sensitivity of land-use forecasts to transportation network conditions and of traffic on the networks to land-use forecasts, and that equilibrium was reached, was the principal result of these initial development stages of the work. In the remaining sections of the chapter, a series of policy tests undertaken with the prototype integrated package will be described and discussed.

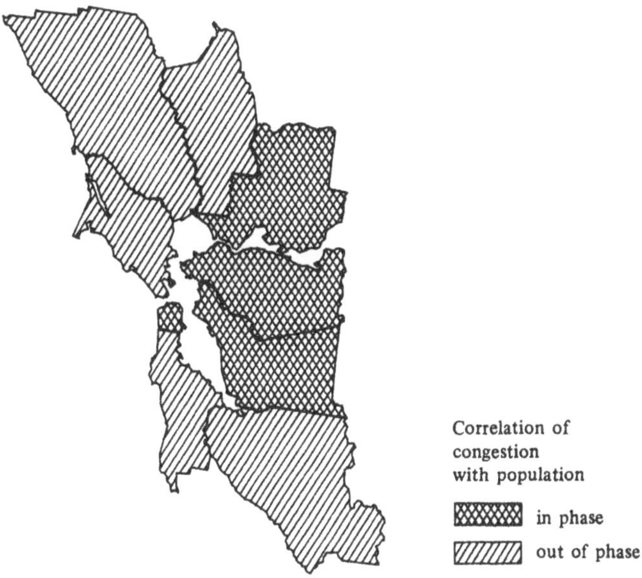

Figure 4.3. San Francisco Bay Area, nine county system: congestion-centralization phasing.

4.3 Policy tests: background

After the completion of the initial development of the prototype integrated package, and after subsequent tests and modifications as described in the previous chapter, several rather simple policy inputs were examined. It was expected that the results of these analyses would provide substantive information of interest to the sponsors of the project (the US Department of Transportation), as well as methodological information useful for guiding further work with the model package. After some consideration, it was decided that three types of policy would be tested. The first involves an arbitrary change in the spatial location of some of the exogenously produced increments in basic (NLMS) employment. The second test involves the imposition of land-use controls by altering land available for development. Finally, the third test examines the results of specific improvements in the transportation network.

Note that the presentation and discussion of these test results cannot be accomplished without reference to the geography and spatial distributions of activities in the San Francisco Bay Area. To assist the reader in understanding these discussions, two maps are provided here. Figure 4.4 shows the full 291 zone system with which the model system operated. Figure 4.5 shows the place names and some of the highway system in the San Francisco Bay Area.

The previous test runs of the package had suggested that the exogenous forecasts of the location of basic employment resulted in the rapid urbanization of the eastern part of Alameda County, in particular the area of Pleasanton and the Crow Creek end of Castro Valley. This effect was so marked, and as this area was indeed experiencing problems of rapid growth, it was decided to concentrate the policy testing and analysis on Alameda County (with the other basic employment forecasts provided by BEMOD left unaltered). Thus these analyses sought to gauge the

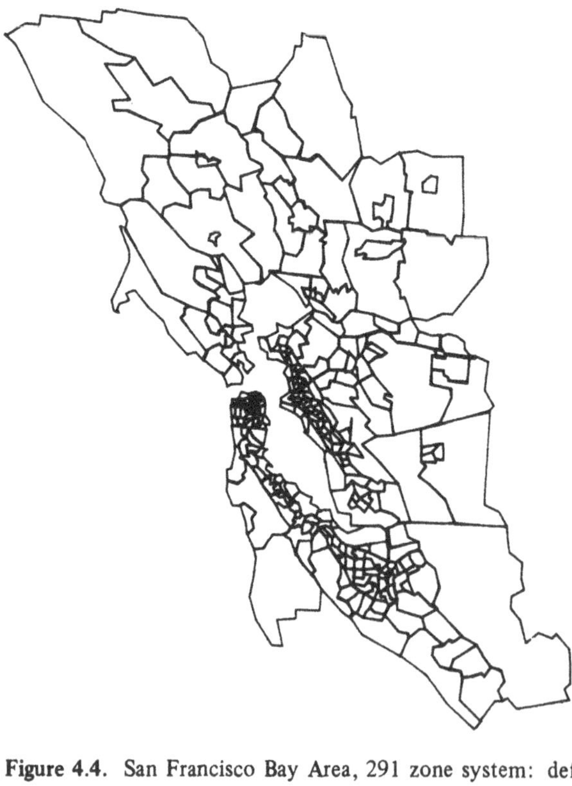

Figure 4.4. San Francisco Bay Area, 291 zone system: definition of analysis zones.

sensitivity of the model to policy input by examining the behavior of these areas in response to different patterns of industrial development. It was felt that in practice it would be difficult to implement policies sufficiently powerful to alter the location of existing industry, but that new industrial development would be more susceptible to attempts at policy manipulation. With this in mind, the forecast 1980 distribution of increments in basic industry was examined. Most of the increments fell in the northwest corner of Alameda County, in the area between San Leandro and Berkeley, with small increments scattered elsewhere.

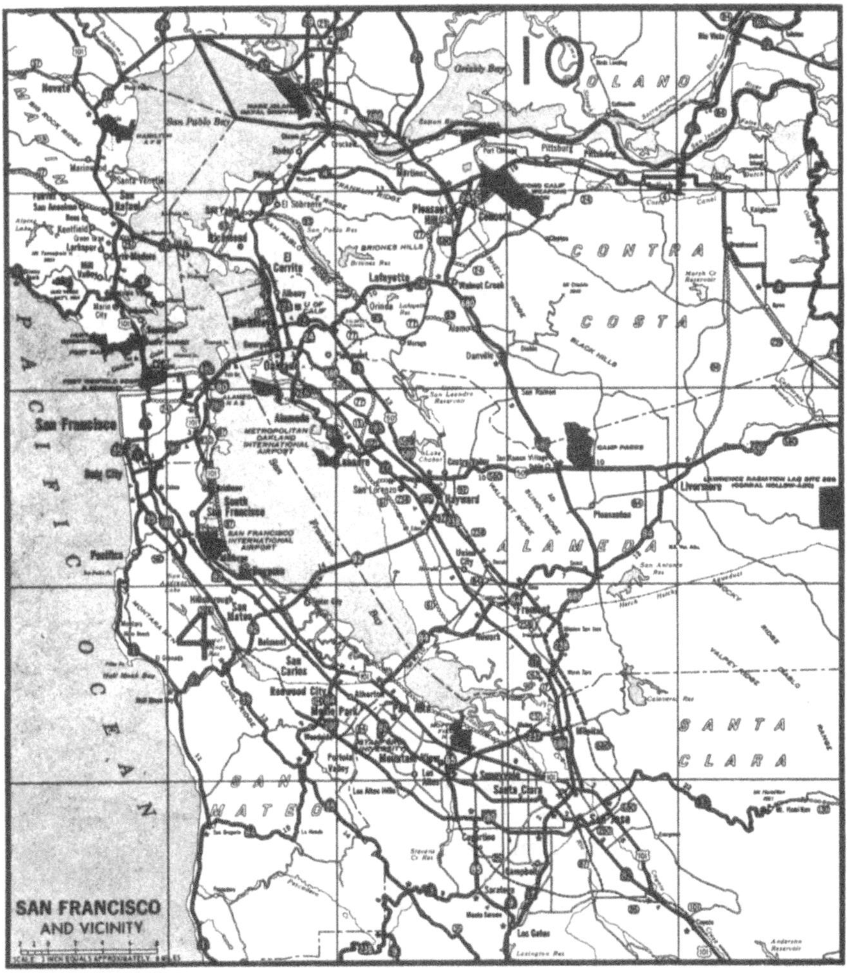

Figure 4.5. San Francisco Bay Area: place names and highways.

Two possible strategies for the location of basic employment seemed worth examining. One strategy might be to attempt to restrict the increases in basic employment to the present heavily urbanized areas of Berkeley and Oakland. In this way employment would be directed away from easy access to the Pleasanton and Castro Valley areas, and so less rapid urbanization could be expected in those areas. In addition, local highway congestion might have resulted from this concentration of employment, and this would have led to longer journey times and possibly to a more centralized urban pattern in the county. An alternative strategy might be to attempt to concentrate the growth of industry in several scattered main centers, or activity clusters. With this type of policy, however, there would probably be strong tendencies towards increased suburban sprawl.

It was also desired to investigate several other possible policy inputs. Consequently an overall strategy for policy testing with the integrated package was developed, consisting of three components:

(1) movement of basic employment in Alameda County to its northwestern corner, as described above;

(2) imposition of land-use controls to restrict land development in the zones running along the mountains from Lafayette to Fremont;

(3) construction of two high-speed highways with very limited access, one from Oakland to Livermore and the other from Fremont to Livermore. These roads would have no intermediate access points and would be superior to existing links.

The overall effect desired from this policy was the establishment of a green belt along the area from Lafayette to Fremont. Furthermore, it was hoped that these policy inputs would yield a development at Livermore with the elimination (or at least some reduction) of sprawl development southeastward along the Berkeley-Fremont axis. This would also require very stringent control of land use in the Livermore area in order to avoid sprawl emanating from that nucleus. Finally, it was hoped that these policies would result in the centralization of development along the eastern shore of the San Francisco Bay in the already developed Berkeley-San Leandro strip.

4.4 Policy test 1: redistribution of basic employment

A full round of model runs (round 4) was done to establish the effects of the redistribution of basic employment in Alameda County as described above. The results of this set of runs are shown in table 4.4. If one bears in mind that the total basic employment in Alameda County was unchanged as a result of this policy, and that the 1965-1980 change represented a little less than 30% of the 1980 total, the effects on the distribution of population appear quite significant. Iteration 3.6AVG of round 3 which was tabulated in table 3.18 is taken to be the base, or benchmark, run. Comparisons between iterations 4.5 and 3.6AVG are given in table 4.5.

The general tendency of the result was that population shifted away from Alameda, and to a lesser extent from Contra Costa and Santa Clara, into the counties to the north of the region. Of the more northern counties, Sonoma and Solano showed the greatest increases, followed by Marin and Napa. San Mateo and San Francisco showed only very small changes.

Table 4.4. Population distributions for the nine counties of the San Francisco Bay Area: round 4.

County	1965	1980 iteration[a]				
		4.2	4.3	4.4	4.5	4.6
San Francisco	754754	762464 (1·02)	763373 (1·14)	765262 (1·39)	764778 (1·33)	765224 (1·39)
Marin	192523	256409 (32·2)	369969 (92·2)	306671 (59·3)	340685 (76·9)	337421 (75·3)
San Mateo	553271	716094 (29·4)	742466 (34·2)	793707 (43·5)	773802 (39·9)	799310 (44·5)
Sonoma	167497	256690 (53·3)	497316 (196·9)	266828 (59·3)	333490 (199·1)	269962 (61·2)
Napa	69378	174451 (151·5)	168534 (142·9)	108524 (56·4)	146955 (111·8)	122767 (76·9)
Solano	149386	366950 (145·6)	329164 (120·3)	248165 (66·1)	356274 (138·5)	274330 (83·6)
Contra Costa	534056	778539 (45·8)	733951 (37·4)	794572 (48·8)	760834 (42·5)	796591 (49·2)
Santa Clara	922357	1334145 (44·7)	1315736 (42·6)	1433150 (55·4)	1379884 (49·6)	1388694 (50·6)
Alameda	1095236	1761796 (60·9)	1487035 (35·8)	1690663 (54·4)	1558833 (42·3)	1653243 (50·9)
Total	4438358	6407469 (44·4)	6407467 (44·4)	6407469 (44·4)	6407470 (44·4)	6407471 (44·4)

[a] Percentage changes 1965–1980 given in brackets beneath each value.

Table 4.5. County-level comparison of 1980 population forecasts: implementation of policy test 1.

County	1965	1980 iteration		Difference	
		3.6AVG	4.5	percentage	absolute
San Francisco	754754	764950	764778	−0·0	−172
Marin	192523	320451	340685	+6·3	+20234
San Mateo	553271	778837	773802	−0·7	−5035
Sonoma	167497	288333	333490	+15·7	+45157
Napa	69378	130566	146955	+12·6	+19389
Solano	149386	307813	356274	+15·7	+48461
Contra Costa	534056	788455	760834	−3·5	−27621
Santa Clara	922357	1401803	1379884	−1·4	−21919
Alameda	1095236	1626328	1558833	−4·2	−67495

It appeared from these results that the 'natural' areas for the residence of workers in northern Alameda County are in the next most northern counties, particularly since the northern section of Alameda and closer sections of Contra Costa are already substantially developed. The concentration of additional population in Sonoma and Solano suggests that commuters find most residential location opportunities available by using the Richmond–San Rafael Bridge (route 17) and Benicia–Martinez Bridge (route 680), both of which connect Contra Costa to the northern counties. In figure 4.6 the broad arrow represents the policy-determined movement of basic employment.

A more detailed examination of the zonal totals, given in table 4.6 and figure 4.7, confirmed the overall impression given by the county totals. In general, the zonal totals changed in the same way as the county totals. There was one significant exception to this in Contra Costa County where the Briones Hills area, zone 135, showed an increase in population. The other zones in Contra Costa County that showed noticeable changes (that is, of above or around 10%) all showed decreases. However, closer inspection showed this result to be in line with the overall pattern, as this zone was close to major roads running north and east, and contained substantial quantities of vacant developable land. In Marin, Napa, and Sonoma the significant zonal changes were found, as might be expected, at the southern edges of the counties in zones close to the Bay. There was one exception to this, in Sonoma, where zone 92, corresponding to the

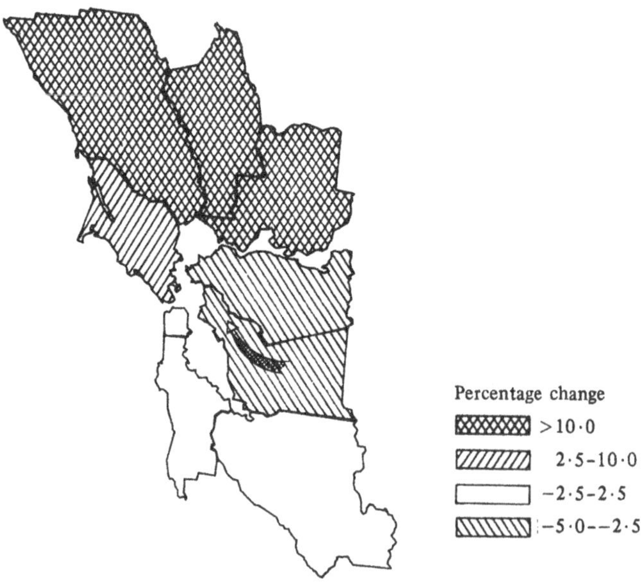

Figure 4.6. San Francisco Bay Area, nine zone system: change in county population due to redistribution of basic employment.

Santa Rosa area, showed large increases. This was to be expected as examination of the zone's attributes clearly suggests that this was one of the more accessible areas in the county, and was already developing.

Almost all the zones in Solano County showed substantial changes in resident population. The exceptions were zones 111, 112, and 125 which

Table 4.6. Round 4 (policy test 1): significant zonal population changes.

County	Zone	Iteration		Percentage difference
		4.5	3.6AVG	
San Francisco		no large changes		
Marin	41	8106	7262	11·6
	49	32909	29274	12·6
	51	17235	14317	20·4
San Mateo		no large changes		
Sonoma	92	92953	60736	50·6
	93	38890	34116	35·9
	100	31385	28609	9·7
Napa	104	7043	5606	25·6
	109	37542	31924	17·6
	110	7325	6638	10·3
Solano	113	37033	34122	8·5
	115	22528	20061	12·3
	117	10262	8252	24·6
	118	79018	65559	20·5
	119	3633	3163	14·9
	120	4239	3628	16·8
	121	8651	7469	15·8
	122	35106	24443	43·6
	123	51779	39825	30·1
	124	5175	4741	9·2
Contra Costa	135	32027	28538	12·2
	137	32906	35289	−6·8
	139	10911	11901	−8·3
	150	13813	15053	−8·2
	151	43636	48347	−9·8
	154	9081	11241	−19·2
	155	2963	4514	−34·3
	156	10910	13238	−17·6
	157	18184	21192	−14·2
	158	29667	33880	−12·4
Santa Clara		no large changes		
Alameda	223	7619	8469	−10·0
	226	115658	137291	−15·8
	227	30807	35264	−12·6
	228	53607	63047	−15·0
	236	84028	100466	−16·4
	252	10637	12649	−15·9

correspond roughly to the Vallejo area. Examination of the land available in those zones suggested that they were already well developed. In contradistinction, the counties Santa Clara, San Mateo, and San Francisco showed little change on a zone-by-zone basis. The zonal changes in Contra Costa and Alameda were concentrated in the inland sections. This suggested that those areas lying to the east of the coastal hills were the 'natural' commuting areas for people working in southern Alameda. Examination of figure 4.5 seems to confirm this, as the main access routes from northern Alameda County into the hinterland were through built-up areas and thus likely to be congested. The routes from the south, although not of such high capacity, were less likely to carry local traffic and were therefore less likely to be congested. The congestion measures by county, as described in the latter part of section 4.2, are shown in table 4.7. It can be seen, as anticipated in the discussion above, that by comparison with iteration 3.6AVG, which is included in table 4.7 for convenience, there were significant increases in congestion both in Napa and in Alameda as a result of this policy.

The weighted average lengths of trips originating in the counties are given in table 4.8. Note that the trips (predominantly work trips) from

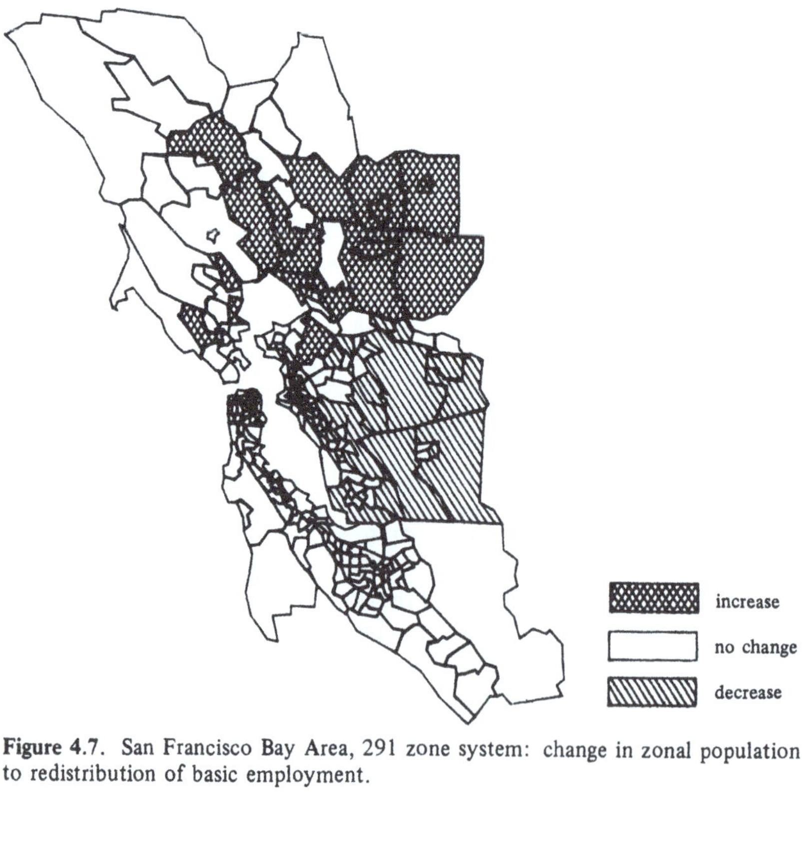

Figure 4.7. San Francisco Bay Area, 291 zone system: change in zonal population due to redistribution of basic employment.

Marin, Sonoma, Napa, and Solano were all somewhat shorter in iteration 4.5, corresponding to the northward shift of the employment locations to which many of the residents of these counties commute. The trips from San Mateo, Santa Clara, and Alameda were, as one would expect, somewhat longer; this was also due to the northward shift of employment locations to which at least some, even in the southern end of the region, of their residents commute. Finally, the total time spent by tripmakers on the network increased from 31 338 000 min in the base run (iteration 3.6AVG) to 33 523 220 min in the 'policy 1' run (iteration 4.5). It is worth noting here how an attempt to limit sprawl and preserve open space, although achieving some degree of success, seemed also to increase trip lengths regionwide.

Table 4.7. County-level measures of congestion v/v_0 on input to iterations of round 4.

County	Iteration				
	4.3	4.4	4.5	4.6	3.6AVG
San Francisco	1·890	1·935	1·925	1·926	1·928
Marin	1·398	1·307	1·075	1·067	1·090
San Mateo	2·243	2·166	2·133	2·115	2·102
Sonoma	2·440	2·503	2·413	2·419	2·407
Napa	1·059	1·591	1·216	1·422	1·262
Solano	1·859	1·584	1·514	1·556	1·551
Contra Costa	1·578	1·757	1·457	1·604	1·566
Santa Clara	1·716	1·675	1·683	1·674	1·677
Alameda	1·974	2·006	1·925	1·933	1·766

Table 4.8. Weighted average length, $\bar{t}$(min), of trips originating in each county: effect of policy test 1.

County	Iteration		County	Iteration	
	3.6AVG	4.5		3.6AVG	4.5
San Francisco	25·30	25·05	Solano	38·01	35·70
Marin	45·46	43·26	Contra Costa	44·43	44·40
San Mateo	38·62	39·52	Santa Clara	30·81	32·24
Sonoma	35·10	34·47	Alameda	34·91	36·57
Napa	35·74	34·37			

4.5 Policy test 2: high-speed road links

The second full round of runs for policy testing, round 5, was an attempt to assess the effect of changes in the highway network. These changes involved the construction of high-speed high-capacity links (shown in figure 4.8) between Oakland (northern Alameda) and Livermore Valley (central Alameda), and between Livermore and an area just northwest of San Jose (northern Santa Clara). Note that the redistributions of basic employment in policy test 1 were not part of the inputs to policy test 2.

The results of round 5 are shown in table 4.9. Again with iteration 3.6AVG as the base, a comparison with iteration 5.5 is shown in table 4.10. The overall regional pattern had some expected features, and some rather unexpected ones. First, as might be expected, the three most northern counties lost population, some of which moved into the southern counties of Santa Clara and San Mateo. The unexpected results came in Alameda, which showed little change in total population, and in Marin, which showed a large increase. The result for Alameda may be attributed to the redistribution of population between zones within the county, which will be discussed below. The increase in Marin was more difficult to explain. It is possible that Marin and the other northern counties were almost equally accessible, but with Marin just that much more so that it only required small changes in the network to tip the balance. On figure 4.8 the bold lines indicate the new network links.

At the zonal level, the zones to the east and south of Santa Clara County showed gains, as shown in figure 4.9 and table 4.11. The largest increases were in zones near the end of the newly constructed road. This was to be expected, since those were the areas rendered more accessible by the changes in the network. San Mateo County showed mixed changes. The largest change was population increase in zone 75 which was large, empty, at the end of the Dumbarton Bridge (route 84) near the city of San Mateo, and thus likely to benefit from the newly constructed road. Large increases took place at the extreme northern end of the county, adjacent

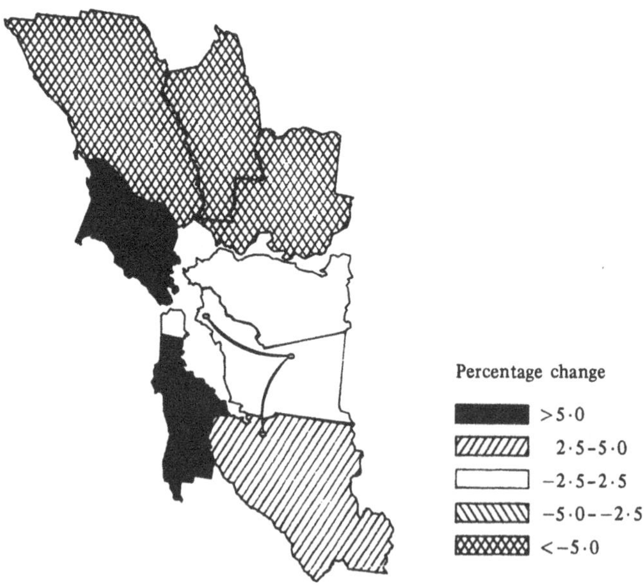

Figure 4.8. San Francisco Bay Area, nine zone system: change in county population due to construction of new high-speed roadway.

to San Francisco, whereas large decreases occurred at the extreme southern and western areas. Zones in Sonoma, Napa, and Solano showed fairly uniform decreases in almost all zones. Marin, however, showed uniform increases, with the largest increase being in the zone adjacent to the end of the Richmond-San Rafael Bridge to Berkeley and Oakland (route 17).

Table 4.9. Population distributions for the nine counties of the San Francisco Bay Area: round 5.

County	1965	1980 iteration [a]				
		5.2	5.3	5.4	5.5	5.6
San Francisco	754754	765562 (1·4)	767672 (1·7)	765841 (1·5)	766898 (1·6)	765946 (1·5)
Marin	192523	265593 (37·9	384865 (99·9)	327091 (69·9)	357235 (85·6)	399042 (76·1)
San Mateo	553271	741227 (33·9)	852372 (54·1)	777238 (40·5)	826875 (49·5)	789550 (42·7)
Sonoma	167497	322346 (92·4)	240455 (43·6)	298937 (78·5)	250551 (49·6)	332285 (98·4)
Napa	69378	166038 (139·3)	111759 (61·1)	130706 (88·4)	114863 (65·6)	136468 (96·7)
Solano	149386	433627 (190·2)	230153 (54·1)	312623 (109·3)	256759 (71·9)	317144 (112·3)
Contra Costa	534056	790064 (47·9)	766631 (43·5)	769153 (44·0)	783668 (46·7)	751502 (40·7)
Santa Clara	922357	1318723 (42·9)	1443655 (56·5)	1331590 (44·4)	1438544 (55·9)	1366135 (48·1)
Alameda	1095236	1604364 (46·5)	1610021 (47·0)	1694371 (54·7)	1612145 (47·2)	1609470 (46·9)
Total	4438358	6407477 (44·4)	6407470 (44·4)	6407469 (44·4)	6407472 (44·4)	6407475 (44·4)

[a] Percentage changes 1965-1980 given in brackets beneath each value.

Table 4.10. County-level comparison of 1980 population forecasts: implementation of policy test 2.

County	1965	1980 iteration		Difference	
		3.6AVG	5.5	percentage	absolute
San Francisco	754754	764950	766898	+0·0	+1948
Marin	192523	320451	357235	+11·5	+36784
San Mateo	553271	778837	826875	+6·2	+48038
Sonoma	167497	288333	250551	−13·1	−37782
Napa	69378	130566	114863	−12·0	−15703
Solano	149386	307813	256759	−16·6	−51054
Contra Costa	534056	788455	783668	−0·6	−4787
Santa Clara	922357	1401803	1438544	+2·6	+36741
Alameda	1095236	1626328	1612145	−0·9	−14183

Zones in Contra Costa County showed, for the most part, only small changes. The northern zones followed the trend of Solano County and showed decreases. The southern and inland zones followed the trend in the inland areas of Alameda County and also showed decreases. However, zone 142 near Moraga in Contra Costa County, which is rather near one end of the new road, increased in population, along with the nearby zone 251 in Alameda County, which also showed a significant increase. Zones in Alameda County showed mixed behavior. As had been expected, zone 252, east of Livermore, where the new superhighway ends, showed a substantial increase in population. This balanced the loss in population in zones which are bypassed by the new road in the Livermore Valley. In part, this was a desirable result, since it suggested that a limited access highway might limit sprawl development in the areas it crosses but to which it does not give access.

Some of the unexpected zonal behavior was ascribed to the fact that the IPLUM allocation functions do not put people into the residence zone nearest their employment zone, but distribute residential locations over many zones. Further, the IPLUM accessibility measures, via their normalization procedures, operate in terms of relative accessibility of one

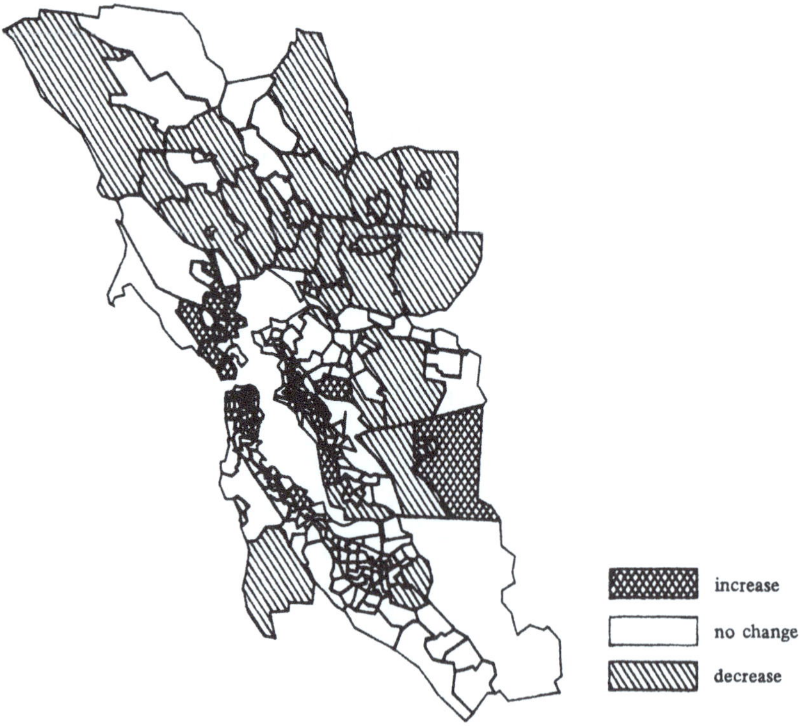

Figure 4.9. San Francisco Bay Area, 291 zone system: change in zonal population due to construction of high-speed roadway.

Table 4.11. Round 5 (policy test 2): significant zonal population changes.

County	Zone	Iteration 5.5	Iteration 3.6AVG	Percentage difference
San Francisco		no large changes		
Marin	41	10506	7262	+44·67
	44	23845	20199	+18·05
	45	21126	18574	+13·74
	48	47381	42148	+12·41
	49	33675	29274	+15·03
	50	26270	21854	+20·20
	52	14428	12157	+18·68
	55	18136	15383	+17·90
	56	19874	16289	+22·00
San Mateo	75	93469	68604	+36·24
	88	6534	10224	−36·09
	89	6930	8136	−14·82
	90	14120	11445	+23·37
	91	13751	10459	+31·48
Sonoma	92	47813	60736	−21·28
	93	29287	34116	−14·15
	96	23288	28988	−19·66
	99	11237	13833	−18·77
	100	18773	28609	−34·38
Napa	102	51998	57740	−9·94
	104	4437	5606	−20·85
	105	923	1268	−27·21
	109	25106	31924	−21·36
	110	5836	6638	−12·08
Solano	113	28491	34122	−16·50
	115	17618	20061	−12·17
	116	1937	2435	−20·45
	117	5162	8252	−37·45
	118	50345	65559	−23·21
	119	2385	3163	−24·59
	120	2690	3628	−25·85
	121	5996	7469	−19·72
	122	16694	24443	−31·70
	123	25258	39825	−36·58
Contra Costa	136	25924	29609	−12·44
	142	27184	23157	+15·14
	156	11323	13238	−14·47
	158	30264	33880	−10·67
Santa Clara	166	81884	71139	+15·10
	169	48004	36519	+31·44
Alameda	223	6681	8469	−21·08
	224	11291	15128	−25·36
	225	15791	20359	−22·44
	226	89023	137291	−35·15
	228	71938	63047	−16·26
	238	107725	85910	+25·39
	251	18846	16958	+11·13
	252	20429	12649	+61·15
	260	20021	16894	+18·51

zone vis-à-vis all others. This is probably representative, at least to some extent, of reality. In any case, when the trip-length measures, shown in table 4.12, were examined, some support for these conclusions became apparent. From these results, it appeared that the regional effect of the new road had been to shift the concentration of employment and population of the entire San Francisco Bay region towards the south. Marin County was 'swamped' with new residents shifting south from Sonoma and Napa, and San Mateo and Santa Clara took the rest of the shift in population from the centrally located counties. Those residents remaining in the northern half of the region appeared to be making much longer work trips. Even the work trips of residents in Santa Clara and Alameda increased in length, as they probably made use of the new road to make the somewhat longer trip to the Berkeley–Oakland area in northeastern Alameda and Contra Costa. Total travel times for the 1980 forecast year increased from 31338000 min in the base run (iteration 3.6AVG) to 32494953 in the 'policy 2' run (iteration 5.5). The congestion measures which resulted from these trips are shown in table 4.13. The principal point to be noted here is that even if the model package had overresponded to this policy input, it has still demonstrated the importance of the regionwide, interrelated effects which may be overlooked in a traditional transportation impact analysis.

Table 4.12. Weighted average length, $\bar{t}$(min) of trips originating in each county: effect of policy test 2.

County	Iteration		County	Iteration	
	3.6AVG	5.5		3.6AVG	5.5
San Francisco	25·30	26·01	Solano	38·01	46·27
Marin	45·46	45·02	Contra Costa	44·43	49·51
San Mateo	38·62	39·61	Santa Clara	30·81	34·78
Sonoma	35·10	42·69	Alameda	34·91	36·46
Napa	35·74	41·65			

Table 4.13. County-level measures of congestion v/v_0 on input to iterations of round 5.

County	Iteration				
	5.3	5.4	5.5	5.6	3.6AVG
San Francisco	1·713	1·925	1·908	1·916	1·928
Marin	0·724	0·935	0·902	0·920	1·090
San Mateo	1·861	2·066	2·056	2·052	2·102
Sonoma	1·856	1·856	1·910	1·863	2·407
Napa	1·368	1·306	1·308	1·268	1·262
Solano	1·500	1·468	1·428	1·367	1·551
Contra Costa	1·280	1·456	1·415	1·440	1·566
Santa Clara	1·368	1·590	1·587	1·588	1·677
Alameda	1·497	1·629	1·610	1·635	1·766

4.6 Policy test 3: land-use controls

Owing to the shortage of time and resources during the final months of the original project, a full round of runs was not done for the land-use control policy test. The response of the model system to land-use control was tested by rerunning iteration 3.6AVG with the land-use controls policy imposed by defining additional land in certain zones as being unavailable for development. These changes were accomplished by transferring land from the 'vacant' category to the 'unusable' category on input to the model. The changes were as shown in table 4.14. The county-level results of this run, called iteration 6.1, are shown compared with the results of iteration 3.6AVG in table 4.15. Figure 4.10 shows the county-level shifts in population, and the areas in which stringent land-use controls were applied are shown shaded.

As can be seen from figure 4.10, the effect of these controls was to reduce relative population growth in the counties (Contra Costa and Alameda) where the controls were applied. Note that this was a reduction from the 'no-controls' base values (iteration 3.6AVG), but that there was still substantial absolute growth forecasted in these counties over the

Table 4.14. Land-use change: inputs to policy test 3.

Zone	Base		Policy test 3	
	L^v (acres)	L^u (acres)	L^v (acres)	L^u (acres)
140	2765	2741	277	5229
141	2780	1883	278	4385
142	8104	6571	810	13865
158	20255	2243	2026	20477
227	9696	8400	970	17126
236	20267	22110	2027	40350

L^v, vacant acreage; L^u, unusable acreage [see equation (3.10)].

Table 4.15. County-level comparison of 1980 population forecasts: implementation of policy test 3.

County	1965	1980 iteration		Difference	
		3.6AVG	6.1	percentage	absolute
San Francisco	754754	764950	763622	−0·17	−1328
Marin	192523	320451	402148	+25·49	+81697
San Mateo	553271	778837	764426	−1·85	−14411
Sonoma	167497	288333	498057	+72·74	+209724
Napa	69378	130566	184445	+41·27	+53879
Solano	149386	307813	446421	+45·03	+138608
Contra Costa	534056	788455	682986	−13·38	−105469
Santa Clara	922357	1401803	1294710	−7·64	−107093
Alameda	1095236	1626328	1370732	−15·72	−255596

Table 4.16. Round 6 (policy test 3): significant zonal population changes.

County	Zone	Iteration		Percentage difference
		3.6AVG	6.1	
San Francisco	— no large changes —			
Marin	41	7262	9741	+34
	43	5415	6434	+12
	44	20199	22793	+13
	45	18574	23006	+24
	47	21121	23417	+11
	48	42148	53482	+27
	49	29274	47306	+61
	50	21854	31656	+45
	51	14317	25599	+80
	52	12157	17923	+47
	53	13312	17078	+28
	54	8599	12344	+42
	56	16289	20436	+25
San Mateo	84	37924	31382	−17
	87	17715	15104	−15
	88	10224	7892	−23
	89	8136	6704	−18
Sonoma	92	60736	160479	+164
	93	34116	54837	+61
	94	29299	38191	+30
	95	36824	54422	+52
	96	28988	53638	+86
	97	19574	31833	+63
	98	10575	12025	+14
	99	13833	17024	+23
	100	28609	46551	+63
	101	19979	23255	+16
Napa	102	57740	79398	+37
	103	2448	3006	+23
	104	5606	9385	+67
	105	1268	2250	+78
	106	5118	10189	+97
	107	9109	13062	+43
	108	3238	4036	+24
	109	31924	46941	+47
	110	6638	8700	+31
Solano	113	34122	41145	+21
	114	29033	36352	+25
	115	20061	23818	+19
	116	2435	3103	+27
	117	8252	12350	+50
	118	65559	102619	+56
	119	3163	4526	+43
	120	3628	7109	+96
	121	7469	12521	+68
	122	24443	54748	+124

Table 4.16 (continued)

County	Zone	Iteration		Percentage difference
		3.6AVG	6.1	
Solano	123	39825	75851	+90
	124	4741	5813	+23
Contra Costa	134	26148	29042	+11
	136	29609	32999	+11
	137	35289	28779	−19
	140	8451	5454	−35
	141	18869	11958	−37
	142	23157	10785	−53
	143	31202	27403	−12
	145	38000	31332	−17
	146	23420	20118	−14
	149	13132	10723	−18
	150	15053	12104	−20
	151	48347	37046	−23
	152	26716	23714	−11
	153	6099	4974	−18
	154	11241	7995	−29
	155	4514	2442	−46
	156	13238	7011	−47
	158	33880	13235	−61
Santa Clara	161	39164	28703	−27
	162	17257	12205	−30
	163	23453	16796	−28
	164	6584	4829	−27
	165	20754	12075	−42
	166	71139	52340	−27
	167	35468	24492	−31
	169	36519	28409	−20
	170	12391	10880	−12
	172	43536	39308	−10
	173	36638	29193	−20
	174	39815	33153	−17
	181	37463	30965	−17
Alameda	224	15128	9674	−36
	225	20359	13897	−38
	226	137291	68461	−50
	227	35264	14820	−58
	228	63047	41888	−34
	229	47553	40219	−16
	230	18148	14677	−19
	232	21035	17155	−19
	234	14982	12963	−18
	235	24218	19557	−19
	236	100466	28925	−79
	238	85910	68097	−21
	241	30582	25976	−15
	252	12649	6884	−46

period 1965–1980. The counties to the south and southeast (San Mateo and Santa Clara) also showed a relative decline in population. This presumably was due to declines in nonbasic (local serving, or LOMS) employment in Contra Costa and Alameda to which employees residing in San Mateo and Santa Clara formerly commuted. Furthermore, indirectly induced declines in population and nonbasic employment took place throughout these four counties. True to the notion that if a growing region is squeezed in one place it bulges out in another, there was substantial relative growth in the four northern counties: Marin, Sonoma, Napa, and Solano. This growth, for the most part, represented the residential location of people employed in San Francisco and Contra Costa (Berkeley, Oakland, and other centers) who were unable to find residential space in Contra Costa and Alameda. Again, the indirect effects of nonbasic employment tend to amplify the trend.

The zonal results are given in table 4.16 and mapped in figure 4.11. There were no results at the zonal level which were not expected from the county-level results. It is, however, worth noting that virtually all the significant zonal changes took place on the urban fringe. This suggested that if appropriate land-use control policies were to be developed and implemented there would be a fair chance of having some effect on the shape of the future development of a region.

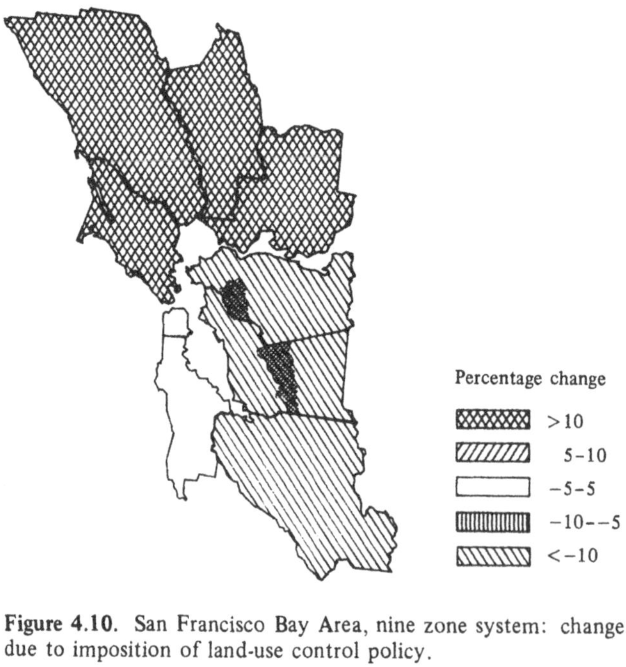

Figure 4.10. San Francisco Bay Area, nine zone system: change in county population due to imposition of land-use control policy.

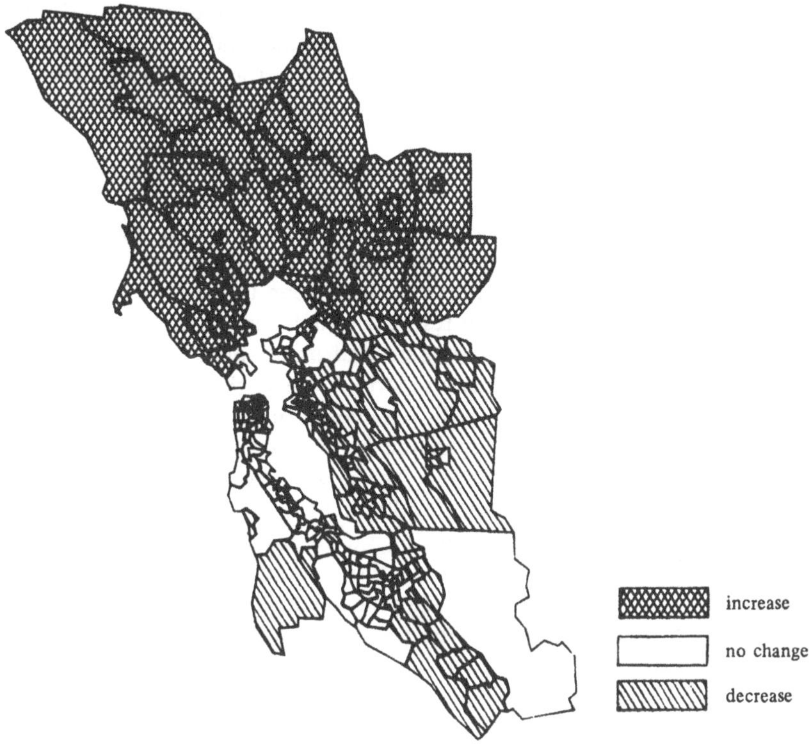

Figure 4.11. San Francisco Bay Area, 291 zone system: change in zonal population due to imposition of land-use control policy.

4.7 Summary of prototype policy tests

In a growing metropolitan region, the actual regional growth rate may be determined by factors over which the regional policymakers have little or no control. This may be even more true for the local policymaker in subareas of the region. To the extent that the overall growth of the region is significant and beyond the control of regional policymakers, they may find themselves in a paradoxical position. A policy which seems perfectly reasonable for one part of the region may have serious undesirable consequences elsewhere in the region. A salient feature of the policy test runs discussed here was the demonstrated fact of this regional interrelatedness. In each of these policy runs, a change in one portion of the region appeared to produce significant changes in other parts of the region.

The results of these policy tests made with the prototype integrated package can be summarized, in comparison with the base run, as follows.

Policy test 1 A policy to inhibit the spatial dispersion of basic employment in Alameda County by prohibiting the location of new basic employment in the south–central and eastern parts of the county did reduce residential and nonbasic (LOMS) employment sprawl both in Alameda and in Contra Costa. In the absence of a comprehensive regional land-use control policy, however, the desirable results in Alameda and Contra Costa were offset by the generation of considerable new sprawl development in Marin, Sonoma, and Solano.

Policy test 2 A policy to centralize impending development in eastern Alameda County by the introduction of a high-speed very limited access highway did not seem to produce the desired results. There was a decline in population and nonbasic employment in the area where it was to be centralized. The new highway seems to have been utilized for commuting over longer distances and to have resulted in a general southerly shift in the location of the region's employment and population. This seems to have had a substantial effect only in Marin County, the southeastern portion of which was swamped with new development of housing, and local-serving employment which resulted from a general southerly shift of the region's population in response to the new highway.

Policy test 3 An attempt to limit development in western Contra Costa County and central Alameda County by the imposition of stringent land-use controls was generally successful in those counties. Again, the lack of a regionwide policy allowed the development which was controlled in those counties to run rampant in the northern counties of Marin, Sonoma, Napa, and Solano.

These policy tests are presented here as examples of possible policy alternatives which could demonstrate the operation of the prototype integrated model package. As they are quite realistic examples of possible policy, they demonstrate, to a limited extent, the types of phenomena that are likely to occur in reality.

Finally, some clear conclusions relating to the overall aims of the early phases of the integrated model package project had emerged from this series of tests. There are strong interrelationships between transportation development and land development which are not adequately accounted for, even in complex computer simulation models, by traditional forecasting procedures. Furthermore, it was clear that as more knowledge of these phenomena was developed, powerful instruments would be identified for possible use by policymakers in shaping the growth of the metropolis. Thus it was also clear that integrated, regionwide planning of both transportation and land use is a necessary condition for the proper design and application of these policies. The next few chapters of this book discuss the many tests and improvements which were made to the components of the integrated transportation and location model package before the additional policy analyses described in chapter 9 were carried out.

5

Modifications and improvements to the transportation component

5.1 Introduction

The development of the first version of the integrated transportation and land-use model package made use of a conventional transportation network analysis procedure. As described in chapter 3, the operation of the integrated package involved the substitution of trips generated implicitly in the residence-location model for the standard procedures for trip generation and distribution used in conventional transportation models. The unconventional use of these implicitly generated trips required that a procedure be developed for conversion of the matrix of trip probabilities into a matrix of vehicle trips which could, in turn, be assigned to the transportation network. The second section of this chapter describes such a conversion procedure.

The development of the disaggregated residential allocation model (DRAM) (see Putman, 1977a) took place after the completion of work with the first version of the integrated package. The rather more straightforward formulation of DRAM makes its generation of trips more explicit, although the conversion of trip probabilities to vehicle trips is still necessary. The third section of this chapter is devoted to a description, including numerical examples, of the trip-generating capability of DRAM. It should be noted that this is the way in which trips are generated and distributed in current versions of the integrated package.

An illustration of a simple network assignment procedure was presented, in the form of a numerical example, in chapter 2. The assignment procedure actually used in the integrated package is of a more sophisticated form and is described in detail, along with the results of several tests conducted to verify the effectiveness of its operation, in the fourth section of this chapter. This ties in with a subsequent series of tests of an alternative assignment procedure involving a multipath assignment algorithm, which is described in the fifth section of this chapter.

The first version of the integrated model package made no attempt to incorporate any form of mode-split model. The reasons for this were lack of data and lack of project resources. The long-term usefulness—and credibility—of any integrated model package will inevitably require the inclusion of a theoretically sound, but practical, mode-split procedure. The sixth section of this chapter contains a description of work done on the development of a mode-split procedure and on a preliminary calibration effort. This continues into the final section of the chapter, where prospects for future work in this vein are discussed.

5.2 Conversions from trip probabilities to vehicle trips

In chapters 2 and 3, a simple formulation of a residence-location model was given as

$$N_i = \sum_j E_j p_{ij} , \tag{5.1}$$

where N_i is the number of persons residing in zone i, E_j the number of persons employed in zone j, and p_{ij} the probability of a work trip between zones i and j.

With this general formulation it was possible to produce simultaneous forecasts of residence location and trip distribution. This is done, if the above model form is used for residence-location forecasts, by scaling up the p_{ij} matrix to a vehicle-trip matrix. To repeat what has been said several times earlier in this book, it was this trip-extraction procedure that made the integrated model package a unique development in transportation and location analysis.

The problem of transforming the p_{ij} term in equation (5.1) into forecasts of actual vehicle trips to be assigned to a capacity-constrained transportation network is a classic illustration of the customary gap between theory and application. One of the underlying hypotheses of both the IPLUM model used in the first versions of the integrated model package and the DRAM model used in current versions of the integrated model package is that place of work is an important (but not exclusive) determinant of place of residence. To reflect this hypothesis, the expanded formulation of p_{ij} contains a function of c_{ij}, the travel time and/or cost between zones i and j. If, for the moment, the rather thorny question of how time and/or cost are to be measured is neglected, one is still faced with the question of when they are to be measured.

An anecdote sometimes told to illustrate high-pressure real estate sales practices clearly illustrates the path to be followed here. The story describes the executive who has been transferred to a branch office (or facility) of the firm, located in a different city. During a preliminary visit to the new city the executive arranges with a real estate agent to be shown a possible new house. A visit to the house is to take place at 2.00 pm (or perhaps on a Saturday or Sunday). On the appointed day a short trip of, say, fifteen to twenty minutes takes the executive from the vicinity of the new office, via an urban expressway, to the prospective new house. Perhaps after several such appointments a purchase is made of a new house for the executive's family. One month later the executive and family relocate to the new city and take up residence in the new house. The first day of work finds the executive finishing breakfast and departing for work. Based on the prior visit(s) to the house, twenty minutes are allowed for the work trip. The executive arrives at work twenty minutes late, owing to the customary, but unexpected, peak hour congestion on the expressway. The rather annoyed executive vows, for the future, to inspect new houses only during the peak hours.

The implication for the development of the model is clear: it is peak hour travel times and/or costs that should be used in the trip-function portion of the probability term in equation (5.1). Further questions remain as to which peak hour should be used, how ratios such as the number of persons per vehicle should be estimated, and the consistency of network specifications of link capacity (daily versus hourly, for example), and so on. By one means or another all of these issues had to be settled before even the first tests of the prototype integrated model package. First, an overview of the necessary transformations will be given.

In most metropolitan areas the afternoon peak ('rush hour') is of greater duration than the morning peak. This is true for the San Francisco Bay region, according to the data collected by the Bay Area Transportation Study Commission (BATSC, 1969). The BATSC home interview and trip surveys taken in 1965 showed approximately 740000 automobile and public transit trips originating during the period from 7.00 am to 8.00 am (weekdays). During the period 4.00 pm to 5.00 pm (weekdays), these surveys showed approximately 900000 trips. Thus the afternoon peak was significantly greater than the morning peak. This would, of course, lead to higher volumes on the network links, higher congestion levels, and consequently greater travel times and costs. A histogram of the hourly trip volumes (weekdays) shown in figure 5.1 shows that if the time period 4.00 pm to 5.00 pm were used, the average congestion levels to be found during the afternoon peak would probably be overestimated. Thus an

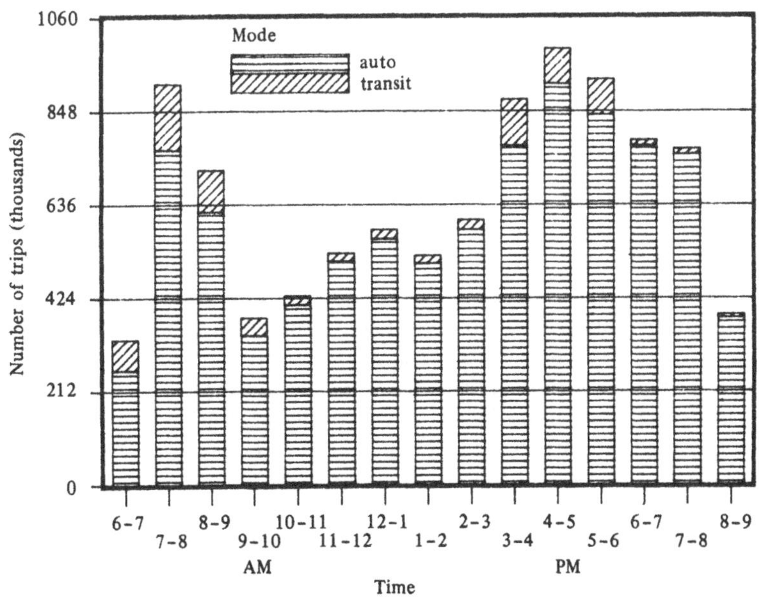

Figure 5.1. San Francisco Bay Area, 1965: hourly person trips by mode. Source: BATSC (1969).

attempt was made to use the traffic volumes for the period 4.30 pm to 5.30 pm, which more closely correspond to the average peak hour congestions.

In the integrated model package, three trip types are generated: (1) work-to-home, (2) work-to-shop, and (3) home-to-shop. Discussion of the 'shopping' trips will be deferred for the moment. The zone-to-zone distribution of the work trips results directly from the p_{ij} matrix as shown in the general form of the residence-location model, equation (5.1). What needs to be done is to convert these probabilities into vehicle trips for loading on the network. It has already been determined that it is afternoon peak vehicle trips that are required. The first step in the conversion was to determine the peak hour ratio of automobile person trips to total person trips. All data used in the following discussion are derived from the BATSC study cited above. The peak hour mode split is as follows (data on peak hour trips both by trip type and by mode were not available):

Time	Auto	Transit
4.00-5.00 pm	820000	100000
5.00-6.00 pm	745000	90000
Total	1565000	190000 .

Thus of the 1755000 trips taking place on an average weekday between 4.00 pm and 6.00 pm, only 190000, or 10·8%, were transit trips.

The next step in the conversion process involves estimating the total daily auto person trips if such trips were taken at the peak hour mode-split percentages. This conversion step is taken in order to obtain, eventually, estimates of auto person trips by trip purpose. The daily average number of auto person trips was 8953000. There were 933000 daily transit trips, and thus a total (not including walking trips) of 9886000 trips. If these trips had been made at the peak hour mode-split ratios, then there would have been 8815721 auto person trips. This figure may then be used to calculate a ratio of the implied number of daily auto person trips (if taken at the peak hour mode-split ratio) to the actual number of trips. This ratio is 8815720/8953000, which is equal to 0·9847.

The next step is to obtain a set of 'auto occupancy' rates to convert person trips into vehicle trips; these rates are shown in table 5.1. Next, the breakdown of peak hour auto person trips by trip purpose must be estimated. The distribution of trips by trip purpose is given for daily trips, and these figures must be prorated to the afternoon peak hour. The average daily data are given in table 5.2. It is now possible to adjust (in an admittedly arbitrary fashion) to compensate for the peak hour mode-split ratio, when there is a slightly greater level of public transit usage. There is a reduction of approximately 3% in the percentage of auto person trips during the peak hour as compared with the rates for the whole day. Thus the proportions of auto person trips are reduced correspondingly, by use of the previously calculated ratio of 0·9847, as shown in the first two columns of table 5.3. If, finally, these adjusted auto person trip factors are divided by the trip-purpose-specific auto occupancy rates, a direct conversion from auto person trips, by trip purpose, to vehicle trips is possible. These figures are shown in table 5.3. The last step in this rather extended discussion is the calculation of peak hour trips as a proportion of total daily trips; these are given in table 5.4.

In the model package there were (and are) only three trip types, work trips, home-to-shop trips, and work-to-shop trips. Thus the various trip purposes shown above were aggregated into these three groups, and a weighted average person–vehicle ratio was used for each of the three types.

All this is given here not because of any theoretical niceties, but to illustrate the rather pragmatic procedures upon which it is sometimes

Table 5.1. San Francisco Bay Area: 1965 data on auto occupancy. Source: BATSC (1969).

Trip purpose	Average auto occupancy
Home-based	
school	2·76
recreational	1·93
social	1·62
comparison shopping	1·46
personal business	1·41
convenience shopping	1·28
work	1·18
other	1·81
Nonhome-based	1·44
Mean	1·44

Table 5.2. San Francisco Bay Area: 1965 data on trip purpose by mode. Source: BATSC (1969).

Trip purpose	Mode		Proportion of auto trips
	auto	transit	
Home-based			
work	2155000	331000	0·867
personal business	1026000	71000	0·935
social	655000	21000	0·969
recreational	456000	25000	0·948
convenience shopping	968000	24000	0·976
comparison shopping	343000	39000	0·898
school	398000	324000	0·551
other	984000	9000	0·991
Nonhome-based	1968000	89000	0·948
Total (mean)	8953000	933000	(0·906)

necessary to fall back in order to fill gaps—that is, to create linkages—between components of a complex model package. Quite possibly some of the assumptions glibly passed by here have significant impacts on the outputs of the model package. This is a risk concomitant with the development of large systems, where the assumptions of the early work become the research topics of subsequent work. For the development work with the integrated package, these conversion procedures were quite adequate, and were subsequently applied to the trip-probability matrices

Table 5.3. San Francisco Bay Area: calculation of adjusted rate of vehicle trips per person trip.

Trip purpose	Proportion of auto trips		Occupancy rate	Vehicle trips/ person trip
	initial	adjusted		
Home-based				
work	0·867	0·8537	1·18	0·7235
personal business	0·935	0·9207	1·41	0·6530
social	0·969	0·9541	1·62	0·5890
recreational	0·948	0·9335	1·93	0·4837
convenience shopping	0·976	0·9610	1·28	0·7508
comparison shopping	0·898	0·8842	1·46	0·6056
school	0·551	0·5426	2·76	0·1966
other	0·991	0·9758	1·81	0·5391
Nonhome-based	0·948	0·9335	1·44	0·6483
Mean	0·906	0·8921	1·44	0·6195

Table 5.4. Derivation of 1965 and 1980 peak hour vehicle trips.

Trip purpose	1965 (BATSC, 1969)			1980 (estimated)	
	mean daily total [a]	evening peak	peak/daily ratio	daily total [a]	evening peak
Home-work	2626	474454	0·181	3952	715312
Home-based					
personal business	1248	110089	0·088	1710	150460
social	795	76484	0·096	1095	105120
recreational	535	25466	0·047	863	40561
convenience shopping	1223	141677	0·155	1704	195960
comparison shopping	409	36056	0·089	585	52065
school	–	29317	–	–	–
other	1095	76413	0·069	1519	104811
Nonhome-based	2447	224219	0·091	3879	352989
Total	10378	1194175		15307	1717278

[a] Values in thousands.

to yield the vehicle trips used in the assignment models. The next section of this chapter describes the theory and application of the procedures used to generate the trip-probability matrices.

5.3 Simultaneous location and trip-generation procedures

In chapters 2 and 3, reference has been made to the fact that forecasts of trips could be extracted from the calculations in the location model. In chapter 2, this point was made during the presentation of the example residence-location models. In chapter 3, the point was made at the end of the discussion of the details and numerical examples of the prototype package. In this section, these earlier discussions are brought together to complete the coverage of this concept of the simultaneous location and trip-generation—and trip-distribution—process.

To begin this discussion it is helpful to refer back to the simple residence-location model given in chapter 2 and again at the beginning of this chapter in equation (5.1). It is there stated that the p_{ij} matrix may be scaled up (that is, multiplied by the appropriate values) to produce a trip matrix. Equation (5.1) is

$$N_i = g\sum_j E_j p_{ij} \,.$$

In the numerical example of chapter 2 it was assumed that

$$p_{ij} = c_{ij}^{-2} \,. \tag{5.2}$$

A c_{ij} matrix was given, and the p_{ij} matrix which resulted was the same as is given here in table 5.5. Now, equation (5.1) results from, or may be considered as resulting from:

$$N_i = \sum_j T_{ij} = \sum_j gE_j p_{ij} \,, \tag{5.3}$$

where T_{ij} is the number of trips from zone i to zone j, thus,

$$T_{ij} = gE_j p_{ij} \,. \tag{5.4}$$

The values of E_j in the example were

$$E_1 = 100 \,, \quad E_2 = 200 \,, \quad E_3 = 300 \,, \quad E_4 = 0 \,, \quad E_5 = 0 \,.$$

Table 5.5. Illustrative p_{ij} matrix from c_{ij}^{-2}.

Zone	1	2	3	4	5
1	1·00	0·44	0·25	0·44	0·16
2	0·44	1·00	0·44	0·25	0·25
3	0·25	0·44	1·00	0·25	0·16
4	0·44	0·25	0·25	1·00	0·44
5	0·16	0·25	0·16	0·44	1·00

The value of g was calculated to be 0·4494. The matrix of trips may then be calculated, element by element, to yield the trip matrix in table 5.6. Recalling from chapter 2 that equation (5.3) can be expanded to

$$N_1 = gE_1p_{11} + gE_2p_{12} + gE_3p_{13} + gE_4p_{14} + gE_5p_{15} , \tag{5.5}$$

we can see that each term of equation (5.5) represents the scale factor g, multiplied by the product of the number of persons working in zone j, with the probability of a trip between zone i and zone j. For each N_i there will be an expanded equation like equation (5.5). The first terms of each of these, in this example, will be

$$gE_1p_{11} , \quad gE_1p_{21} , \quad gE_1p_{31} , \quad gE_1p_{41} , \quad gE_1p_{51} ,$$

and these terms represent all the possible residence locations for employees who work in zone 1. The sum of these terms should be equal to the total employment of zone 1. Reference to the trip matrix given in table 5.6 makes it clear that these column sums do *not* equal the zonal employments. This inconsistency was common to all Lowry derivative models in application in the USA, including PLUM, before the work on the prototype integrated package described in chapter 3. In this work, the solution to the inconsistency was simple: the probabilities were normalized to sum, down columns, to unity. Obviously if

$$\sum_i p_{ij} = 1 \cdot 0 , \tag{5.6}$$

then

$$\sum_i gE_jp_{ij} = gE_j . \tag{5.7}$$

Furthermore, if equation (5.6) holds, then the value of g in these equations also becomes unity; this was also discovered to be true empirically when the experiments were being done on the correction factors in the early versions of PLUM, as described in chapter 3.

Table 5.6. Trip matrix from numerical example and equation (5.3).

Zone	1	2	3	4	5	$\sum_j T_{ij}$
1	44·94	39·55	33·71	0	0	118·20
2	19·77	89·88	59·32	0	0	168·97
3	11·24	39·55	134·82	0	0	185·61
4	19·72	22·47	33·71	0	0	75·95
5	7·19	22·47	21·57	0	0	51·23
$\sum_i T_{ij}$	102·91	213·92	283·13	0	0	599·96

Normalization of the column sums of the matrix may most easily be done by dividing each element p_{ij} by the column sum:

$$p'_{ij} = \frac{p_{ij}}{\sum_i p_{ij}} . \tag{5.8}$$

This is, of course, exactly the same equation as equation (2.6) which results when a trip-end balancing factor B_j is added to the original model of equation (2.3). If the p_{ij} matrix given in table 5.6 is normalized in this fashion, we obtain the matrix given in table 5.7. The trip matrix which results from this normalized probability matrix is given in table 5.8. Now, of course, the column sums equal the zonal employments. With the probabilities normalized in this fashion, the model is known as a singly constrained spatial interaction model. It is in precisely this manner that the version of IPLUM used in the ITLUP prototype produced trip distributions (zone-to-zone trip matrices) simultaneously with its calculation of residence location.

One may often find in the literature reference to production-constrained or attraction-constrained spatial interaction models, or to origin-constrained or destination-constrained models. All of these are singly constrained spatial interaction models. The different designations refer to trip directions and to which end of the trip is constrained. Suppose, for example, that the T_{ij} matrix in table 5.8 had been defined as a matrix of home-to-work

Table 5.7. Normalized p_{ij} values from table 5.5.

Zone	1	2	3	4	5
1	0·437	0·185	0·119	0·185	0·079
2	0·192	0·420	0·210	0·105	0·124
3	0·109	0·185	0·476	0·105	0·079
4	0·192	0·105	0·119	0·420	0·219
5	0·070	0·105	0·076	0·185	0·499
$\sum_i p_{ij}$	1·000	1·000	1·000	1·000	1·000

Table 5.8. Normalized trip matrix (numerical example).

Zone	1	2	3	4	5	$\sum_j T_{ij}$
1	43·70	37·0	35·70	0	0	116·40
2	19·20	84·0	63·00	0	0	166·20
3	10·90	37·0	142·80	0	0	190·70
4	19·20	21·0	35·70	0	0	75·90
5	7·00	21·0	22·80	0	0	50·80
$\sum_i T_{ij}$	100·00	200·0	300·00	0	0	600·00

trips, where zone i was the zone of residence and zone j was the zone of employment. By constraining a home-to-work trip matrix at the work end of the trip, one adopts an attraction-constrained or destination-constrained procedure. Similarly, constraint at the home end of the trip implies a production-constrained or origin-constrained procedure.

It is possible, and sometimes necessary, to constrain both ends of the trip. This is done with a doubly constrained model, and is illustrated here to complete the exposition. The problem in a doubly constrained spatial interaction model is to estimate the T_{ij} matrix, given exogenously determined row and column totals and an equation which describes the nature of the interactions.

We shall assume that, in the example used above, the locations of population are known to be

$$N_1 = 110\,, \quad N_2 = 170\,, \quad N_3 = 200\,, \quad N_4 = 60\,, \quad N_5 = 60\,.$$

The traditional gravity model has the form

$$T_{ij} = gO_iD_j\,\mathrm{f}(c_{ij})\,, \tag{5.9}$$

where O_i is the number of trips originating in zone i, and D_j the number of trips terminating in zone j.

If, for the purpose of illustration, it is assumed that

$$O_i = N_i\,, \tag{5.10}$$

$$D_j = E_j\,, \tag{5.11}$$

$$\sum_i N_i = \sum_j D_j\,, \tag{5.12}$$

$$\mathrm{f}(c_{ij}) = c_{ij}^{-2}\,, \tag{5.13}$$

then a trip matrix can be calculated by using the regionally constrained formulation in equation (5.9). The data from the earlier simple example in chapter 2 may be used, and to begin, with g set equal to unity, we obtain

$$T_{11} = (110)(100)(1{\cdot}00) = 11000, \quad T_{12} = (110)(200)(0{\cdot}44) = 9680,$$

$$T_{13} = (110)(300)(0{\cdot}25) = 8250, \quad T_{14} = (110)(0)(0{\cdot}44) = 0,$$

$$T_{15} = (110)(0)(0{\cdot}16) = 0.$$

These and the other, similarly calculated values of T_{ij} are listed in table 5.9. The total number of trips in this first round of calculations is 192430. Thus the value of the normalized scaling factor, g', must be

$$g' = 600/192430 = 0{\cdot}003118\,.$$

The matrix given in table 5.9 is then multiplied by this value to give the final trip matrix, shown in table 5.10. Note, again, how a regionally constrained model produces the correct number of trips for the entire region, but has neither row (origins) nor column (destinations) totals that correspond to the input values O_i, D_j. The example given earlier in this section illustrated the operation of a trip-end-constrained (the work end) model. What remains to be done here is to go through the rather tedious calculations of a doubly constrained model which will insure the consistency both of the origin and of the destination inputs.

The equation to be used is

$$T_{ij} = A_i B_i O_i D_j \mathrm{f}(c_{ij}) \,, \tag{5.14}$$

where the balancing factors A_i and B_j are defined as

$$A_i = \left[\sum_j B_j D_j \mathrm{f}(c_{ij})\right]^{-1} , \tag{5.15}$$

$$B_j = \left[\sum_i A_i O_i \mathrm{f}(c_{ij})\right]^{-1} . \tag{5.16}$$

The problem is that neither A_i nor B_j are known in advance; in fact, they must be calculated iteratively.

The process begins by assuming an initial value of 1·00 for all B_j. Then, in the example, equation (5.15) yields

$$A_1 = [(100)(1{\cdot}00) + (200)(0{\cdot}44) + (300)(0{\cdot}25)]^{-1} = 0{\cdot}00380 \,;$$

similarly,

$$A_2 = 0{\cdot}00266 \,, \quad A_3 = 0{\cdot}00242 \,, \quad A_4 = 0{\cdot}00592 \,, \quad A_5 = 0{\cdot}00877 \,.$$

Table 5.9. Trip matrix calculated from the traditional gravity model [equations (5.9)-(5.13)], with $g = 1$.

Zone	1	2	3	4	5
1	11000	9680	8250	0	0
2	7480	34000	22440	0	0
3	5000	17600	60000	0	0
4	2640	3000	4500	0	0
5	960	3000	2880	0	0

Table 5.10. Normalized trip matrix resulting from the traditional gravity model example.

Zone	1	2	3	4	5	$\sum_j T_{ij}$
1	34·30	30·18	25·72	0	0	90·20
2	23·32	106·02	69·97	0	0	199·31
3	15·59	54·88	187·08	0	0	257·55
4	8·23	9·35	14·04	0	0	31·62
5	2·99	9·35	8·98	0	0	21·32
$\sum_i T_{ij}$	84·43	209·78	305·79	0	0	600·00

Note the fourth and fifth terms in equation (5.15) can be ignored in the above calculations as $D_j = E_j$, $E_4 = E_5 = 0$.

Now a trial set of B_j values may be calculated by using the above trial set of A_i values in equation (5.16):

$$B_1 = [0{\cdot}00380(110)(1{\cdot}00)+0{\cdot}00266(170)(0{\cdot}44)+0{\cdot}00242(200)(0{\cdot}25) +0{\cdot}00592(60)(0{\cdot}44)+0{\cdot}00877(60)(0{\cdot}16)]^{-1} = 1{\cdot}02203\ ;$$

similarly,

$$B_2 = 0{\cdot}93508, \qquad B_3 = 1{\cdot}04117,$$

$$B_4 = 0{\cdot}99532, \qquad B_5 = 1{\cdot}06400.$$

These new values of B_j may then be used to calculate a second set of values for A_i. These in turn may be used to calculate yet another set of B_j values. The process is continued until successive iterations in the calculation of A_i and B_j show no further change. These iterations are tabulated in table 5.11. Note that the values of B_4 and B_5 need not be calculated for this example, as E_4 and E_5 are zero.

Having now found equilibrium values of all the A_i and B_j, it is possible to calculate the values of T_{ij} from equation (5.14):

$$T_{11} = A_1 B_1 O_1 D_1 c_{11}^{-2} = 0{\cdot}00382(1{\cdot}01758)(110)(100)(1{\cdot}00) = 42{\cdot}76\ ,$$

$$T_{12} = A_1 B_2 O_1 D_2 c_{12}^{-2} = 0{\cdot}00382(0{\cdot}92900)(110)(200)(0{\cdot}44) = 34{\cdot}35\ ,$$

and so on. This yields the trip matrix given in table 5.12. Here it is readily seen that (within rounding errors) *both* trip-end constraints (origin and destination) are indeed satisfied.

Table 5.11. Iterative calculation of balancing factors A_i, B_j.

	Trial						
	1	2	3	4	5	6	7[a]
B_1	1·0[b]	1·02203	1·01919	1·01909	1·01795	1·01758	
B_2	1·0[b]	0·93508	0·92981	0·92968	0·92926	0·92900	
B_3	1·0[b]	1·04117	1·04526	1·04516	1·04486	1·04459	
B_4	1·0[b]	0·99532					
B_5	1·0[b]	1·06400					
A_1	0·00380	0·00381	0·00381	0·00382	0·00382	0·00382	
A_2	0·00266	0·00271	0·00271	0·00271	0·00271	0·00271	
A_3	0·00242	0·00238	0·00238	0·00238	0·00238	0·00238	
A_4	0·00592	0·00589	0·00589	0·00589	0·00590	0·00590	
A_5	0·00877	0·00884	0·00885	0·00885	0·00886	0·00886	

[a] All values identical to those in trial 6: equilibrium attained.

[b] Assumed initial value.

Whether one uses a regionally constrained, singly constrained, or doubly constrained model depends upon the situation being analyzed. The important point in this discussion is that the calculation of residence location requires the calculation of a trip matrix (or at least a trip-probability matrix in some forms of the models). This matrix may then be adjusted to exogenously provided regional trip totals and further adjusted to yield vehicle trips.

A particular virtue of this joint calculation process is that the trips matrix is, of course, consistent with the spatial distribution of tripmakers, their trip origins, and trip destinations. This is a rather different situation from the more traditional procedure: first, estimating the spatial distributions of activities; then, by use of multiple regression analysis, separately estimating trip origins and destinations; and finally, by use of a gravity model, producing a trip matrix. A trip matrix so produced might or might not be consistent with the spatial distribution of the activities whose trips are being estimated—that is, row or column sums may not equal zonal totals of population or employment.

After the trip matrices so generated have been modified by some procedure for conversion into vehicle trips (such as the procedure detailed in the previous section of this chapter), the next step is to assign those trips to the regional transportation network. The next section of this chapter describes the assignment procedure currently used in the integrated transportation and land-use model package.

Table 5.12. Trip matrix resulting from the doubly constrained model example.

Zone	1	2	3	4	5	$\sum_j T_{ij}$
1	42·76	34·35	32·92	0	0	110·03
2	20·63	85·60	63·52	0	0	169·75
3	12·11	38·91	149·17	0	0	200·19
4	15·85	16·44	27·73	0	0	60·02
5	8·66	24·69	26·65	0	0	60·00
$\sum_i T_{ij}$	100·01	199·99	299·99	0	0	599·99

5.4 Trip assignment: incremental tree-by-tree

Regardless of what techniques for trip generation and distribution are used, the next step in the process of transportation network analysis is trip assignment. The trip-assignment procedure involves taking an exogenously calculated origin–destination trip matrix (which gives the numbers of trips from each zone to each other zone) and determining the paths that these trips take over the transportation network in going from their origins to their destinations. In typical situations the transportation network used in the model is an abstraction of the actual 'on-the-ground' transportation system of the region being analyzed. The process of preparing such 'coded' networks is both painstaking and tedious. The details of such work are beyond the scope of this book, but may be found in any of several works on transportation planning (for example, Black, 1981; USDoT, 1973; 1977). The network may contain one or several modes, though the most common practice is to use single-mode networks, for example, one for highways and one for public transit. In the remainder of this discussion, attention will be confined to a single-mode network, where the mode is highways. The question of mode split and additional networks will be taken up later in this chapter.

The coded network which is input to the assignment process will contain all the connections within the network in terms of network links, as well as the connections, via load nodes, of the tripmakers in each of the region's zones (that is, residents and employees) to the network. For each link there will be information on the free flow travel time and the link capacity. Each link with two directions of traffic flow will normally be coded as two separate one-way links. This coded network, together with a trip matrix (converted from trip probabilities to vehicle trips by a procedure such as the one described in section 5.2) are the inputs required for trip assignment.

Trip assignment, as one of the three basic components of an integrated transportation and location analysis, was described in terms of a small numerical example in chapter 2. There are, of course, numerous complexities of all types which can be added to this simple form already described. Regardless of the particular assignment procedure used, three tasks must be accomplished. First, a decision rule must be defined to describe the basis on which each tripmaker selects a path through the network. Second, an algorithm must be developed to trace paths through the network according to this decision rule. Third, a procedure must be selected for allocating the trips to the paths.

The customary decision rule is that the tripmaker will select the path of minimum generalized cost through the network. Data describing generalized cost, including components such as travel time, vehicle operating costs (both direct and indirect), and highway user charges, are often difficult to obtain; in practice a single measure such as travel time or approximate travel cost is used instead.

The algorithm for tracing paths through the network depends in part both upon the decision rule and the trips-to-paths allocation rule. The most frequently used trips-to-paths allocation rule is the 'all-or-nothing' rule (to be discussed shortly). There have been quite a number of path-tracing algorithms proposed and tested and which are described in the literature (Bureau of Public Roads, 1964; Jewel, 1967; Martin and Manheim, 1965). The specific workings of these algorithms are not of concern here, but the question of which trips-to-paths allocation procedure is of concern here and consists, for the most part, of deciding between all-or-nothing or some other trip-allocation procedure. The bulk of current applications involve the use of an all-or-nothing allocation procedure where the trips being allocated to a particular path at a particular point in the overall assignment process are simply all assigned (allocated) to the then current best path between the given pair of zones $[ij]$. A probabilistic assignment algorithm is discussed in the next section of this chapter.

Most of the variation between the alternative procedures in current use for network assignment results from the ways in which capacity restraint is applied to the loaded network links. Four such procedures were tested during the initial development of the ITLUP package.

The first assignment procedure tested was the form originally used in transportation network analysis, and was perhaps the most commonly applied procedure during the early 1970s. This procedure might best be described as *iterative full network loading.* The procedure calls first for all trips between each pair of zones $[ij]$ to be loaded on the shortest (minimum time or cost) path between zone i and zone j. After all the trips from all zones to all other zones have been loaded on the links of the network, then the volume-delay function is evaluated for each link. This results in a new set of 'congested' travel times (or costs) for the network links.

The iterative part of this procedure has varied from application to application. In essence, the congested link times are used for a complete reloading of the network. The earlier trip assignments are discarded, and from the congested link times, an entire new set of minimum paths and trip assignments (that is, link loadings) is completed. Then, of course, the congested link times are recomputed by use of the volume-delay function and the process is begun again. In the early applications of this procedure (circa 1960–1965) it was discovered that severe oscillations could (and usually did) result from this iterative scheme. The solution which finally evolved was to average the congested link times over the successive iterations. There were two main drawbacks to this procedure. First, it often took a number of iterations (sometimes ten or more) for the process to converge. Second, the overall process was still basically an all-or-nothing loading of the network links; that is, all trips between a specific origin–destination (OD) pair were loaded on the single best (albeit congested) path.

A subsequent development attempted to avoid the oscillation problems associated with full network loading by loading trips in several increments in successive passes through the assignment procedure. This *incremental full network loading* procedure is thereby able to achieve some overall improvement of the situation. This is done by beginning with, say, 40% of the trips between each OD pair. These 40% volumes are then assigned to the network links in the standard way. After they have all been assigned, the volume-delay function is used to compute the congested time on each link, given the volumes resulting from the 40% of all the trips then loaded on the network.

A second increment of, say, 30% of the trips between each OD pair (note that this is one-half of the trips remaining after the first 40% increment is loaded) is then loaded onto the partially congested network.

After these trips are assigned to the minimum (40%-congested) paths, the volume-delay function is again evaluated, now with 70% of the total trips loaded on the network. The last increment of 30% of the total trips is then assigned to the network, which is already congested with the 70% of the trips previously loaded. This process results in a much more steady, nonoscillatory increase of trips and congestion levels on the network. Of course, trip increments of different sizes could be used in this process. One might, for example, have five increments of 20% each, or four increments of 20% plus two increments of 10%, and so on.

Another possible loading procedure involves a compromise between the first two described. In *incremental full network reloading*, the procedure again begins with the assigning of, say, 40% of the trips between each OD pair. When these trips have been assigned to the network links, the volume-delay function is used to determine the congested travel times (or costs) on these links due to the first 40% of the trips. Then the trip volumes on the links are discarded and a new portion of the trips, equal to the original 40% plus a second increment of, say, 30%, are assigned to the 40%-congested network.

Now, this new increment of 70% of the trips between each OD pair is loaded onto the network. Again note that this is an all-or-nothing loading onto the minimum paths of the 40%-congested network. After this assignment is completed the volume-delay function is recomputed for each 70%-loaded link. The process is then completed by one last iteration where, the 70% trip volumes having been discarded, the full 100% of the trips for each OD pair are loaded onto the network which is, at this point, characterized by 70%-loaded link times. This overall process, too, results in a more gradual 'filling-up' of the network, with the intent of avoiding the oscillations associated with the full network loading procedure. It was not clear, before these experiments, whether there were any advantages of this procedure over the incremental full network loading procedure.

In all three of the assignment procedures discussed above, all trips being loaded at any given iteration in the process 'see' the same network characteristics. So, for example, in full network loading, the trips from zone 14 to zone 37 see (or encounter) the same link characteristics as the trips from zone 14 to zone 92 or from zone 85 to zone 163. After the volume-delay function has been applied, the same situation is true for the second portion of trips being loaded in the incremental full network loading procedure, and so too for the third pass through the incremental full network reloading procedure.

A rather different view of the situation is taken in *tree-by-tree loading*. In this case, as soon as the trips between a particular OD pair are assigned to a network link, the volume-delay function is calculated for that link. Thus, at the end of each assignment of a set of trips for a given OD pair, all the links traversed by those trips are partially congested by the trips just assigned. If 100% of the trips are being assigned, then the trips between each subsequent OD pair see a progressively more congested network. This tree-by-tree loading provides a much smoother loading of the network, but does suffer from a bias introduced by the order in which zonal origins are selected for tree tracing and assignment. If, for example, the zone 3–zone 56 path is traced and loaded first, then the links of this path (minimum time or cost) will be different from those on the zone 3–zone 56 path if it were, say, the twenty-second path to be traced and loaded. Thus the link volumes at the end of an assignment run will be a function of the order (sequence) in which the OD pairs are selected for loading.

One proposed solution to the problem of this origin-sequence selection bias is to select the origins at random, rather than taking them in simple numerical order. This solution does not actually eliminate the bias, it simply applies it randomly, the distribution of congestion depending upon which paths are selected for loading at which points in the random sequence. Furthermore, in its original form the tree-by-tree procedure still used 100% of the trips between each OD pair in an all-or-nothing assignment procedure.

A procedure which minimizes these last problems, and the one which was actually used in the integrated transportation, location, and land-use analyses described in this book, is the *incremental tree-by-tree loading* procedure. Here, an initial increment of perhaps 40% of the trips between each OD pair is assigned to the network by means of a tree-by-tree method of capacity restraint, and a random selection sequence for OD pairs. This is followed by the loading of a second increment of an additional, say, 30% of the trips between each OD pair. Again, a tree-by-tree loading procedure is followed and a new randomly selected sequence of OD pairs is used. This step may then be followed by one or more additional increments also loaded tree by tree with randomly selected OD pairs. By using a procedure which is both incremental and randomly selected tree by tree, a rather smooth and unbiased network loading should result.

A series of computer test runs was conducted to compare the various assignment procedures described above. The volume-delay function given in chapter 2 during the presentation of the simple numerical example of trip assignment, equation (2.30), was used in these tests as well. It is interesting to note in passing that the second part of the assignment problem example in chapter 2 may now be identified as having been tree-by-tree loading. In the first test runs, several minor problems emerged which required attention. The most important of these was, particularly with the small test network first used, that the original volume-delay function had no upper bound. Thus with an all-or-nothing loading, and with the implicit discontinuities present in a small test network, some links could become congested, rather easily, to fifty or one hundred times their design capacity. An arbitrary limit of $2 \cdot 5$ was placed on v/v_0, the volume/capacity ratio. From equation (2.30),

$$t_1 = t_0\left[1{\cdot}0+0{\cdot}15\left(\frac{v}{v_0}\right)^4\right],$$

we recall that t_1 is the congested link travel time, t_0 the uncongested link travel time, v the link volume (trips) on the link, and v_0 the link design capacity (trips). If the maximum value of v/v_0 is $2{\cdot}5$, then

$$t_{1\,\max} = t_0[1{\cdot}0+0{\cdot}15(2{\cdot}5)^4] = 6{\cdot}86t_0 . \tag{5.17}$$

Thus, a link which during off-peak travel takes ten minutes to traverse could be congested to just under an hour and ten minutes. Obviously other values could have been selected for the cutoff, but, in practice, the value selected seems to work reasonably well.

The first series of tests[(4)] compared the results from a full network loading to the results from incremental network loadings. Several measures were used to compare the results of these runs. The principal statistic used was total travel time, which is of course an implicit measure of the degree of minimization of disutility that is subsumed in the best path criterion. Secondary statistics were: (a) the number of links loaded beyond 2·5 times capacity; (b) the mean and standard deviation of actual link volume/capacity ratios; (c) restrained volume/capacity ratios; (d) the number of unused links in the network (Huber et al, 1968). The clearest indication of network congestion is given by the mean and standard deviation of the volume/capacity ratio. The other statistics reflect the design of the network as well as the quality of the particular assignment technique. The presence of many restrained links, for example, can indicate poor network design, overestimated traffic generation, or simply an overloaded network. The effects of restraining links are shown by comparing the mean of the actual volume/capacity ratio to the mean of the constrained volume/capacity ratio.

The first tests were run on the small test network, and confirmed that, within limits, all the statistics moved together. For example, even the best possible loading onto a given network will leave some links with zero volume. However, up to this best point, the number of links with zero volume will decrease as better solutions, that is, those with lower total network travel time, are found. Similarly, each improvement in a solution will be accompanied by lower mean v/v_0 ratios, and fewer capacity-constrained links.

Some test runs were done on a full-scale version of the 1965 highway network for the San Francisco Bay Area. This network contained 5107 links and 2022 nodes (of which 291 were load nodes, connecting the network to the 291 zones of the region). To save computer costs, some

[(4)] The computer runs described in this section were performed by Peter Kuner (see Kuner, 1973).

tests were run on a much smaller network containing only 184 links and 45 nodes (of which nine were load nodes connecting this prototype network to a prototype region).

First a series of tests was run on the large network. These tests were to compare a single loading of 100% (full network loading, without iterations) with an incremental loading consisting of a set of five increments of 20% each (incremental full network loading). A single increment of 40% was also done, as will be discussed shortly. The results of these runs are given in table 5.13.

Several interesting results are to be noted. First, the full network loading using 100% of the trips yields a slightly smaller total travel time (−8·01%) than is found after the last increment in the incremental full network loading. This is an artifact of the arbitrary maximum constraint imposed on the volume/capacity ratio. The total travel time statistic is the sum, over all links, of the final trip volume multiplied by the final (congested) travel time for each link. On the majority of the high-volume links the congested travel time is based on the constrained value of the volume/capacity ratio ($v/v_0 = 2{\cdot}5$). Because their congested travel time is arbitrarily constrained, some of these links are assigned more volume than they might otherwise carry. Thus when there are a large number of constrained links in the loaded network, the total travel time statistic will be somewhat less accurate.

A more indicative measure is the number of unused links. A greater number of unused links implies less even loading of the trips on the links of the network. From table 5.13 it can be seen that there are almost two-thirds fewer unused links after the incremental loading than after the

Table 5.13. Comparison of assignment runs for the large network: full versus incremental full network loading.

	Full network loading		Incremental full network loading				
Trip increments (%)	40	100	20	20	20	20	20
Cumulative trips (%)	40	100	20	40	60	80	100
Total time [a]	11322	58246	2959	10767	23188	41070	63319
Unused links	974	974	983	885	675	471	363
Constrained links [b]	65	622	2	38	187	401	686
v/v_0 (unconstrained)							
mean	0·412	1·018	0·206	0·410	0·729	1·047	1·363
SD	0·601	1·508	0·301	0·542	0·919	1·131	1·290
v/v_0 (constrained)							
mean	0·401	0·800	0·206	0·408	0·627	0·943	1·191
SD	0·548	0·893	0·300	0·531	0·710	0·810	0·861
Total trips assigned	214065	547066	107033	214065	332994	428136	547066

[a] Measured in thousands of minutes. [b] Links where $v/v_0 > 2{\cdot}5$ cutoff applied.

full network loading. This means that the trips were more evenly spread over the network links after the incremental loading, as is also indicated by the increase of about 10% in the number of constrained links for the incremental process, compared with that for the full network loading procedure. A greater spread of traffic over the links is a more realistic result, as in reality tripmakers spread themselves out over many alternative paths between points in the region.

The mean volume/capacity ratio is also distorted by the presence of large numbers of constrained links. Suppose, for example, that a great many trips are using a particular network link which is constrained to the maximum volume/capacity ratio of 2·5. For this example assume that there are forty other network links and that the sum of the volume/capacity ratios on the other forty links is equal to 50. The mean value v/v_0 is then equal to 50/40 = 1·25. If one additional link, constrained to a maximum v/v_0 of 2·5, is included, then the mean value of v/v_0 is 52·5/41 = 1·28. If, instead, those trips which were congesting the single links in excess of the maximum constraint value were loaded onto three other links where the resulting v/v_0 was 2·0, then the mean v/v_0 would be 56/43 = 1·30. On a congested network a more even spread of the trips over the links can give these misleading results. In fact, even for an unconstrained value v/v_0 on our offending link, so long as that value was less than 3·3, the three links loaded at v/v_0 = 2·0 would yield a greater mean v/v_0 in the example.

The single 40%-increment run was made to compare with the results after two 20% increments. In this situation the comparison shows the expected results. The two 20% increments yield, in every respect, a better network loading (utilization) than the single 40% increment.

With the incremental full network loading shown to be superior to full network loading, the next step was to compare incremental full network loading with incremental tree-by-tree loading. Preliminary comparisons were done on the small network, and the results appear in table 5.14. These results show that there is a dramatic reduction in total travel time when the incremental tree-by-tree method is used. Whether two or three increments are used, the results for the tree-by-tree procedure are vastly superior to five increments of the incremental full network loading procedure. All the other statistics presented also show improvement when the tree-by-tree method is used.

Finally, three more runs were made with the large network. All three runs were tree-by-tree assignments, and the intent was to compare the effects of different numbers of increments. The first of these runs loaded 100% of the trips in a single increment, by using the tree-by-tree assignment procedure. This run may first be compared with the results of full network loading and of the last increment of the incremental full network loading given in table 5.13. Again, however, the large number of restrained links limits the accuracy of a comparison between the runs. The numbers of unused links show a clear improvement in the tree-by-tree loading compared with the full network loading. In fact the tree-by-tree loading gives as good results on a 100% increment as the incremental full network loading did for five 20% increments.

The tree-by-tree loading of two 50% increments shows some improvement over the tree-by-tree loading of a single 100% increment. The last results, shown in table 5.15, are for a tree-by-tree loading of three increments of 40%, 40%, and 20%, respectively. Again, there is a modest improvement; in particular, the unused-link statistic is the lowest for any of the procedures run on the large network. In fact, all of the network statistics were best on this run, though the improvements over, say, the two-increment tree-by-tree run were rather modest.

Table 5.14. Comparison of incremental assignment runs for the small network: full network versus tree by tree.

	Incremental full network loading					Incremental tree-by-tree loading	
Trip increments (%)	20	20	20	20	20	40, 40, 20	50, 50
Cumulative trips (%)	20	40	60	80	100	100	100
Total time [a]	105	372	1662	8876	50816	1237	1252
Unused links	20	8	6	5	3	5	5
Constrained links [b]	3	10	16	22	36	29	27
v/v_0 (unconstrained)							
mean	0·316	0·742	1·177	1·603	2·076	1·996	1·883
SD	0·641	0·941	1·445	1·989	2·382	2·367	2·207
v/v_0 (constrained)							
mean	0·340	0·670	0·959	1·169	1·360	1·342	1·273
SD	0·548	0·737	0·814	0·841	0·825	0·892	0·888
Total trips	3041	6082	9122	12163	15203	15203	15203

[a] Measured in thousands of minutes. [b] Links where $v/v_0 > 2{\cdot}5$ cutoff applied.

Table 5.15. Incremental tree-by-tree method: comparison of three large-network assignments.

Trip increments (%)	100	50, 50	40, 40, 20
Cumulative trips (%)	100	100	100
Total time [a]	59555	54468	53197
Unused links	374	325	287
Constrained links [b]	574	456	422
v/v_0 (unconstrained)			
mean	1·328	1·276	1·259
SD	1·257	1·111	1·074
v/v_0 (constrained)			
mean	1·193	1·178	1·169
SD	0·839	0·807	0·796
Total trips	547066	547066	547066

[a] Measured in thousands of minutes. [b] Links where $v/v_0 > 2{\cdot}5$ cutoff applied.

Comparison of the best incremental full network loading with the best incremental tree-by-tree loading shows that the tree-by-tree procedure produces substantially better results. Total travel time was reduced by 16%; the number of unused links was reduced by 21%; the number of maximally constrained links was reduced by 38%; the mean volume/capacity ratio was reduced by just under 8%. As the computation cost of three tree-by-tree increments was approximately equal to the computation cost of five full network loading increments, the tree-by-tree procedure was subsequently adopted for use in future analyses with the integrated model package.

Even after this series of tests, and after the decision had been made to use the incremental tree-by-tree assignment procedure, questions remained about the assignment problem. It was believed that the superiority of the method may have depended, in part, on the form of the volume-delay equation used and on the actual magnitude of the total trips being loaded on the network in relation to some unspecified measure of the overall carrying capacity of the network. For these reasons, as well as the fact of increasing numbers of reports in the literature of new work with various multipath assignment procedures, a new examination of the overall problem of traffic assignment was undertaken, and its results (which post-date the results described above by about four years) are described in the next section of this chapter.

5.5 Trip assignment: multipath with capacity constraint[5]

In chapter 3, the description of the work done on and with the prototype versions of the integrated transportation and land-use package drew attention to certain problems of system instability. As described in chapter 3, a series of investigations of the detailed functioning of the elements of the model package were undertaken, in an attempt to ascertain the source of these oscillations. The assignment routine was found to be a significant source of the problem. This matter will be discussed again in a later chapter, but the problem may be briefly described here. In essence, despite the use of incremental tree-by-tree assignments, the trips being loaded at any instant were loaded by an all-or-nothing procedure. It is true that with the incremental tree-by-tree method the trips are being loaded on a partially congested network, and that in the final increment only the last 20% of the trips are being assigned. Yet, when the actual assignment of these trips to specific links is made, it is to what is then the minimum path according to an all-or-nothing criterion. It was found that even with the use of incremental tree-by-tree loading, and only a 20% increment, there were still oscillations: a particular link might be highly congested on one iteration and much less congested on the next. This was due to the subsequent increase in travel time and rearrangement of

(5) Portions of this material have appeared, in a somewhat different form, in Soberanes (1978).

activity locations which would yield an altered trip matrix and thus different network loadings (trip volumes assigned to links).

An investigation was then undertaken into the possibilities for the use of multipath assignment procedures. These dispense with the all-or-nothing rule by, in effect, assignment of trips between a particular OD pair to several possible paths. The trips are assigned to these paths in proportion to their relative travel times (or costs). The investigation of this entire matter did, however, begin at the beginning, by reanalyzing the assignment problem. Then, after examining the several aspects of the problem, a set of comparison tests between the incremental tree-by-tree method and several forms of multipath assignment was conducted. The results of this investigation are presented here.

To begin, the objectives of traffic-assignment analyses will vary considerably according to the purpose of the study project within which the work is being done. The principal objectives are usually the determination (or forecasting) of traffic volumes on selected portions of the transportation networks of a region. These portions may be links, routes, or transport corridors. The choice of an assignment procedure obviously depends on the required level of detail of the results. Van Vliet (1976a; 1976c) has divided the objectives into primary and secondary according to the level of detail required, as follows.

Primary objectives (less detailed):

- prediction of the volume of traffic on each network link;
- determination of the minimum travel cost (impedances) between each OD pair.

Secondary objectives (more detailed):

- determination of the route or routes used by each OD pair;
- analysis of the origins and destinations of traffic that passes through given links;
- isolation of external and through traffic for a subnetwork to be studied in greater detail;
- analysis of turning movements at intersections;
- visual presentation of data.

We note first that it is the second of the two primary objectives given above which is of greatest concern in the integrated transportation and land-use modelling work (and in ITLUP, per se); it should also be noted that constraints such as computer expense and computer capacity will, for large networks, also play a major role in the choice of an assignment procedure.

Most traffic-assignment methods are based on the basic principle formulated by Wardrop (1952). This principle can be stated as:

> "The journey times on all routes actually used are equal, and less than those which would be experienced by a single vehicle on any unused route".

According to this principle the traffic pattern on the network(s) converges to an equilibrium situation in which no driver (tripmaker) can reduce travel time by switching to another route. The main assumptions on which this principle is based are:

1. Individuals select the route of minimum perceived cost.
2. All individuals perceive cost in the same way.

The first assumption represents a particular difficulty in solving the traffic-assignment problem, as there is no general agreement on a proper definition of perceived cost. In many transportation studies, perceived cost is defined by such concepts as travel time, travel distance, disutility, and sometimes as weighted combinations of all of these. Since the problem of defining perceived cost was beyond the scope of this work, it was assumed that costs on the links were to be established according to some definition of average travel cost. Thus from Wardrop's principle and the two basic assumptions, it is possible to distinguish two cases:

(a) *Assumptions 1 and 2 hold* A multiplicity of route choices arises from congestion effects: as the best routes under free flow conditions rapidly become congested, then alternative routes must also be used.

(b) *Assumption 2 does not hold* A multiplicity of route choices arises from the fact that not all individuals perceive cost in the same way. This leads to a spread of (that is, to several) selected routes.

In reality, congestion effects and variations in perceived cost both lead to a multiplicity of route choices. Consequently, a major consideration is to define when, and under what circumstances, one factor is more relevant than the other, or when both are relevant to such choices. Clearly, different approaches for solving the traffic-assignment problem will best be applied to different situations.

In order to define the mathematical structure of the problem several variables need to be defined first. The network is represented by a set of links $\{\ell\}$ corresponding to the highways or roads, and a set of nodes $\{\eta\}$ which represent either zones or intersection of roads. Origin and destination zones are considered as a subset of the nodes, generally referred to as 'load nodes' or 'centroids'. The cost on each link is assumed to be a function of the total flow of traffic along the link, thus:

$$c_\ell = \mathrm{F}(v_\ell)\ , \tag{5.18}$$

where c_ℓ is the cost of traversing link ℓ, and v_ℓ the volume, of flow, of trips on link ℓ. Furthermore, let T_{ij} be the total demand for travel (trips) between origin zone i and destination zone j. The T_{ij} matrix is, for the purposes of this analysis, exogenously determined and remains fixed. This assumption will be discussed further in the final section of this chapter.

Given all this, a feasible flow pattern is defined as a pattern of flows such that, for each OD pair $[ij]$:

(a) a definite number of routes exist between origin and destination;

(b) the total number of trips along each route are determined;

(c) the sum of the trips following all routes between a given $[ij]$ pair is equal to T_{ij}.

Now, according to Wardrop's basic principle, a feasible flow pattern is also an equilibrium solution to the assignment problem if, of all routes between any particular OD pair, any route with greater than zero trips is a minimum cost route. This of course implies that if any tripmaker chooses a new route, that there will be an increase in that tripmaker's travel cost and thus in the systemwide travel cost as well.

This equilibrium solution may also be expressed as a nonlinear mathematical programming problem. The dimensions of the problem are quite staggering, so that for example a fifty-node, two-hundred-link network yields a problem with 2450 flow constraints, and 10000 variables and nonnegativity constraints. Algorithms have been developed to deal with small-scale problems (Nguyen, 1974; LeBlanc et al, 1975). For practical applications such as the data for the San Francisco Bay Area used in the incremental tree-by-tree tests and policy analyses described above, these nonlinear programming algorithms are still in the experimental stages of development. Consequently the only currently practical solution is in the use of various heuristic methods.

When the assignment problem is presented as above, with its central objective being the finding of a solution that satisfies Wardrop's principle, then it is the congestion on the network's routes which is the major factor leading to the multiplicity of route choices. Thus those traffic-assignment procedures which attempt to find an equilibrium assignment solution according to Wardrop's principle are the most appropriate for use in cases where proper representation of and response to congestion is of paramount importance. Capacity-constraint procedures are included in this group.

It is of course possible that in some instances variations in perception of travel cost outweigh the effects of congestion. In such cases it is more appropriate to consider the use of stochastic assignment procedures (Dial, 1971; McLaughlin, 1966; Murchland, 1968; Robillard, 1974). These methods assume that the costs on the links are predetermined and fixed and thus independent of link flows, in contradistinction to the case in the discussion above. Considered in terms of volume-delay functions relating link costs to link flows, stochastic assignment methods are most appropriate for rather flat curves, whereas capacity restraint methods (based on Wardrop's principle) are most appropriate for steeply sloping curves.

Several capacity-constraint methods were described in section 5.4, together with the results of operational tests of the methods. In the years since those tests were conducted, the tree-by-tree methods have seen increasing application. Experience with these techniques has shown that while these methods are capable of reaching convergence to an equilibrium solution, several inherent difficulties remain (Van Vliet, 1976b). In particular, the sizes of the increments are arbitrarily selected, yet they determine the procedure's ability to converge. The smaller the increment, the closer will the procedure come to an equilibrium solution. However, in general the method does not guarantee an equilibrium solution (Ferland et al, 1974; Florian and Nguyen, 1974). This is because once an increment of flow is assigned, it cannot be reassigned to another route. Thus some routes selected for the early increments may be congested and unreasonably expensive in terms of the resulting travel-cost pattern. However, Van Vliet (1976b; 1976c) argues that very good convergence to an equilibrium solution can be achieved by using the incremental tree-by-tree method and eliminating the worst routes by reassigning trips to another route as an extra step to the above method. Such procedures for reassigning trips are described by Van Vliet (1976b; 1976c) and by Gleave (1973). Further, Van Vliet (1976c) has developed an expression, ξ, which quantifies the degree of convergence to an equilibrium solution:

$$\xi = \frac{\sum\limits_{i,j} v_{ij} c_{ij}(v_{ij}) - \sum\limits_{i,j} T_{ij} c_{ij}^{\min}}{\sum\limits_{i,j} T_{ij} c_{ij}^{\min}}, \tag{5.19}$$

where v_{ij} is the volume of trips on links between i and j, and $c_{ij}(v_{ij})$ the cost on links between i and j, given v_{ij}.

Thus, ξ is a measure of the total excess travel cost expressed as a fraction of the total minimum travel cost. The first term in the numerator is the total travel cost actually incurred on the network given by the resulting flow pattern of the assignment, and the second is the total travel cost if all trips take the current minimum-cost route, then $c_{ij}^{\min}$ is the minimum cost from i to j given link costs $c_{ij}(v_{ij})$. Furthermore, an equilibrium solution according to Wardrop's principle is achieved if ξ equals zero.

A small value of ξ does not, however, necessarily imply a good solution to the congested assignment problem, given the inherent errors that exist in the input data. For practical purposes it would be necessary to define an acceptable value of ξ according to the level of errors inherent elsewhere (that is, in aggregation of the network characteristics, in the input data, and so on).

Stochastic assignment procedures assume that, owing to variations in tripmakers' perceived travel cost, of all trips arriving at a node, not all of them will have travelled along minimum cost routes, and of all trips leaving a node, not all of them will take minimum cost routes. As a consequence of this assumption, stochastic methods never explicitly define routes. Thus, in order to represent the 'nondeterministic' behavior of drivers, these methods divert trips at each network node, in a probabilistic manner, independently of past and future link-choice decisions. A basic difference between various stochastic assignment methods is the process by which they divert trips at each node.

Trips at each node are assigned to the different links entering the node according to a splitting function. Several forms of splitting functions have been proposed (McLaughlin, 1966; Murchland, 1968). One such splitting function was suggested by Dial (1971):

$$F(c_{(i,j)}) = \begin{cases} \exp(\theta \Delta c_{(i,j)}) & (c_i < c_j) \\ 0 & (c_i > c_j) \end{cases} ; \tag{5.20}$$

here

$$\Delta c_{(i,j)} = c_j - c_i - c_{(i,j)} , \tag{5.21}$$

where c_j is the minimum travel cost from the origin to node j, c_i the minimum travel cost from the origin to node i, $c_{(i,j)}$ the travel cost of link (i, j), and θ an arbitrary trip-dispersion parameter. This function is evaluated for each link (i, j) in the network. Thus $\Delta c_{(i,j)}$ represents the extra cost incurred in travelling from the origin to node j, through node i rather than travelling along the minimum cost route to j. It may be confirmed that if node i is on the minimum cost route to j, then $\Delta c_{(i,j)} = 0$. The parameter θ is used to control the degree of diversion of trips away from the minimum cost routes at each node. Then, for large values of θ there is less dispersion, and for low values of θ more dispersion.

Two main types of stochastic assignment method or procedure have been developed for network loading: the single-pass procedure and the double-pass procedure. To load all trips from one origin to all destinations, the *single-pass* procedure consists of the following steps.

1. Assign to each destination node, j, all trips coming from the origin 0, T_{0j}, and set the volume of trips at all other nodes to zero.
2. Perform the following steps at each node j (one at a time) in descending order of travel cost from the origin (that is, starting with the most 'distant' node from the origin 0, and working back until the origin is reached):
   (a) for each link $\ell = (i, j)$, entering node j, calculate the value, F_ℓ, of the splitting function described above (or any other functional form);
   (b) divide the trips T_j at node j among the links ℓ (entering node j) according to the expression

$$T_j^\ell = T_j \left(\frac{F_\ell}{\sum_\ell F_\ell} \right) ; \tag{5.22}$$

   (c) increment the accumulated trips at each node i by T_j^ℓ.

This procedure is repeated once for every origin in order to load the whole trip matrix $\{T_{ij}\}$.

The *double-pass* procedure is so called because it requires two passes through the whole network, a forward pass and a backward pass, in order to load all trips from an origin 0 to all destinations j. The forward pass has the following steps:

1. Examine each node i in the network, in ascending order of travel cost from the origin.

2. For all links ℓ, $\ell = (i, j)$, leaving node i, calculate a weight w_ℓ according to the expression

$$w_\ell = F_\ell w_i \ , \tag{5.23}$$

where F_ℓ is the splitting function evaluated for link ℓ, and w_i the sum of the weights of all links entering node i (for the origin node $w_0 = 1$).
When the destination node most 'distant' from the origin is reached, the 'forward' pass ends. Next the 'backward' pass is performed. The backward pass is almost identical to the single-pass procedure described above, and has the following steps:
1. Assign to each destination node, j, all trips that come from the origin 0, T_{0j}, and set the volume of trips at all other nodes to zero.
2. Starting with the most 'distant' node from the origin, divide the trips T_j at each node j among the links l (entering node j) according to the expression

$$T_j^\ell = T_j \left(\frac{w_\ell}{\sum_\ell w_\ell} \right) . \tag{5.24}$$

3. Increment the accumulated trips at each node i by T_j^ℓ.
Once all the nodes have been examined in a descending sequence of travel cost from the origin, and the origin is reached, then the backward pass ends.

In general, single-pass procedures seem to perform better than double-pass procedures. Furthermore, single-pass procedures use less computer core storage and less computer processing time. Double-pass procedures also seem to have a tendency to load too many trips onto secondary roads because the procedure places greater emphasis on the number of possible routes by using weights, w, in the backward pass. This tendency could have effects which are both significant and incorrect in assignment, as route choice could thus depend on whether or not various secondary roads are included in the network coding.

Another important issue concerning stochastic assignment procedures relates to the formulation of the splitting function. For example, in the formulation proposed in equation (5.20), the division of trips is a function of the absolute value of extra cost, which implies that drivers choose their route according to differences in absolute values of the excess (above minimum) travel cost incurred. However, it seems more reasonable to assume that a driver will be more aware of differences of two minutes in ten than two minutes in one hundred. In order to reflect this behavior the apportioning of trips should perhaps be a function of the relative excess of travel cost (Murchland, 1968). The formulation given in equation (5.20) does, nonetheless, possess several desirable properties because of its exponential form (Dial, 1971). Perhaps an even better formulation would result from a combination of the exponential form and the relative excess of travel cost.

It has been proved (Van Vliet, 1976b) that stochastic assignment methods are inconsistent when flow-dependent costs in the links are considered. These methods would, however, be consistent if the link costs as calculated from the final link flows, via a volume-delay curve, were equal to the initial costs. Thus, as implied above, stochastic assignment methods are most appropriately applied in situations where congestion effects are minimal.

In the context of work on and with the integrated package, it is necessary to determine the routes of minimum travel cost from each zone to each other zone on the network after the loading of all trips has been done, for input to the location models. Preliminary tests of the ITLUP package, as described above, resulted in the selection of an incremental tree-by-tree method for traffic assignment, with the trips loaded in three increments of 40%, 40%, and 20%. Subsequently it was decided to investigate a method that could take into account variations in the perception of travel cost, and at the same time simulate congestion effects. It was hoped that these two phenomena taken together would yield a more realistic description of driver behavior. Thus it was decided to attempt to implement a stochastic method in an incremental fashion, allowing for capacity-constraint adjustment of link costs. The method actually implemented was a combination of the incremental tree-by-tree method and a stochastic double-pass procedure. This method was thus similar to the incremental tree-by-tree method described in section 5.4. The main difference was that instead of the trips being loaded onto the minimum paths by an all-or-nothing procedure, the loading of trips from the origin to all destinations was done by the double-pass procedure just described.

A series of test runs of this new assignment procedure was performed to compare its performance with that of the incremental tree-by-tree assignment method. Several tests were run, for a small network as well as for a larger network. The small network consisted of 100 links and 38 nodes, of which 9 were load nodes. The larger network corresponded to the San Francisco Bay Area highway network, used in the previously described tests of assignment procedures. As several minor modifications had also been made to the original incremental tree-by-tree procedure, several tests were performed with the incremental tree-by-tree method as well, and these results were then compared with the results of the multipath method. The volume-delay function used here was the same as was used above, and the maximum permissible volume/capacity ratio was again set to $2 \cdot 5$ in all tests.

In table 5.16(a) the results of the incremental tree-by-tree assignment procedure on the small test network are shown. There are no unexpected results here, as the performance of the incremental tree-by-tree method has been described previously. In table 5.16(b) the first results from the incremental multipath procedure are presented. Note that the trip-dispersion parameter θ was assigned a value of one for these first tests.

Table 5.16. Comparison of assignment runs for the small network.

	Run 1	Run 2		Run 3			Run 4				
(a) *Incremental tree-by-tree assignment: reruns of incremental tree-by-tree loading (modified procedure)*											
Trip increments (%)	100	50, 50		40, 40, 20			20, 20, 20, 20, 20				
Cumulative trips (%)	100	50	100	40	80	100	20	40	60	80	100
Total time[a]	73324	27851	82877	18432	60883	80848	6494	18390	32539	54250	80993
Unused links	5	7	6	8	5	1	8	8	5	4	2
Constrained links[b]	28	13	26	12	23	27	5	12	16	24	28
v/v_0 (unconstrained)											
mean	2·737	1·367	2·821	1·094	2·231	2·801	0·547	1·093	1·659	2·222	2·801
SD	4·952	2·479	4·903	1·985	3·932	4·903	0·992	1·984	2·951	3·926	4·895
v/v_0 (constrained)											
mean	1·167	0·848	1·308	0·741	1·151	1·294	0·476	0·741	0·968	1·142	1·308
SD	0·928	0·895	0·932	0·844	0·950	0·920	0·684	0·843	0·899	0·926	0·919
Total trips assigned	1405936	702968	1405936	562374	1124749	1405936	281187	562374	843562	1124749	1405936
(b) *Incremental multipath algorithm* ($\theta = 1{\cdot}0$)											
Trip increments (%)	100	50, 50		40, 40, 20			20, 20, 20, 20, 20				
Cumulative trips (%)	100	50	100	40	80	100	20	40	60	80	100
Total time[a]	73348	27836	82850	18424	60871	80951	6495	18373	32561	54597	79980
Unused links	2	2	2	3	1	0	4	3	1	1	0
Constrained links[b]	28	13	26	12	23	27	5	12	16	24	27
v/v_0 (unconstrained)											
mean	2·737	1·367	2·822	1·094	2·231	2·803	0·547	1·093	1·656	2·213	2·799
SD	4·952	2·479	4·903	1·984	3·931	4·901	0·992	1·984	2·952	3·931	4·899
v/v_0 (constrained)											
mean	1·167	0·848	1·308	0·741	1·150	1·298	0·476	0·741	0·965	1·133	1·304
SD	0·926	0·894	0·931	0·844	0·949	0·917	0·684	0·843	0·900	0·933	0·918
Total trips assigned	1405936	702968	1405936	562374	1124749	1405936	281187	562374	843562	1124749	1405936

[a] Measured in thousands of minutes. [b] Links where $v/v_0 > 2{\cdot}5$ cutoff applied.

It is interesting to note that in these first runs on the small test network the only significant difference between the incremental tree-by-tree and incremental multipath assignment procedures is that the latter consistently has fewer unused network links. Again, a warning should be given about the evaluation of the total-travel-time statistic. This is calculated by adding the products of link travel time with link volume. The accuracy of the resulting value depends on the number of maximally constrained links, as the travel times for these links are calculated on the basis of constrained volume/capacity ratios. Thus, the fewer the restrained links, the more accurate the value of the total-travel-time statistic.

It was felt necessary to examine the consequences of changing values of θ for a fixed network and set of trip increments, thus two further small network run results are presented. In these two runs θ was assigned the values of one-half and two, respectively. These results are given in table 5.17. A comparison of the three 'forty, forty, twenty' runs of the incremental multipath procedure for different values of θ suggests that, in the range tested, the lower the value of θ, the fewer the unused links. The remainder of the network statistics did not show significant differences. It seems, however, that the use of $\theta = 0{\cdot}5$, which results in greater dispersion of trips to links than does the use of $\theta = 2{\cdot}0$, is perhaps somewhat better. It was not known whether the response of loadings as a function of θ was linear or nonlinear, and there were inadequate resources at the time of these tests to complete the examination.

Finally, several tests were attempted for the full-scale San Francisco Bay Area network; table 5.18 shows the results. Two runs of the incremental tree-by-tree procedure are presented, one for a single 100% increment and one for the 40%, 40%, and 20% increments. As expected the three-increment run shows better results than the single-increment run.

Table 5.17. Comparison of assignment runs for the small network: incremental multipath algorithm with two values of θ.

	$\theta = 0{\cdot}5$			$\theta = 2{\cdot}0$		
Trip increments (%)	40, 40, 20			40, 40, 20		
Cumulative trips (%)	40	80	100	40	80	100
Total time [a]	18449	60121	80432	18432	60976	80961
Unused links	1	0	0	4	2	0
Constrained links [b]	12	23	27	12	23	27
v/v_0 (unconstrained)						
mean	1·094	2·230	2·817	1·094	2·231	2·801
SD	1·984	3·931	4·894	1·985	3·932	4·903
v/v_0 (constrained)						
mean	0·742	1·148	1·317	0·742	1·151	1·295
SD	0·842	0·942	0·903	0·844	0·950	0·920
Total trips assigned	562374	1124749	1405936	562374	1124749	1405936

[a] Measured in thousands of minutes. [b] Links where $v/v_0 > 2{\cdot}5$ cutoff applied.

From the difference between incremental tree-by-tree for the single 100% increment and incremental multipath for the single 100% increment, it is obvious that the latter procedure yields the best results. For the two incremental tests, it again appears as if the incremental multipath procedure shows better results than the incremental tree-by-tree procedure. The question remains as to how much better these results are.

Though the incremental multipath procedure gave improved network statistics, it required three times as much computer execution time as the incremental tree-by-tree method. Because of this, especially if it is noted that the incremental tree-by-tree method was considered to be an 'expensive' assignment procedure in any case, further work with the incremental multipath procedure was not pursued. What is required, in addition to resources, is a definitive criterion for the excellence of a particular assignment pattern. Further work to examine the consequences of the various practical problems raised might then be usefully undertaken.

There also remain some theoretical questions. In particular, the multipath procedures as originally proposed assumed predetermined link travel costs, which are therefore independent of link volumes. The capacity-constraint notion is that link travel costs are explicitly a function of link volumes. The full theoretical implications of attempting to combine these approaches are not discussed here. There is, however, some new theoretical work which attempts to resolve some of these problems. After the discussion of mode-split procedures in the next section, the final section of this chapter will take up these questions again.

Table 5.18. Comparison of assignment runs for the large network: modified incremental tree-by-tree loading versus incremental multipath algorithm.

	Tree-by-tree loading				Multipath algorithm ($\theta = 1{\cdot}0$)		
Trip increments (%)	100	40, 40, 20			100	40, 40	
Cumulative trips (%)	100	40	80	100	100	40	80
Total time[a]	246427	54453	164061	233536	220588	51933	150311
Unused links	868	1160	867	674	730	961	742
Constrained links[b]	166	20	96	138	155	20	97
v/v_0 (unconstrained)							
mean	0·7098	0·2734	0·5604	0·7186	0·6797	0·2656	0·5426
SD	0·9022	0·3822	0·6643	0·7861	0·8749	0·3705	0·6487
v/v_0 (constrained)							
mean	0·6562	0·2714	0·5439	0·6876	0·6295	0·2637	0·5268
SD	0·6651	0·368	0·5871	0·6553	0·6445	0·3569	0·5730
Total trips assigned	547066	214065	428136	547066	547066	214065	428136

[a] Measured in thousands of minutes. [b] Links where $v/v_0 > 2{\cdot}5$ cutoff applied.

5.6 Development of a prototype mode-split procedure[6]

It is perfectly obvious that, in practice, integrated transportation, location, and land-use analyses will need to take proper account of the question of mode split. The proportions of travellers who are using, or who can be persuaded to use, nonautomobile transportation facilities will be of increasing importance in the coming decades. The original sponsorship of work on the integrated model package, coming as it did just before the massive energy price adjustments of the mid to late 1970s, did not provide for any effort at all to be devoted to the study of how mode split might be included in the model package. Despite the superficially myopic appearance of this approach, the fact was that very few urban areas in the USA had, at that time, any really substantial portion of their trips by public transit modes. Further, several studies done in the same era showed that transit usage in many European cities seemed to be declining in direct proportion to increases in personal income. Yet it is difficult to believe, with human populations still increasing and traditional energy sources declining, that the last decades of the twentieth century will not see a substantial increase in transit usage.

In light of this belief that energy price increases will, inevitably lead to some increased transit usage, especially in those urban areas with existing systems (or those where federal sponsorship is allowing the development of new systems) it seemed wise at least to investigate the prospects for inclusion of mode-split procedures in the integrated model package. This section describes the development and trial calibration of a prototype mode-split model which could be used as an integral part of integrated transportation and land-use analyses. The procedure was, in fact, also subjected to limited testing (on a small prototype data set) as a subprogram in the computer model package.

Through the mid 1950s very little formal attention was paid to mode-split analysis. The transportation analyses of the era used three principal steps: trip generation, trip distribution, and trip assignment. With the subsequent emergence of interest in mode-split analysis, several options for inclusion of mode split in these procedures were tested. The two most evident options were trip-end models and trip-interchange models. This general state of affairs persisted for some time, as is illustrated by one of the standard guides to mode-split application (Fertal et al, 1970). Trip-end mode-split models apportion the total trips originating and terminating in the various zones to the several modes. The mode-specific origins and destinations are then distributed—for example, the OD trip matrix is then estimated—for each mode. Trip-interchange models, on the other hand, calculate the mode splits after the trip-distribution step. Thus a matrix of the total trips between each OD pair is taken as input. The mode-split proportions are then calculated for each of the elements of that matrix.

(6) Portions of this material have appeared, in a somewhat different form, in Frank (1978).

Since the early 1970s, increasing research and application efforts have been devoted to refinement of behavioral models of mode choice (Atherton and Ben-Akiva, 1976; Domencich and McFadden, 1974; Koppelman, 1976). Both the trip-end and trip-interchange models as initially developed (and regularly applied) are aggregate deterministic models. Their actual mode-split calculation procedures are based on empirically derived (usually by multiple regression) functions often with rather vague theoretical underpinnings. The behavioral models, usually referred to as disaggregate models, are often derived from the notions of individual choice theory. It is also possible to derive models which jointly estimate mode split and trip distribution. In this section, efforts tending in this direction are discussed. After several possible mode-split formulations have been derived, one such model is applied to the San Francisco Bay Area data.

Of the many alternative transportation systems which exist—automobiles, buses, subways, taxicabs, and so on—most can be classified into two major categories: private transport, where the vehicles are owned and operated by individuals, and public transport, which is typified by bus or rail. In the San Francisco Bay Area, for example, over 70% of a sample of the 1965 population journeyed between place of residence and place of employment by auto. Slightly more than 5% made the same trip by bus, and the remainder were split between pedestrian, rail, taxicab, and trolley. Given the fact that the major proportion of all trips are made in autos or buses, the analysis presented here deals only with two transport modes, though the addition of more modes to this analysis is not precluded by any of its aspects. The general statements and conclusions should hold true for any number of modal alternatives. Two alternative formulations for addition of modal choice to the integrated transportation and land-use model perhaps are worth considering here. In one case choice of destination and of mode is combined into a single model. The other case, a more typical implementation, a modal choice submodel is inserted into the analysis procedure after the trip-distribution submodel.

As initially specified, DRAM, the residence-location submodel of the integrated package, may be described as being a production-constrained spatial interaction model. Though this form is customarily referred to as singly constrained, there are two basic constraints necessary to its implied entropy-maximizing derivation. The first constraint states that the total number of persons living in all zones i and working in a particular zone j, summed over all residence zones i, is equal to the total number of persons employed in zone j. Thus, we can write

$$\sum_i \sum_m T_{ij}^{mn} = E_j^n , \tag{5.24}$$

where T_{ij}^m is the number of person trips by transportation mode m. Note that this production constraint is stratified by n income categories.

The second constraint states that the total transportation cost of the region, C, is equal to the sum, over all residence zones i, all employment zones j, and all modes of transport, of the product of trips and costs:

$$\sum_i \sum_j \sum_m T_{ij}^{mn} c_{ij}^m = C^n \,. \tag{5.25}$$

Note that the cost constraint is also stratified by n income categories.

Subject to constraints (5.24) and (5.25), the entropy-maximizing derivation leads to the following joint distribution and mode-split model equation (Wilson, 1974)

$$T_{ij}^{mn} = B_j^n W_i^n E_j^n \exp(-\beta^n c_{ij}^m) \,, \tag{5.26}$$

where

T_{ij}^{mn} is the number of persons of type n living in zone i, working in zone j, and using mode m to travel between zones i and j;

B_j^n is a balancing factor,

$$B_j^n = \left[\sum_i W_i^n \exp(-\beta^n c_{ij}^m)\right]^{-1} ;$$

W_i^n is the attractiveness of zone i to type n tripmakers;

E_j^n is the number of persons of income group n employed in zone j;

β^n is an empirically derived, person-type-specific, trip-function parameter;

c_{ij}^m is the cost on mode m of travelling between zones i and j.

The mode-split proportions may then be obtained by forming the ratio of trips by one mode over total trips:

$$M_{ij}^{mn} = \frac{\sum\limits_n E_j^n \exp(-\beta^n c_{ij}^m)}{\sum\limits_m \sum\limits_n E_j^n \exp(-\beta^n c_{ij}^m)} \,, \tag{5.27}$$

where m_{ij}^{mn} is the proportion of trips by type n persons between zones i and j which take place by mode m. If the destination constraint is stratified by income categories, the E_j^n factors will not cancel out. However, one frequently finds nonhome-based trip ends characterized by a zone attribute of trip land-use activity, rather than by characteristics of tripmakers such as income. If the destination constraint is, in fact, not stratified by income categories, then the mode-split proportions may be stated as

$$M_{ij}^{mn} = \frac{\exp(-\beta^n c_{ij}^m)}{\sum\limits_m \exp(-\beta^n c_{ij}^m)} \,. \tag{5.28}$$

Finally, if there are only two possible modes of transport the mode-split proportion for mode 1 can be restated as:

$$M_{ij}^{1n} = \frac{\exp(-\beta^n c_{ij}^1)}{\exp(-\beta^n c_{ij}^1) + \exp(-\beta^n c_{ij}^2)} = \frac{1}{1 + \exp[-\beta^n (c_{ij}^2 - c_{ij}^1)]} \,. \tag{5.29}$$

This is a statement of mode split in terms of modal cost differences. It is a theoretical derivation of the typical empirical calculations of modal choice which when plotted as a function of cost differences yields a logistic curve. It is worth noting that the entropy-maximizing approach yields the same equation form as the behavioral choice theory approach. If the total cost constraint is stratified by income type, the parameter β^n must be estimated for each income class. The value of β^n, for each income type n, determines the average distance travelled by that person type, and also, in the mode-split equation, it determines the sensitivity of that person type to differences in costs between modes.

It is unfortunately the case that the data necessary to estimate such models are usually unavailable. This was indeed true of the San Francisco Bay Area data set available at the time of the study being discussed here. Thus the second, post-distribution mode-split procedure was the only real application possibility. In practice, if locational factors are deemed to be separate from the choice of mode, a post-distribution (or post-residential-allocation) modal choice submodel is a reasonable selection. This involves the calibration of two submodels, as two different parameters are to be estimated. The β parameter in the trip-distribution submodel shows the sensitivity of tripmakers to the mean distance between residence and place of work. A second parameter in the mode-split submodel shows the sensitivity of people to differences in modal costs.

By maximizing a function of all trips, T_{ij}^{mn}, subject to constraints, it is possible to derive, in a fashion analogous to that of a production-constrained trip-distribution model, a residential allocation model quite similar to DRAM. The destination constraint is that of equation (5.24). In place of equation (5.25), the total cost constraint is specified in terms of a generalized (that is, not mode-specific) cost:

$$\sum_i \sum_j T_{ij}^n c_{ij} = \hat{C}^n . \tag{5.30}$$

The precise form of the generalized cost variable to be used is determined by the assumptions about behavior which have been incorporated into the model. One possible assumption is that people behave for residential allocation as if they perceive only the minimum cost of all modes available to them:

$$c_{ij}^n = \min c_{ij}^m . \tag{5.31}$$

It is then possible, via the entropy-maximizing approach, to derive the following residential allocation model:

$$T_{ij}^n = B_j^n W_i^n E_j^n \exp(-\beta^n c_{ij}^n) . \tag{5.32}$$

A separate modal choice submodel can be derived in a similar fashion. The cost of travelling between each pair of zones is divided into the amount spent on travel by the minimum cost mode, plus the additional amount necessary to travel by other than the minimum cost mode, a notion similar to that expressed in equation (5.21). Then, if Z is the cost of travel on the minimum cost mode, and z^n is the extra cost to a type n traveller of travel on other than the minimum cost mode, then

$$c^n = Z + z^n . \tag{5.33}$$

If we represent the minimum cost between zone i and zone j by Z_{ij}, an extra cost constraint may be stated:

$$\sum_m (c_{ij}^m - Z_{ij}) T_{ij}^{mn} = z_{ij}^n . \tag{5.34}$$

The total number of trips on all modes is equal to the total number of trips:

$$\sum_m T_{ij}^{mn} = T_{ij}^{*n} . \tag{5.35}$$

Maximizing, per the entropy-maximizing approach,

$$\ln \frac{T_{ij}^{*n}!}{\prod_m T_{ij}^{mn}!} ,$$

subject to equations (5.34) and (5.35), we obtain

$$\frac{T_{ij}^{mn}}{T_{ij}^{*n}} = \frac{\exp[-\lambda_{ij}^n (c_{ij}^m - Z_{ij})]}{\sum_m \exp[-\lambda_{ij}^n (c_{ij}^m - Z_{ij})]} . \tag{5.36}$$

If it is assumed that the parameter λ^n is independent of zones i or j, the Z_{ij} cancel out and the following mode-split equation is obtained:

$$\frac{T_{ij}^{mn}}{T_{ij}^{*n}} = \frac{\exp(-\lambda^n c_{ij}^m)}{\sum_m \exp(-\lambda^n c_{ij}^m)} . \tag{5.37}$$

Interestingly enough, equation (5.37) is of the same form as equation (5.28), with a different set of parameters from that of the allocation model.

This model structure of distinct submodels is far more practical and easier to implement as the data requirements are far less stringent. The residential allocation submodel is calibrated for the β parameter. The modal choice submodel can be calibrated for the λ parameter, from cost data for all competing modes of transport. These costs can be stated in terms of travel time, travel cost, or some combination of the two. Having more or less settled on such a model, the next task was to attempt empirical estimates of the λ parameters.

The calibration of a modal choice submodel such as equation (5.37) involves the specification of a statistical procedure for obtaining accurate estimates of the parameter(s). The particular statistical procedure to be used depends upon the implicit theoretical background employed in the derivation of the modal choice submodel. The theoretical underpinnings of mode-split analysis are those of the theory of choice behavior.

The theory of choice behavior is predicated on the idea of probabilities of choices that are dependent upon attributes of items. If the aggregate choice behavior (for example, the total number of trips taken on each mode) of a large number of people is observed, then the proportions of the population choosing the various alternatives provide the best estimate of probabilities of each possible choice.

Let Q_1 be the probability of travelling by mode 1; it is possible to define a probability function

$$Q_1 = g(X) \; , \tag{5.38}$$

where X is a vector of attributes of the alternative modes. These attributes include such items as travel time or travel costs. These items are assumed to have associated λ parameters, which transform the vector of attributes into a linear function. For our purposes, let us assume that

$$X = c^1 - c^2 \; , \tag{5.39}$$

where c^1, c^2 are the costs of travelling by mode 1 (auto) and mode 2 (bus), respectively.

Because Q_1 is defined to be a probability, it must have the characteristics of probabilities; primarily, it must conform to the range between zero and unity. The function g can take on any form which satisfies this requirement. Two commonly used functions are the cumulative normal distribution and the cumulative logistic distribution. The normal distribution produces the probability function:

$$Q_1 = \phi(\lambda X) = \frac{1}{(2\pi)^{1/2}} \int_{-\infty}^{X} \exp\left(\frac{-\lambda X^2}{2}\right) \mathrm{d}X \; , \tag{5.40}$$

where ϕ is the standard cumulative normal distribution. This equation is termed the probit probability model. The logistic distribution gives the probability function:

$$Q_1 = \frac{1}{(1+\exp\lambda X)} \; , \tag{5.41}$$

which is termed the logit probability model. Both of these models give rise to ogive shapes, differing only at probabilities close to zero or one, where the probit model approaches the limiting values more rapidly. The logit model is more commonly used because of the computational advantage of dealing with a closed functional form. The probit model, on the other hand, has as its argument the limit of an integral which cannot easily be expressed in its closed form. The logit function is graphed in figure 5.2.

A number of computer programs were available to calibrate either logit or probit models. However, for data which is highly skewed (for example, where 90% travel by auto and 10% travel by bus), most of these computer programs are unable to fit reasonable parameter values. Several such programs typically predicted, with 90% confidence, that *all* trips would be by automobile. This was a serious obstacle, as such highly skewed mode splits are common in American urban areas. Much of the early work done on calibration of modal choice models was done in the UK, where there was more use of public transport than in the USA. A typical UK mode split would be 60% by auto and 40% by public transit. The standard computer programs can do a reasonably good job of calibration with such data.

An approach utilizing a program written by Southworth (1975) was attempted in an effort to overcome this problem. This procedure fits a logit curve to a set of derived points. These points are the percentage of people who travel by car when the cost differential between the two modes, X, is within a given interval. The cost difference for the data used here was the difference in travel times for people travelling by each mode on an uncongested network. This program has the added advantage of only using those OD pairs for which there are data on the number of people travelling by each mode. Since the data source in the trial calibration, a survey of trips in the San Francisco Bay Area nine county region, created a very sparse (empty) trip table for calibration purposes, this attribute of the program was a most decided advantage. The procedure uses a gradient-search technique to attempt to fit a logit curve to this constructed set of points. For the fitted curve, the program outputs two parameters, slope and intercept. The slope gives the sensitivity of tripmakers to cost differences between modes, and the intercept gives the penalty (or bonus) ascribed to taking a public transport mode.

The trial calibration was accomplished with data for trips and travel times for the 291 zone system of the San Francisco Bay Area. Forty-four cost intervals, each two minutes in length, were used. Each interval contained over 2000 trips, with anywhere from 59–100% by automobile.

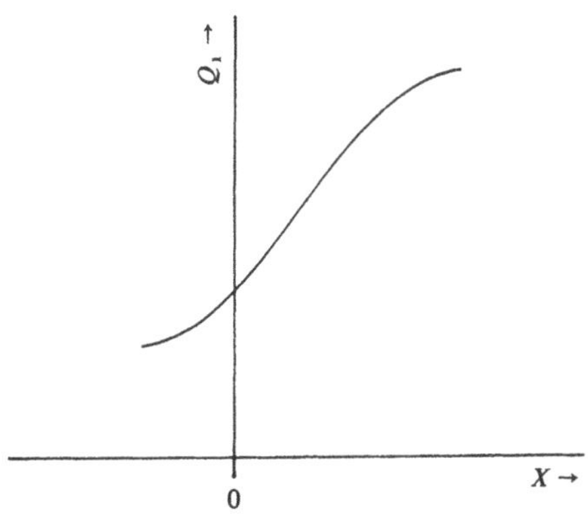

Figure 5.2. Logit curve for modal choice probability Q_1 [see equations (5.38)-(5.40)].

The values of the parameters obtained with this procedure yielded an explicit form of the logit function:

$$Q_1 = \frac{1}{1+\exp(-0{\cdot}041X+18{\cdot}928)} \cdot \tag{5.42}$$

The total number of trips for the entire region in the sample data set is 664030, with 571925 (or 86%) by auto. The model estimated 578799 trips by auto, an overestimate of 6874 trips or 1% of the total.

A second equation was then estimated, with the same data regrouped into one-minute intervals. With one exception (908 trips), all the one-minute intervals contained in excess of 1000 trips. The second logit function, estimated with the trip intervals reduced to one-minute widths, was

$$Q_1 = \frac{1}{1+\exp(-0{\cdot}046X+14{\cdot}785)} \cdot \tag{5.43}$$

In this redefined data set there were 512665 trips, of which 432675 were auto trips. The model estimated 434441 auto trips, an error of 0·3%. This function yielded similar trip estimates to those of equation (5.42) and, by its rather similar parameters, suggested a desirable lack of sensitivity to the size of the trip interval of the constructed data.

Yet a third function was tested, with the two-minute cost intervals of equation (5.42), but with perceived travel times as reported in the sample survey rather than calculated travel times. Here, the results were

$$Q_1 = \frac{1}{1+\exp(-0{\cdot}001X+52{\cdot}427)} \cdot \tag{5.44}$$

These rather different parameter values also yielded less-satisfactory trip estimates which reflect what are apparently considerable differences between individually perceived and agency-calculated estimates of travel cost.

By way of comparison it is interesting to note that models of the form

$$Q_1 = \frac{1}{1+\exp(aX+b)+\exp(cY+dZ+eS+f)} , \tag{5.45}$$

where

X is the travel time by car minus travel time by bus,

Y is the income of traveller,

Z is the travel cost by car minus travel cost by bus,

S is a service attribute, and

a, b, c, d, e, f are empirically derived parameters,

have been fitted by Peat, Marwick, Mitchell and Co. (1972) to data for San Diego, San Francisco (Marin County), and Boston. Coefficients of the travel-time-difference variable are all of the same order of magnitude. The coefficients for San Francisco determined above were roughly equal to 0·04 when differences in travel time less than 45 minutes were used.

The coefficients determined by Peat, Marwick, Mitchell and Co. for San Diego were 0·01 for CBD-oriented trips, and 0·04 for non-CBD-oriented trips; and their coefficients for San Francisco were 0·05 for CBD-oriented trips, and 0·06 for non-CBD-oriented trips.

Finally, having calculated the best-fit parameters for the logit function, in an effort to test the fit of these parameters, several statistical tests were to be calculated. The typical test of linear least-squares models, the coefficient of determination, R^2, is not valid in this case, as it embodies the assumption that the mean value of the predicted data is forced by the modelling procedure to be equal to the mean value of the observed data. Not only is this assumption not met by the logit model, but the mean values for the data used, as computed afterwards, are not even close. Therefore, an R^2 statistic cannot usefully be computed.

Of the several other statistics which are often used to test the results of forecasting models, two seemed appropriate here. The mean absolute percentage error, MAPE, is computed from the observed data, Q_i, and the predicted data, $\hat{Q}_i$:

$$\mathrm{MAPE} = \frac{100}{N}\left[\sum_{i}^{N}\left(\frac{Q_i - \hat{Q}_i}{Q_i}\right)\right] . \tag{5.46}$$

MAPE values of less than 3% are accepted as giving reasonable standards. The mean squared error, MSE, is computed as follows:

$$\mathrm{MSE} = \left[\frac{1}{N}\sum_{i}^{N}(\hat{Q}_i - Q_i)^2\right]^{1/2} . \tag{5.47}$$

MSE values of less than 1% are accepted as showing good estimating capability for the model.

These two performance statistics were then calculated for a selected set of zones from the San Francisco data set. As the sample data yielded a rather sparse matrix of observations, there were but fifteen zones for which MAPE and MSE could usefully be calculated. In each of these zones the sample included residents travelling to at least six different employment zones. The results of these calculations are given in table 5.19. For this rather limited analysis, these results seemed quite good.

It may be helpful here to try to tie together the implications of these mode-split explorations. First, two types of mode-split analysis were discussed as being of potential interest in the context of integrated transportation and land-use analysis. The first of these was a combined trip-distribution and mode-split model, implicitly a combined location and mode-split model. Although elegant in theory, the application of such joint models (which will be mentioned again in the next section of this chapter) requires more data for the purposes of parameter estimation than are normally available for US cities. Thus attention shifted to a post-distribution mode-split procedure. The logit function, equation (5.41), yielded by the derivation of this model was then empirically estimated with survey data from the San Francisco Bay Area. After a brief discussion of the estimation procedure, the results were presented. Based on this somewhat limited analysis, these results seemed to confirm the usefulness of a data-grouping approach to parameter estimation. The need for this approach resulted from the rather small percentage of actual transit trips in the data base.

Were such a post-distribution mode-split procedure to be integrated with the existing model package, the following general procedure would be followed for forecasting purposes. The residence location would be based not on auto–highway travel cost alone, as is currently the case, but on a generalized cost involving some weighted combination of transit cost and auto cost. As part of the residence-location procedure a trip distribution would be calculated. The output of this trip-distribution process is the $\{T_{ij}^{n}\}$ matrix of trips by type n tripmakers between each OD zonal pair. Then, based on the estimated logit function this matrix would be split into mode-specific $\{T_{ij}^{mn}\}$ matrices. Each such matrix of type n travellers using mode m between each OD pair would be separately assigned to the appropriate network. After the assignment process a new set of minimum paths through congested transit and highway networks would be calculated. These would be combined and the residence location recalculated. The process would probably need to be iterated two or three times to reach a relatively stable solution.

In point of fact, several prototype tests of this expanded version of the integrated package have been tested with good results. A full-scale practical application has yet to be attempted, but is probably not too far off. Even so, there are still questions, both theoretical and empirical, to resolve. The final section of this chapter takes up some of these issues.

Table 5.19. Performance statistics for the logit mode-split model: selected zones of the San Francisco Bay Area.

Origin zone	Sample size, N	MAPE	MSE	Origin zone	Sample size, N	MAPE	MSE
1	33	0·48	0·35	262	8	0·36	0·28
2	59	0·76	0·28	275	6	0·25	0·23
3	31	0·47	0·21	276	30	0·31	0·28
10	8	0·32	0·27	278	7	0·58	0·47
11	8	1·01	0·31	279	7	0·21	0·15
39	45	0·56	0·26	283	8	0·37	0·28
40	8	0·42	0·30	287	7	0·31	0·27
66	8	0·70	0·35				

5.7 Future prospects: applied and theoretical

The preceding five sections of this chapter have covered material which ranges from pragmatic to theoretical. Section 5.2 on conversion from trip probabilities to vehicle trips exemplifies the need in any actual application for resolution of problems which, although rather mundane, may well be critical to the success of the project. The part of section 5.6 on simultaneous trip-distribution and mode-split models discusses some work which, although of great interest from a theoretical point of view, is so demanding of data as to be unlikely soon to be applied. In this final section, a selection of further points will be taken up across the entire range from theory to application. The purpose is to suggest the direction in which applied investigations as well as theoretical conjectures on the transportation component of integrated analyses should go. The discussions begin with the more practical and move, more or less, towards the more theoretical.

One of the practical aspects of transportation analysis which sometimes becomes a critical problem in integrated transportation and location analyses, is the determination of levels of spatial detail. This topic is also taken up in the discussion of implementation of location models, and thus the comments here are confined to those specifically relevant to the transportation component of these model systems. The main problem here stems from the difficulty of producing a proper mathematical representation of a transportation network. The actual on-the-ground network contains, for any metropolitan area, tens, if not hundreds, of thousands of components. Streets, their lanes, their intersections and interconnections to land uses, entry and exit ramps, and so on are all part of the system. Clearly a one-to-one representation of all this is infeasible and, quite probably, undesirable. The question then becomes one of selection of aggregation procedures. The transportation planners have, in the main, traditionally favored great detail in network representation. Yet it is far from clear that the generation, distribution, and assignment procedures in current use are able to perform adequately at ultrafine levels of detail. The literature is virtually devoid of organized discussion of this topic.

In many metropolitan areas, transportation planners eventually settle for five hundred to one thousand 'traffic zones' and an abstract transportation network to connect them. Even this compromise on the part of the transportation analysis is much too detailed for the location analysis. Again, aggregation procedures become important, and again the literature is virtually devoid of help. Yet some aggregation method is clearly needed in order to bring the transportation and location models together. The little that has been published on this problem (Chan, 1976) is but a start towards the resolution of this perhaps mundane, but certainly critical, problem.

Before moving to the assignment process per se, note should be made of the obvious need for general work on the procedure for conversion from trip probabilities to vehicle trips. Presumably a survey of the many reports on trip-generation models would yield much valuable information on trip-generation rates. This material could then be combined with the notion of extracting trip distributions from the location models to yield a more satisfactory conversion procedure. The importance of a proper resolution of this problem becomes more clear when it is considered as follows. The vehicle-trip conversion ultimately yields the numbers of vehicle trips to be assigned to the network. As the assignment is a capacity-constrained assignment, the numbers of vehicle trips will determine the degree to which the network is congested. The extent of network congestion determines the revised zone-to-zone travel times and/or costs. These, in turn, determine the spatial pattern of activities in the region, and so on. Thus the consequences will be felt throughout the model system.

Another potentially critical set of questions revolves around the determinations of intrazonal travel times and/or costs, and of 'terminal' times. Frequently one finds that intrazonal times are produced by educated guesswork in the transportation agency. This, in itself, is no great problem. However, when these intrazonal times are used in the context of capacity-constrained assignment, an obvious inconsistency appears. How are congestion effects to be represented on the intrazonal connections of the transportation systems? A similar problem arises with terminal times. These are the times taken to represent a traveller gaining access to the network from within a particular zone to the load node of that zone. Terminal times, too, should be responsive to (that is, be functions of) network trip volumes. Yet when they are exogenously provided there is no sensible way to have them vary with congestion levels.

Again, the problems are made more serious by the interconnectedness of the model system. The intrazonal times or costs will usually tend to be the smallest values in the zone-to-zone cost matrix. Indeed many early location-model applications arbitrarily assumed that the difficult-to-obtain c_{ii} values were either zero or one. To adopt this practice would, without question, seriously bias the estimates of location-model parameters as well as the subsequent forecasts from the model.

The network assignment procedure used in the ITLUP analyses could include a combined intrazonal and terminal time, if the load node had only one link. The impedance of this link would be automatically added to every path connecting to the load node. This impedance would then also serve as the intrazonal time. To some extent this would still be an inaccurate representation, as some components of terminal times (parking fees, for example) should be invariant under loading of the network, whereas others should vary with varying loads. Unfortunately, no information on the determination of the terminal and intrazonal times was provided with the data sets delivered. The San Francisco Bay Area networks provided had load nodes imbedded in the network, usually with three or four links per node. This caused errors in the final impedance matrix insofar as the diagonal elements did not represent an intrazonal time, merely an arbitrary link time, and the matrix as a whole does not include terminal times. However, these 'errors' were corrected in a later program that inserted estimates of the appropriate values.

To utilize the ITLUP network assignment program in this way, dummy load links would have to have been coded for each network. The link impedances would then have been some combination of the exogenously supplied intrazonal and terminal times. Furthermore, assumptions would have to have been made about the capacities of these links. If the links were given too large a capacity value, there would never have been any volume delays. If the capacity had been too small, the volume delays would have been excessive. In the wide middle ground, there was no particular criterion for choosing one capacity as opposed to any other. Coding these links would have required substantial map work, rerunning of the network coding program, and retracing of the free flow minimum paths. It should be noted that this practice of imbedding load nodes into the network was described in the US Department of Transportation's FHWA network package manual (USDoT, 1972). The FHWA package per se has no provision for intrazonal times, which was one of the reasons it was not used for ITLUP. The standard solution to this difficulty, as suggested in the FHWA manual, would be to use as the intrazonal time 50% (or some other percentage) of the average impedance from each zone to all contiguous zones. The implementation of this solution would have required the preparation of a zone contiguity table. A second solution is to weight the free flow intrazonal impedances by the average percentage change due to volume delays in the load links. This solution was implemented and is now an integral part of the model package. Thus, when the assignment procedure, including the capacity-constraint calculations, is complete, a post-processing procedure is performed. This procedure calculates the new intrazonal times as modified according to the above and inserts this vector of times into the diagonal of the zone-to-zone travel-time matrix.

A further area of attention is that of whether the trip-assignment configuration which results from the last increment of the incremental tree-by-tree assignment process is at, or near, an assignment equilibrium. It is likely that this solution is not an equilibrium assignment and that, furthermore, on some links at least, it is quite far from being an equilibrium assignment. A palliative measure would be to have a further pass through the assignment process. On this pass the first step would be to remove some trips from the most heavily overloaded links. These trips would subsequently be reassigned to the network as congested after the last trip increment had been assigned. By removing these trips from the worst routes through the network and then reassigning them, the resulting assignment would be more likely to approximate more closely an equilibrium assignment.

Finally, there is the peculiar matter of negative trip volumes. As work with the integrated package progressed, it became clear that after the first time period in a forecast for several time periods, there was what could be considered a stable load of trip volumes on the network. Thus, it was reasoned, substantial computer time could be saved by not completely reloading the network after the location and land-use forecasts for each time period. Thus the link volumes were saved after, say, 60% or 80% of the trips for the first time period were loaded. The 100%-loaded volumes were used to calculate the link congestions which ultimately yielded the zone-to-zone travel times used as input to the location and land-use forecasts for the next (second) time period. The trip matrix for the second time period is output from the location forecasts. From this trip matrix is subtracted, say, the 80% trip matrix from the first time period. This leaves an incremental trip matrix to be assigned to the network (which is congested with the volumes from the 80% trip matrix for the first time period). With this procedure it is possible that the spatial redistribution of activities from one iteration to the next will result in a redistribution of trips which will result, on some links, in negative trips when the one trip matrix is subtracted from the other. Consequently, the volume of negative trips thus to be unloaded from a link can exceed the existing link volume because of the nature of the network assignment procedure.

Generally speaking, in the assignment procedure, some number of trips originating at zone A and terminating at zone B is to be loaded onto the network. To do this, the shortest path through the network from A to B is found and the trips are added to each link in this path. Each time trips are loaded on a given link, the time required to traverse that link increases owing to congestion of the link. This is dealt with operationally in the calculations of the volume-delay function. Because of this, trips traversing a certain portion of the network will find that a link that was on the minimum (that is, shortest time) path at one time, will not be on it at a later time.

When negative trips from A to B are to be (un)loaded, the assignment procedure is not able to deal with them appropriately. Any link on the minimum path from A to B may have been included in the path by virtue of congestion on a 'parallel', or alternative, link and may thus have just a few trips which were loaded on it when the previous 'best' link became overloaded. If the negative trips being (un)loaded are in excess of the existing trips on the link, there is no way for the assignment program to determine what other link had been best before the loading of the link in question. Consequently there is no way to determine where (from which link) the excess negative trips should be (un)loaded. These excess negative trips therefore end up on the most recent best link which, in turn, is left with a negative trip volume. Operationally this is dealt with by restricting the volume/capacity ratio to be greater than zero, but the negative trip volume remains either to the end of the network run, or if positive trips (from another OD pair) are loaded on that link they reduce (or eliminate) the negative volume. Although in practice this does not occur frequently enough to cause any real difficulties, it does illustrate the kinds of problem that may arise with procedures which under normal circumstances are perfectly satisfactory. Problems of this sort appeared regularly during the development of the integrated package, pointing out many areas where the paucity of previous integrative attempts has allowed important considerations to 'fall between the cracks'. This leads into the brief mention of some theoretical developments in these areas which will undoubtedly come to bear on future development efforts with integrated models.

The first item here centers on the basic question of network equilibrium. We recall that the earliest work with network assignment procedures soon came up against the equilibrium problem, and it has been a focus of attention ever since. It was not until the mid 1970s that precise definitions of network equilibrium resulted in the development of reasonably efficient mathematical programming techniques for calculating network equilibria (Florian and Nguyen, 1975; Florian, 1976). This work was extended to multimodal automobile and transit equilibria (Florian, 1977). It remains, to this author's knowledge, to determine the practical efficiency of these procedures. Do they yield more realistic network loadings, and if so, how much better are they and at what cost, as compared with some of the procedures now in use? Although some results have recently become available, they are not yet definitive (Eash et al, 1979).

The logical next step was the development of combined distribution and assignment models. Procedures for the solution of this problem soon emerged (Evans, 1976; Florian et al, 1975). Finally, it was shown that distribution, mode split, and assignment could, in theory all be solved simultaneously (Florian and Nguyen, 1978).

As has been frequently mentioned elsewhere in this book, if one has a trip-distribution model of the proper form then one has a residence-location model. Thus once it has been demonstrated that distribution, mode split, and assignment could all be solved simultaneously it was to be expected that various location, mode-split, and assignment models would soon follow. Such models have recently been described, though again, their practical applicability remains to be demonstrated (Boyce, 1978; Boyce and Southworth, 1978; Los, 1979). Inevitably these theoretical developments will affect the use of transportation models and location models in practice, and a watch should be kept for further development in this direction.

6

Development of employment-location models

6.1 Introduction

Chapter 3 has given an overview of the early development of an integrated transportation and land-use model package; these next few chapters are devoted to elaborations in each of the main subareas of the overall model system. This chapter is devoted to a description of the ways in which analysis of employment location was developed.

At the time of the initial development of the prototype form of the models, one of the most commonly accepted strategies for dealing with intraurban employment location was to assume away the most difficult part of the problem. As such, employment was classified as either 'basic' or 'nonbasic'. This latter was also known as 'local-serving' or 'population-serving' (LOMS) employment. Although this was in some sense an arbitrary dichotomy, the intent was that the basic (NLMS) employment types were export oriented with respect to the particular urban area being analyzed, and thus not too dependent upon the intraurban spatial distribution of population. In contrast, the intraurban location of population-serving employment was assumed to be highly dependent on the intraurban location of population. In most studies the location of basic employment was made external to the location models, and the location of nonbasic employment was calculated simultaneously with the location of population.

Having defined the location of basic employment as being exogenous to the model system, many studies then proceeded to ignore it entirely, though clearly it was a driving force behind the location of other activities. The simultaneous location of population and population-serving employment became one of the central features of the Lowry model (and many of the Lowry derivative models). The next section of this chapter describes such a linked model structure for population and population-serving employment, as this was used in the early versions of the integrated model package. This is followed in section 6.3 by a presentation of two procedures developed for estimating the intraurban location of basic employment, and which were used in early work with the integrated model package. Section 6.4 contains a presentation of a more recently developed model which handles both basic and population-serving employment types. Finally, a discussion of the calibrations of this new employment model and of the equations necessary in calibration is given in section 6.5.

6.2 Models of residence and population-serving employment

It may be recalled from the second section of chapter 2 that the simple residence-location model subsequently extended for use in the integrated package is, in fact, a singly constrained spatial interaction model. By a similar but by no means identical path of reasoning, a model for population-serving employment location may be developed, as follows.

It is first assumed that retail trade will suffice in this example as a representative type of population-serving employment. Then, it is hypothesized that retail employment in a zone is a simple direct function of retail sales in the zone:

$$E_j^{\mathrm{R}} = \mathrm{g}(S_j) \,, \tag{6.1}$$

where E_j^{R} is the retail employment in zone j, and S_j the retail sales in zone j. Then it is assumed that the total retail sales in zone j are a function of the sum, over all zones in the region, of consumers' purchasing power multiplied by the probability that a consumer who lives in zone i will make retail purchases in zone j. In equation form, this is expressed as

$$S_j = a\sum_i Y_i p_{ij}^{\mathrm{s}} \,; \tag{6.2}$$

here a is a constant (empirically determined); Y_i is the purchasing power of consumers living in zone i,

$$Y_i = \mathrm{f}(N_i) \,,$$

so that Y_i is a function of the population of zone i; p_{ij}^{s} is the probability that a consumer living in zone i will make retail purchases in zone j. In most cases the 'shopping' probability p_{ij}^{s} will be taken to be a function of some measure of retail attractiveness in zone j and some measure of the cost of interaction between zones i and j. Thus

$$p_{ij} = \mathrm{k}[W_j^{\mathrm{R}}\mathrm{f}(c_{ij})] \,, \tag{6.3}$$

where W_j^{R} is the retail 'attractiveness' of zone j, f is the trip function, and c_{ij} the generalized travel cost between zone i and zone j.

If, as was the case for the simple residence-location model, the p_{ij}^{s} term is properly constructed as a probability, then any inconsistency between zonal 'sales' and regionwide consumer purchasing power may be resolved. This, not surprisingly, results in a singly (production) constrained spatial interaction model of the form

$$S_{ij} = A_i Y_i W_j^{\mathrm{R}}\mathrm{f}(c_{ij}) \tag{6.4}$$

with

$$A_i = \left[\sum_j W_j^{\mathrm{R}}\mathrm{f}(c_{ij})\right]^{-1} . \tag{6.5}$$

Now, substituting equation (6.5) into equation (6.4), we obtain

$$S_{ij} = Y_i\left[\frac{W_j^{\mathrm{R}}\mathrm{f}(c_{ij})}{\sum_j W_j^{\mathrm{R}}\mathrm{f}(c_{ij})}\right] ; \tag{6.6}$$

this expression can be summed for zone j:

$$S_j = \sum_i S_{ij} = \sum_i Y_i\left[\frac{W_j^{\mathrm{R}}\mathrm{f}(c_{ij})}{\sum_j W_j^{\mathrm{R}}\mathrm{f}(c_{ij})}\right] . \tag{6.7}$$

Now, as an added bit of information, it should be noted that the original nonbasic employment component of the Lowry-type models was a slightly extended version of equation (6.7), with the notion of work-to-shop trips added. Equation (6.7), by representing potential consumers at their places of residence, is in essence dealing only with home-to-shop shopping. Yet it is easily observable in most shopping areas that there are consumers who are coming from their places of work—work-to-shop shopping. These are dealt with in Lowry derivative models by adding in a term to represent potential work-to-shop shoppers:

$$S_j = \sum_i Y_i \left[\frac{W_j^{\mathrm{R}} \mathrm{f}(c_{ij})}{\sum_j W_j^{\mathrm{R}} \mathrm{f}(c_{ij})} \right] + \mathrm{f}(E_j^{\mathrm{T}}) \ . \tag{6.8}$$

The implied assumption here is that all work-to-shop shopping in zone j results from the total employment, E_j^{T}, in zone j. Perhaps a more general form for this term would be a second spatial interaction term such as equation (6.7) itself. In such a formulation Y_i might be redefined as purchasing power of consumers working in zone i. This, in turn, would be a function of the number of employees working in zone i. Similarly the $p_{ij}^{\$}$ term might be redefined as the probability that a consumer working in zone i will make retail purchases in zone j. Further, one would expect the empirically estimated trip function $\mathrm{f}(c_{ij})$ to differ quite considerably between the home-to-shop and work-to-shop cases.

Models of this form for population-serving or nonbasic employment could easily be linked directly to a singly constrained spatial interaction model for residence location. The linked models could then be run, and before that they could even be calibrated as a linked system (Batty, 1976, chapter 5). The IPLUM model (in each of its several versions), as well as the first version of DRAM, had this general configuration. Thus population and population-serving employment were accounted for, but it remained to define some process for producing forecasts of the intraurban location of basic employment. Two of the early models developed to deal with that problem are presented in the next section.

6.3 Models for exogenous forecasts of location of basic employment

During the research work done in developing the prototype of the integrated model package, a decision was taken that, under the circumstances then current, no attempt would be made to develop a model of basic (NLMS) employment location. Yet, it was desired to evaluate several policy options and thus it was necessary to do something about forecasts of basic employment. There had been developed, several years earlier, a model of basic employment for use in the San Francisco Bay Area. This model, known as BEMOD, had been used by the Metropolitan Transportation Commission and the Association of Bay Area Governments in some of their forecasting exercises for the region. The design and development of this model are described only in a 'fugitive' document (Nathanson, 1970) and thus warrant at least brief recapitulation here.

BEMOD was a linear multiple regression model. The level of spatial detail was the census tract, of which there were just under eight hundred in the region. The regression analyses were run for subsets of the eight hundred census tracts. There were six such subsets, five of which were whole counties: San Francisco, San Mateo, Santa Clara, Alameda, Contra Costa. The last subset was an aggregation of Marin, Solano, Napa, and Sonoma Counties. The regressions were thus run separately for each of these groups of census tracts. The independent variables used were:

- average slope
- average elevation
- presence of navigable waterway
- presence of railroad facilities
- generalized accessibility
- density of existing development
- residential land
- unused land
- vacant land
- distribution of employment types.

The results of the regression analyses performed by the Bay Area Transportation Study Commission (BATSC) can only be described as less than spectacular. A separate regression was calculated for each of ten employment sectors for each of the six subareas. F-tests on these regressions showed most of them to be significant, but the details of coefficients and the rather mixed showing of t-tests of coefficients for all sixty regression equations are really not worth repeating here. Results for the goodness of fit, R^2, are given in table 6.1 for use in comparison with subsequent model developments.

Table 6.1. Values of goodness-of-fit parameter, R^2, for BEMOD regression analyses by BATSC, May 1968.

Employment type [a]	Area					
	San Francisco	San Mateo	Santa Clara	Alameda	Contra Costa	North Bay [b]
MFG1	0·7098	0·3107	0·2584	0·1078	0·0697	0·0938
MFG2	0·6376	0·6469	0·1336	0·2091	0·1292	0·1681
MFG3	0·6344	0·7763	0·0607	0·1712	0·0762	0·1622
MFG4	0·5543	0·6772	0·0674	0·4958	0·0953	0·1128
MFG5	0·5636	0·7444	0·0272	0·6459	0·2357	0·1082
TRAN	0·8725	0·6413	0·0281	0·0927	0·0665	0·2361
WHOL	0·8725	0·8993	0·1716	0·1672	0·2680	0·1457
FIN	0·3921	0·0571	0·1754	0·0546	0·1889	0·3522
SERV	0·4710	0·4521	0·3243	0·0565	0·5008	0·1058
GOVT	0·4969	0·7693	0·2502	0·1381	0·2680	0·4412

[a] MFG—manufacturing: 1, new technology; 2, centralized urban; 3, decentralized urban; 4, metal fabrication and machinery; 5, petrochemicals, primary metals; TRAN—transportation; WHOL—trade; FIN—finance and insurance; SERV—services; GOVT—government.

[b] Aggregation of Marin, Solano, Napa, and Sonoma.

Despite the somewhat less than satisfactory performance exhibited by these regression results, the time and resources available to the prototype integrated package project made it impossible to do other than use the BEMOD forecasts. Thus the exogenous forecasts of basic employment used in the preliminary policy tests described in chapter 3 were those originally obtained for the Bay Area Transportation Study Commission by use of BEMOD.

Some modest efforts were devoted to examining the results of alternative spatial and sectoral aggregations. Data sets with twelve and eight employment sectors were analyzed, as were several different levels of zonal aggregation. In general the results of these analyses did not yield any noticeable differences from the original work with BEMOD. Thus it was left to a subsequent research project to attempt to improve upon the BEMOD results.

In 1974, not long after the completion of the work on the prototype integrated package, work on a new project was begun (Putman, 1976).

Table 6.2. Definitions of variables for EMPIRIC equations.

Variable	Definition
LIQ	households in lowest income quartile
LMIQ	households in lower-middle income quartile
UMIQ	households in upper-middle income quartile
HIQ	households in highest income quartile
MISC	construction and other miscellaneous employment
MFGW	manufacturing and wholesale employment
TCU	transportation, communications, utilities employment
RET	retail employment
SVCFIR	service, finance, insurance, real estate employment
LGOVED	local government and education employment
HAHU	highway accessibility to households
TAHU	transit accessibility to households
AHU	composite (sum of highway and transit) accessibility to households
HAEMP	highway accessibility to employment
AEMP	composite accessibility to employment
SEWER	percentage of district 'sewered'
NCA	net commercial area
NIA	net industrial area
NPA	net public and semipublic area
NRA	net residential area
USEDAC	used area (= NCA + NIA + NPA + NRA)
VACAC	vacant or agricultural area
DEVAC	developable area (= USEDAC + VACAC)
TOTAC	total area of the district
TOTHU	total housing units
TOTEMP	total employment
Δ	indicates 'change-in-share' variables; all others are base year shares.

One of the central goals of that project was to compare calibrations of different land-use models on identical data sets. After some amount of sifting and sorting, two models were selected for comparison, the EMPIRIC model and the IPLUM model. Both models were to be calibrated on a data set for the Minneapolis–St Paul metropolitan area. EMPIRIC, being a reasonably well-defined linear-difference-equation model presented no particular calibration problems. This model deals simultaneously with both residence location and employment location[7]. Two-stage least-squares and ordinary least-squares estimation procedures were both used in parameter estimation for EMPIRIC. Variables used in the regression analyses for EMPIRIC are defined in table 6.2. The results from the two different procedures were not very different, and only the two-stage least-squares results are given here, in table 6.3. These results are virtually the

Table 6.3. Two-stage least-squares regression results for EMPIRIC: Minneapolis–St Paul region.

$\Delta\mathrm{LIQ}$	$= 0{\cdot}407\Delta\mathrm{LMIQ} - 0{\cdot}377\Delta\mathrm{HIQ} + 0{\cdot}106\Delta\mathrm{LGOVED} - 0{\cdot}415\mathrm{LIQ} + 0{\cdot}357\mathrm{LMIQ} - 0{\cdot}890\Delta(\mathrm{HAEMP} * \mathrm{USEDAC}) + 0{\cdot}269\Delta\mathrm{SEWER} + 0{\cdot}060(\mathrm{SEWER} * \mathrm{VACAC}) + 0{\cdot}112(\mathrm{TOTEMP}/\mathrm{TOTHU})$
$\Delta\mathrm{LMIQ}$	$= 0{\cdot}299\Delta\mathrm{LIQ} + 0{\cdot}425\Delta\mathrm{UMIQ} + 0{\cdot}092\mathrm{UMIQ} - 0{\cdot}109(\mathrm{AEMP} * \mathrm{USEDAC}) + 0{\cdot}300\Delta(\mathrm{HAEMP} * \mathrm{USEDAC})$
$\Delta\mathrm{UMIQ}$	$= -0{\cdot}144\Delta\mathrm{LIQ} + 0{\cdot}415\Delta\mathrm{LMIQ} + 0{\cdot}261\Delta\mathrm{HIQ} - 0{\cdot}163\mathrm{LIQ} + 0{\cdot}058(\mathrm{SEWER} * \mathrm{TOTAC}) + 0{\cdot}104(\mathrm{UMIQ}/\mathrm{TOTHU})$
$\Delta\mathrm{HIQ}$	$= -0{\cdot}416\Delta\mathrm{LIQ} + 0{\cdot}01\Delta\mathrm{LMIQ} + 0{\cdot}830\Delta\mathrm{UMIQ} + 0{\cdot}248\Delta\mathrm{SEWER} - 0{\cdot}260(\mathrm{HIQ}/\mathrm{TOTHU}) + 0{\cdot}274(\mathrm{INDUS}/\mathrm{TOTEMP})$
$\Delta\mathrm{MISC}$	$= 0{\cdot}44\Delta\mathrm{RET} + 0{\cdot}20\Delta\mathrm{SVCFIR} - 0{\cdot}026(\mathrm{TOTEMP}/\mathrm{TOTHU}) + 0{\cdot}112(\Delta\mathrm{SEWER} * \mathrm{TOTAC}) - 0{\cdot}256\mathrm{MISC} - 0{\cdot}096\mathrm{SVCFIR} + 0{\cdot}109(\mathrm{NIA} * \mathrm{VACAC}/(\mathrm{USEDAC} + \mathrm{VACAC}) + 0{\cdot}094\mathrm{TAHU}))$
$\Delta\mathrm{MFGW}$	$= 0{\cdot}013\Delta\mathrm{SVCFIR} + 0{\cdot}190(\mathrm{SEWER} * \mathrm{TOTAC}) + 0{\cdot}254\mathrm{SVCFIR} - 0{\cdot}189\mathrm{MFGW} - 0{\cdot}268\mathrm{NCA} - 0{\cdot}531(\mathrm{USEDAC}/(\mathrm{USEDAC} + \mathrm{VACAC})) - 0{\cdot}248(\Delta\mathrm{HAHU} * \mathrm{USEDAC}) + 0{\cdot}52\mathrm{HAHU}$
$\Delta\mathrm{TCU}$	$= 0{\cdot}737\Delta\mathrm{RET} + 0{\cdot}919\Delta\mathrm{SEWER} + 0{\cdot}249(\mathrm{NIA} * \mathrm{VACAC}/(\mathrm{USEDAC} + \mathrm{VACAC})) - 0{\cdot}352\mathrm{MFGW} + 0{\cdot}60\mathrm{USEDAC}/(\mathrm{USEDAC} + \mathrm{VACAC}) + 0{\cdot}1827\mathrm{CU} - 0{\cdot}53(\mathrm{TOTEMP})/(\mathrm{NIA} + \mathrm{NCA} + \mathrm{NPA}) + 0{\cdot}31(\mathrm{TOTEMP}/\mathrm{TOTAC}) - 0{\cdot}423(\mathrm{NCA} * \mathrm{VACAC}/(\mathrm{USEDAC} + \mathrm{VACAC}))$
$\Delta\mathrm{RET}$	$= 0{\cdot}473\mathrm{SVCFIR} + 0{\cdot}518\Delta\mathrm{LMIQ} + 0{\cdot}077\mathrm{NCA} * \mathrm{VACAC}/(\mathrm{USEDAC} + \mathrm{VACAC}) - 0{\cdot}32\mathrm{RET} + 0{\cdot}291\mathrm{AHU} * \mathrm{USEDAC}$
$\Delta\mathrm{SVCFIR}$	$= 0{\cdot}169\Delta\mathrm{UMIQ} + 0{\cdot}202\mathrm{MFGW} + 0{\cdot}344\mathrm{RET} - 0{\cdot}154\mathrm{GOVED} - 0{\cdot}228\mathrm{SVCFIR} + 0{\cdot}236\mathrm{RET}$
$\Delta\mathrm{LGOVED}$	$= 0{\cdot}29\Delta\mathrm{LIQ} + 0{\cdot}313\mathrm{TAHU} + 0{\cdot}214\mathrm{NCA} - 0{\cdot}539\mathrm{LGOVED}$

(7) For a review of EMPIRIC and some of its applications, as well as an overall evaluation of the model, see chapter 3 in Putman (1979).

same as those that were obtained in the original application of EMPIRIC in the Minneapolis–St Paul region (Peat, Marwick, Mitchell and Co., 1971).

The precise definitions of the six employment sectors and the goodness-of-fit (R^2) results are given in table 6.4. These results were in fact reasonably good, as they were for dependent variables which were in terms of change in a sector's employment in a zone, over a ten-year period. Nonetheless it was felt that the general form of the model left something to be desired: the ten-year change variable could easily have been obscuring important shorter-term changes.

A new multiple-regression model was then formulated in an attempt to improve somewhat on these results. After a great deal of statistical experimentation with the Minneapolis–St Paul data set, eight employment sectors were defined (goodness of fit, R^2, given here too), as in table 6.5. The variables used in this new model formulation, known as EMPMOD, are defined in table 6.6. Regression results for EMPMOD (statistically significant coefficients only) are given in table 6.7. These results were reasonably good, and were in fact used to generate estimates of basic employment for use in comparing the performance of the two models on the Minneapolis–St Paul data set. They were, however, not sufficiently good to warrant the use of EMPMOD as an integral submodel of the integrated transportation and land-use model package. The development of an employment model of sufficient reliability for this purpose was yet to come.

Table 6.4. EMPIRIC employment-sector definitions and R^2 results.

Employment sector	R^2
1 construction, miscellaneous	0·746
2 manufacturing, wholesale	0·708
3 transportation, communication, utilities	0·464
4 retail	0·790
5 services, finance, insurance, real estate	0·754
6 local government, education	0·545

Table 6.5. EMPMOD employment-sector definitions and R^2 results.

Employment sector	R^2
1 agriculture, forestry, fisheries	0·421
2 heavy manufacturing (durable)	0·758
3 light manufacturing (durable)	0·811
4 manufacturing (nondurable)	0·812
5 wholesale and retail trade	0·866
6 finance, insurance, real estate	0·970
7 services	0·901
8 government	0·960

Table 6.6. EMPMOD: definition of dependent variables, by zone.

Variable	Definition
E(K)	employment in industry group K (K = 1, 8) in 1970
E(K)LAG	employment of type K, in 1960
AVL	mean (average) of TL
TL	total land
MAXTL	maximum value of TL
MINTL	minimum value of TL
TLA	(AVL − TL(I)) ** 2: variance of zonal total land
TLB	(MAXTL − TL(I))/(MAXTL − MINTL): normalized zonal total land
BL	industrial land + available land, 1960, in I
IDENS	industrial density, 1960, in zone I
POP	residential population in I, 1960
RDENS	residential density in I, 1960
DGR1	percentage change in number of households in income quartiles (1+2)
DGR2	percentage change in number of households in income quartiles (3+4)
DACIND	change in composite accessibility to manufacturing employment
DACCM	change in composite accessibility to commercial employment
DACG1	change in composite accessibility to income group 1
DACG2	change in composite accessibility to income group 2
DSWR	change in sewer system land
STE	percentage of total employment in I in 1960
SACTE	percentage of composite accessibility to total employment in I in 1960
SACHH	percentage of composite accessibility to HH in I in 1960
HH	total households

Table 6.7. Estimated employment equations from EMPMOD.

$$E1 = -0{\cdot}0181\,TLA + 0{\cdot}0007\,POP + 1331{\cdot}4\,DGR2 + 18629{\cdot}0\,DACG1 - 15748{\cdot}0\,DACG2 + 8{\cdot}5920$$

$$E2 = 1{\cdot}1745\,E1LAG + 770532{\cdot}0\,DACG1 - 1009028{\cdot}0\,DACG2 + 17931{\cdot}0\,DSWR + 28192{\cdot}0\,SACTE - 29{\cdot}163$$

$$E3 = 1{\cdot}3604\,E3LAG + 15{\cdot}848\,RDENS + 23385{\cdot}0\,DGR1 + 23652{\cdot}0\,DGR2 + 109{\cdot}96$$

$$E4 = 0{\cdot}9491\,E4LAG + 152642{\cdot}0\,DACIND - 107994{\cdot}0\,DACG2 + 166{\cdot}76$$

$$E5 = 1{\cdot}0515\,E5LAG + 0{\cdot}5846\,IDENS + 0{\cdot}0160\,POP + 26253{\cdot}0\,DGR1 + 45314{\cdot}0\,DGR2 - 131{\cdot}62$$

$$E6 = 0{\cdot}8977\,E6LAG - 0{\cdot}1785\,TLA - 5{\cdot}7198\,TLB + 1{\cdot}9488\,IDENS + 0{\cdot}0035\,POP + 4{\cdot}6650\,RDENS + 4471{\cdot}2\,DGR1 + 7411{\cdot}3\,DGR2 + 65497{\cdot}0\,DACG1 - 72701{\cdot}0\,DACG2 - 3969{\cdot}4\,DSWR + 4157{\cdot}8\,STE + 449{\cdot}19$$

$$E7 = 0{\cdot}6957\,E7LAG - 0{\cdot}4636\,TLA - 14{\cdot}782\,TLB - 0{\cdot}6809\,IDENS - 0{\cdot}0089\,POP + 30452{\cdot}0\,DGR1 + 26937{\cdot}0\,DGR2 + 290845\,DACIND - 422560{\cdot}0\,DACCM + 85920{\cdot}0\,DACG1 + 16910{\cdot}0\,STE + 33841{\cdot}0\,SACHH + 1136{\cdot}9$$

$$E8 = 1{\cdot}5057\,E8LAG - 0{\cdot}7470\,IDENS - 0{\cdot}0062\,POP - 28{\cdot}378\,RDENS + 13015{\cdot}0\,STE + 77{\cdot}806.$$

Note. All coefficients significant at 5% or better.

These attempts at developing a model for forecasting basic employment underscored the difficulty of the task, and at the same time, suggested that perhaps a wrong avenue of approach had been taken. In the next section, the exploration of an entirely new approach is described. This eventually led to the development of a rather reliable employment-forecasting model which was eventually called EMPAL.

6.4 Further development of employment models[8]

At this point in the development of employment models, the situation was far from satisfactory. Several attempts at models for forecasting basic employment had been examined in detail. BEMOD, used by the Bay Area Transportation Study Commission, was not satisfactory, nor were the employment equations from EMPIRIC. The more recently developed EMPMOD was an improvement of sorts, but still did not fill the bill. Nonetheless, EMPMOD was being used in the integrated model package to produce the forecasts of basic employment.

The situation for forecasting of nonbasic employment was only slightly better. A simple nonbasic employment model of the type used by Lowry, and discussed in section 6.2, had long been a component of IPLUM. This nonbasic model was taken, intact, and used as an integral part of DRAM to produce the nonbasic employment forecasts. In its general form, this model differs from the nonbasic employment portion of the Lowry model, as given in equation (6.8), only in the second term (shopping trips originating from place of work). In equation (6.8), these trips are assumed to originate only in the zone in which the shopping takes place, that is, the zone for which the nonbasic employment is being estimated. In the nonbasic employment section of IPLUM, and in the various early versions of DRAM, these work-to-shop trips are allowed to originate anywhere in the region. Thus a second probability term is required to represent the probability of an individual who works in zone j making a shopping trip to zone i.

Initially, the regional ratio of nonbasic employees to total population, K, is calculated. Then, two potential demand values are calculated: first demand due to the nonworking (resident) population, D^{r}, is given by

$$D_j^{\mathrm{r}} = KP_j^{\mathrm{r}} , \tag{6.9}$$

where P_j^{r} is the nonworking (resident) population in zone j; second, the demand due to the working population, D_j^{w}, is

$$D_j^{\mathrm{w}} = KP_j^{\mathrm{w}} , \tag{6.10}$$

where P_j^{w} is the working population (employees) employed in zone j.

[8] Portions of this material have appeared, in a somewhat different form, in Ducca (1978).

Then, nonbasic employment in zone i, E_i^{nb}, is calculated

$$E_i^{nb} = K\left(\sum_j P_j^r p_{ij}^r + \sum_j P_j^w p_{ij}^w\right) . \tag{6.11}$$

where p_{ij}^r is the probability that a (nonworking) resident of zone j will shop in zone i, and p_{ij}^w the probability that an employee employed in zone j will shop in zone i.

Both p_{ij}^r and p_{ij}^w consist of normalized attractiveness functions multiplied by a trip function. The attractiveness measure at time t was simply the nonbasic employment in zone i at a previous time, $t-1$. The trip function was a function of travel time (or cost):

$$f(c_{ij}) = c_{ij}^{-2} \exp\left(\frac{-\alpha}{c_{ij}}\right) . \tag{6.12}$$

The value of α was different for p_{ij}^r and p_{ij}^w.

We can therefore write

$$p_{ij}^r = \frac{E_{i,t-1}^{nb} c_{ij}^{-2} \exp(-\alpha^r/c_{ij})}{\sum_k E_{k,t-1}^{nb} c_{kj}^{-2} \exp(-\alpha^r/c_{kj})} ; \tag{6.13}$$

and p_{ij}^w is similar, except that α^w is substituted for α^r. The actual forecasting equation for $E_{i,t}^{nb}$ results from substitution of equation (6.13) into equation (6.11), to yield

$$E_{i,t}^{nb} = K\left\{\sum_j P_j^r \left[\frac{E_{i,t-1}^{nb} c_{ij}^{-2} \exp(-\alpha^r/c_{ij})}{\sum_k E_{k,t-1}^{nb} c_{kj}^{-2} \exp(-\alpha^r/c_{kj})}\right] + \sum_j P_j^w \left[\frac{E_{k,t-1}^{nb} c_{ij}^{-2} \exp(-\alpha^w/c_{ij})}{\sum_k E_{k,t-1}^{nb} c_{kj}^{-2} \exp(-\alpha^w/c_{kj})}\right]\right\} . \tag{6.14}$$

In point of fact, the actual algorithm (computer code) used in IPLUM (as adapted from the original computer code for PLUM) was somewhat different from this. These differences are, however, unimportant for the purposes of this discussion. Furthermore, it should be noted that only one type of nonbasic employment was involved here, even though later versions of PLUM attempted to forecast as many as eight different types of nonbasic employment. As these procedures were rather ad hoc in nature and not essential to the various projects undertaken, this later version of the nonbasic employment model from PLUM was never incorporated into IPLUM or DRAM.

Although these equations were more or less contemporary with then current practice they were really not on a par with those of the residence-location model DRAM (to be described in chapter 7). This was especially true with regard to the estimation of their parameters which was, at that time, strictly an ad hoc procedure.

At this same time, however, there was some new work being done on retail trade models (Ducca and Wilson, 1976). Ducca and Wilson's model represented an attempt to modify, for use as a shopping-center-location model, the equilibrium-model approach to retail trade as originally proposed by Harris (1964). The basic structure of the revised model was as follows:

$$D_{ij}^{R} = D_i^{R} B_i E_j^{R} (q_j + \delta A_j)^{\gamma} c_{ij}^{\beta} ; \tag{6.15}$$

here

$$B_i = \left[\sum_k E_k^{R} (q_k - \delta A_k)^{\gamma} c_{ik}^{\beta} \right]^{-1} , \tag{6.16}$$

and

E_j^R is the retail employment in zone j,
D_i^R is the demand for retail employment originating in zone i,
D_{ij}^R is the demand for retail employment originating in zone i and arriving in zone j,
c_{ij} is the time, cost, or distance between zone i and zone j,
q_j is the density of retail trade (employees per unit land area) in zone j,
A_j is the planned floor space for a shopping center in zone j, and
β, γ, δ are empirically derived parameters.

The model, equation (6.15), was then solved subject to the constraint

$$\sum_i D_i^{R} = \sum_j E_j^{R} , \tag{6.17}$$

this being a statement of the equilibrium requirement that the amount of demand for retail trade attracted is equal to the amount of retail employment.

The calibrations (parameter estimation to base year data sets) of this model were reasonably good. Yet, when an attempt was made to extend the model to deal with location of all nonbasic employment, the results were somewhat disappointing. This led to a reexamination of the general performance of the then current nonbasic employment component of DRAM. There it was found that a simple extrapolation of retail employment (using the Minneapolis–St Paul metropolitan area data base) from 1960 to 1970 yielded a value of R^2 of 0·88 between observed 1970 retail employment and the extrapolated 1970 retail employment. Yet, the R^2 between observed 1970 retail employment and the 1970 retail employment estimated by the model was only 0·77—significantly lower than the extrapolation results. Clearly some further development work was necessary.

If the variable E_j^R is removed from the right-hand side of equation (6.15) the model no longer needs to be considered to be an equilibrium model requiring an iterative solution. In fact, the resulting formulation is quite similar to the singly constrained model presented earlier in equation (6.7). Furthermore, note that equation (6.14) is also of this form, but has two sources of demand for nonbasic employment, residence-based demand (home-to-shop) and workplace-based demand (work-to-shop).

These modifications yield a model of the form

$$D_{ij}^{R} = \sum_{n} W_n [D_i^{R,n} B_i (q_j + \delta A_j)^{\gamma} c_{ij}^{\beta}] , \tag{6.17}$$

and

$$E_j^{R} = \sum_{i} D_{ij}^{R} = \sum_{n} \psi_n \left\{ \sum_{i} [D_i^{R,n} B_i (q_j + \delta A_j)^{\gamma} c_{ij}^{\beta}] \right\} , \tag{6.18}$$

where ψ_n is a weighting factor associated with each type n source of demand, such that $\sum_{n} \psi_n = 1 \cdot 00$, and where B_i is defined by equation (6.16).

An extensive series of calibration tests was undertaken for this model, all based on the Minneapolis–St Paul metropolitan area data, but using different definitions of D_i^{R}, q_j, and A_j, as well as different sectoral (employment-type) definitions.

In order to examine, empirically, the notion of residence-based demand vis-à-vis workplace-based demand a set of separate calibrations was done for each of two employment types. The specific variable definitions used for the terms in equation (6.18) were:

- q_i retail employment/zonal land area, in zone j;
- A_j shopping center floorspace in zone j;
- c_{ij} travel time in minutes between zone i and zone j;
- $D_i^{R,r}$ demand due to nonworking population residing in zone i;
- $D_i^{R,w}$ demand due to employed persons employed in zone i;
- ψ defined equal to unity in these tests, as each demand source was tested individually.

The calibrations were done for two types of nonbasic employment, retail employment and service employment. For retail employment, the following best-fit parameters were obtained:

residence demand	workplace demand
$\delta = 0 \cdot 24$	$\delta = 0 \cdot 08$
$\gamma = 0 \cdot 99$	$\gamma = 0 \cdot 61$
$\beta = -8 \cdot 09$	$\beta = -8 \cdot 34$
	$R^2 = 0 \cdot 58$.

For service employment, the results were:

residence demand	workplace demand
$\delta = 0 \cdot 002$	$\delta = 0 \cdot 007$
$\gamma = 1 \cdot 21$	$\gamma = 1 \cdot 00$
$\beta = -9 \cdot 07$	$\beta = -9 \cdot 09$
	$R^2 = 0 \cdot 78$.

These results were not too surprising, and it was gratifying to see that service employment was less sensitive to the presence of shopping center floorspace than was retail employment.

A subsequent calibration test combined the residence and workplace demands, allowing the value of ψ_n to be determined in the calibration process. For retail employment, the results were:

residence demand	workplace demand
$\delta = 0{\cdot}00$	$\delta = 0{\cdot}20$
$\gamma = 0{\cdot}61$	$\gamma = 1{\cdot}03$
$\beta = -8{\cdot}03$	$\beta = -9{\cdot}45$
$\psi_r = 0{\cdot}49$	$\psi_w = 0{\cdot}51$
	$R^2 = 0{\cdot}73$.

For service employment, the results were:

residence demand	workplace demand
$\delta = 0{\cdot}00$	$\delta = 0{\cdot}01$
$\gamma = 1{\cdot}28$	$\gamma = 0{\cdot}99$
$\beta = -9{\cdot}00$	$\beta = -9{\cdot}50$
$\psi_r = 0{\cdot}38$	$\psi_w = 0{\cdot}62$
	$R^2 = 0{\cdot}78$.

In yet a further set of calibrations, a different trip function was used, with a modified gamma function

$$f(c_{ij}) = c_{ij}^{\alpha} \exp(\beta c_{ij}) \tag{6.19}$$

substituted for the simple power function c_{ij}^{β} used in equation (6.18). Again the parameters were first estimated for each demand source alone and then for both taken together, with the weighting factor. The results of these parameter estimations were, for retail employment:

residence demand	workplace demand
$\delta = 0{\cdot}38$	$\delta = 0{\cdot}00$
$\gamma = 0{\cdot}85$	$\gamma = 0{\cdot}49$
$\alpha = 0{\cdot}05$	$\alpha = -0{\cdot}10$
$\beta = -0{\cdot}18$	$\beta = -0{\cdot}78$
$\psi_r = 0{\cdot}35$	$\psi_w = 0{\cdot}65$
	$R^2 = 0{\cdot}77$;

and, for service employment,

residence demand	workplace demand
$\delta = 0{\cdot}00$	$\delta = 0{\cdot}00$
$\gamma = 0{\cdot}75$	$\gamma = 1{\cdot}55$
$\alpha = -0{\cdot}03$	$\alpha = 0{\cdot}46$
$\beta = -0{\cdot}66$	$\beta = -1{\cdot}13$
$\psi_r = 0{\cdot}07$	$\psi_w = 0{\cdot}93$
	$R^2 = 0{\cdot}79$.

These results actually appear to make a good deal of sense. First, for retail employment, one finds that residence demand is sensitive to shopping-center floorspace, whereas workplace demand is not. The positive value of β for residence demand implies a skewed, peaked, shopping-trip distribution which is consistent with other empirical evidence showing travel to nearby, but not immediately adjacent, retail shopping opportunities. The negative value of α for workplace demand, along with the more negative value of β, implies rather shorter trips, more consistent with the notion of lunch-hour or just-after-work shopping trips from place of work.

The service-employment results, too, are quite sensible. First note the rather small weight ($\psi_r = 0 \cdot 07$) placed on residence-based demand. The few home-to-service trips that are implied here are likely to be rather short, as suggested by the negative value of α. On the other hand, the workplace demand, which accounts for most of the demand for service employment, shows (in the form of a positive value of α) the skewed, peaked, trip distribution; although the peak will be sharper and skewed nearer the origin than was the case for residence-based retail demand. All in all, these results were quite promising. However, a further set of calibrations attempted for the San Francisco data base did not yield consistent results, and so the problem of a reliable generalizable employment model was not yet solved.

6.5 A new employment model—EMPAL[9]

In most of the work which has been done on forecasting the spatial distribution of employment in metropolitan (urban) regions, the employment being located has been split into at least two types, basic and nonbasic. It was then commonly assumed that basic employment responded a good deal more slowly to changes in the intraurban location of other activities than did nonbasic employment. It was often further assumed that the location of nonbasic employment was most appropriately calculated simultaneously with residence location. In addition, it was also the custom to assume that the intraurban location of basic employment was not sensitive to the intraurban location of population (residence). Yet, there exists empirical evidence which is contrary to the implications of all these assumptions.

It is not unusual to observe any of the following phenomena. Often, nonbasic employment seems to persist in certain urban locations even after some of its markets have shifted their locations. Furthermore, in newly developing locations, the locations of nonbasic employment, with the exception of a certain amount of speculative activity, seem mostly to follow population location rather than to locate simultaneously. Similarly, although a significant portion of basic employment appears to be reluctant to shift locations, the suburbanization of basic employment has become increasingly evident in the past two decades.

If, in the face of such observations, all of the earlier assumptions are discarded, one may make a rather modest revision of equations (6.17) and (6.18) which appears, on the basis of the analyses to be described below, to yield a simple and reliable model for forecasting the intraurban location of employment of all types. The revision involves first making the assertion

[9] Portions of this material have appeared, in a somewhat different form, in Ducca (1979).

that, owing to a kind of inertia which results from the costs of relocation as well as to possible agglomeration economies, the past and/or present location of employment exerts a significant influence on the future location of employment. Furthermore, it makes good sense to use this information in forecasting. Second, the location of resident population, whether in its role as consumer or in its role as labor force, also has a significant influence on the future intraurban location of employment of all types. With these, plus the arguments typically used to construct singly constrained spatial interaction models of retail trade, a simple model emerges, combining a spatial interaction term with a lagged-variable term.

Consider the following equation:

$$E_j = \sum_i P_i A_i W_j \mathrm{f}(c_{ij}) , \tag{6.20}$$

where

$$A_i = \left[\sum_k W_k \mathrm{f}(c_{ik})\right]^{-1} , \tag{6.21}$$

and E_j is the employment in zone j, P_i the population (or some function of population) in zone i, W_j an attractiveness measure for zone j, and c_{ij} the travel time (cost) from zone i to zone j.

This is the standard spatial interaction formulation and, as such, has no explicit time dependence, as it is assumed to be a cross-sectional model. If a time subscript is added, then a lagged term may also be added, to yield

$$E_{j,t} = \lambda_1 \sum_i P_{i,t} A_{i,t} W_{j,t} \mathrm{f}(c_{i,j,t}) + \lambda_2 E_{j,t-1} , \tag{6.22}$$

where λ_1 and λ_2 are weighting factors related to the extent to which the employment type is oriented to past locational determinants or to current location of population. It is of course true that a model formulation such as this glosses over the original determinants of the employment at time $t-1$, $E_{j,t-1}$. These determinants may have been anything from inexpensive land fifty years earlier, to proximity to railroad facilities, or some form of intersectoral aggregation economies. Thus if any of these variables continue to play a role in the future location of a particular employment type, their appearance in zones of the urban area where they previously did not exist will be unrecognized by this model formulation. On the other hand, many of these determinants are very stable in their location (proximity to port facilities, for example), and thus are represented by the $E_{j,t-1}$ variable. Others of these prior determinants may well have ceased to be of importance. Finally, for many urban areas the major economic growth sectors are indeed those sectors whose location is most responsive to the current location of population. Over time, the location of employment in response to these variables in fact provides new agglomerations whose effect then shows up in subsequent time periods in the $E_{j,t-1}$ term.

There remain several questions of specification for this model. First, it is helpful in interpreting the model performance to require that the two weighting factors sum to unity. Thus a preferable equation form would be

$$E_{j,t} = \lambda \sum_i P_{i,t} A_{i,t} W_{j,t} \mathrm{f}(c_{i,j,t}) + (1-\lambda) E_{j,t-1} \, . \tag{6.23}$$

Then there is the question of a form for the trip function. Since the two-parameter function used in DRAM [equation (6.19)] has the ability, depending on the signs and magnitude of its parameters, to take either a 'declining exponential' or the 'skewed-peaked' shape, this function is used as the trip function, to yield

$$E_{j,t} = \lambda \sum_i P_{i,t} A_{i,t} W_{j,t} c_{i,j,t}^{\alpha} \exp(\beta c_{i,j,t}) + (1-\lambda) E_{j,t-1} \, . \tag{6.24}$$

The definitions of $P_{i,t}$ and $W_{j,t}$ remain to be resolved. For the preliminary tests of this new formulation several different possible variable definitions were examined. One of the first forms tested was

$$P_{i,t} = (\text{population of zone } i \text{ at time } t) \; ;$$

$$W_{j,t} = (E_{j,t-1}^{\mathrm{T}})^a L_i^b \, , \tag{6.25}$$

where $E_{j,t-1}^{\mathrm{T}}$ is the total employment in zone j at time $t-1$, L_i the total area of zone i, and a, b are empirically derived parameters.

For the Minneapolis–St Paul data set, with time t 1969 and time $t-1$ 1962, and for the retail employment sector, the resulting parameters were

$a = -0{\cdot}05$ $b = 0{\cdot}74$ $\alpha = 0{\cdot}51$

$\beta = -0{\cdot}83$ $\lambda = 0{\cdot}30$ $R^2 = 0{\cdot}79$.

The negative sign of the parameter a was not what was desired, indicating that total employment as part of the attractiveness measure was not the proper form. The attractiveness variable was subsequently redefined:

$$W_{j,t} = (E_{j,t-1}^{\mathrm{R}})^a L_i^b \, , \tag{6.26}$$

where $E_{j,t-1}^{\mathrm{R}}$ is the retail employment in zone j at time $t-1$.

With retail employment at time t as the locating variable, the calibration results were

$a = 0{\cdot}52$ $b = 0{\cdot}49$ $\alpha = 0{\cdot}50$

$\beta = -0{\cdot}18$ $\lambda = 0{\cdot}50$ $R^2 = 0{\cdot}86$.

Here all the signs of the parameters are in accord with expectations. The weighting parameter, λ, shows the equal importance of prior retail employment location and spatial interaction. The trip-function parameters, α and β, show the skewed-peaked trip distribution for shopping trips.

To test the assumption that it is not necessary to locate retail employment and population simultaneously the model was recalibrated by using the definition of attractiveness given in equation (6.25), but substituting $P_{i,t-1}$ for $P_{i,t}$. This gives the following equation

$$E_{j,t}^{R} = \lambda \sum_{i} P_{i,t-1} A_{i,t-1}^{R} W_{j,t-1}^{R} c_{i,j,t}^{\alpha} \exp(\beta c_{i,j,t}) + (1-\lambda) E_{j,t-1}^{R} , \tag{6.27}$$

where

$$W_{j,t-1}^{R} = (E_{j,t-1}^{R})^{a} L_{i}^{b} , \tag{6.28}$$

and

$$A_{i,t-1}^{R} = \left[\sum_{i'} (E_{i',t-1}^{R})^{a} L_{i'}^{b} c_{i',j,t}^{\alpha} \exp(\beta c_{i',j,t}) \right]^{-1} . \tag{6.29}$$

The calibration results for this version of the model were

$a = 0 \cdot 51$	$b = 0 \cdot 56$	$\alpha = 0 \cdot 51$
$\beta = -0 \cdot 14$	$\lambda = 0 \cdot 48$	$R^2 = 0 \cdot 85$.

As these differ only slightly from the results obtained by using $P_{i,t}$, the assertion that it makes little difference whether or not retail employment and resident population are located simultaneously seemed to be supported.

The same model formulation was then calibrated with two variables and an additional weighting parameter substituted for $P_{i,t-1}$ in equation (6.27). The resulting form was

$$E_{j,t}^{R} = \sum_{i} (\lambda_1 P_{i,t-1} + \lambda_2 E_{i,t-1}^{T}) A_{i,t-1}^{R} W_{j,t-1}^{R} c_{i,j,t}^{\alpha} \exp(\beta c_{i,j,t}) + \lambda_3 E_{j,t-1}^{R} , \tag{6.30}$$

subject to the requirement that

$$\lambda_1 + \lambda_2 + \lambda_3 = 1 \cdot 00 . \tag{6.31}$$

The results of this calibration yielded only marginally different results, and thus this equation form, analogous to the multiple-demand-source retail model described in section 6.4, was not pursued further.

The model of equation (6.27) was then calibrated for five employment sectors with the Minneapolis–St Paul metropolitan area data set. The results of these calibrations are given, along with the sectoral definitions, in table 6.8. These results were quite reasonable overall and, as will be shown below, agree quite closely with a subsequent set of calibrations done with data from the Washington DC metropolitan area.

Four employment sectors were available in the Washington DC data set. These were

1. finance, insurance, and real estate;
2. wholesale and retail trade;
3. manufacturing, transportation, communication, and public utilities;
4. government.

The time points were 1968 for t and 1960 for $t-1$. The calibration results and sectoral definitions for this data set are given in table 6.9. A comparison of the results for the two different metropolitan areas shows a reassuring number of similarities. All values of α are positive and all values of β are negative. With the exception of 'miscellaneous' employment in Minneapolis–St Paul, retail employment has the highest value of λ, and wholesale-retail has the highest value of λ in Washington DC. This confirms the intuition that these sectors should be the most sensitive to the intraurban location of population. Manufacturing employment in Minneapolis–St Paul also shows a rather high value of λ; this suggests that a substantial quantity of manufacturing employment in that area is decentralized, and perhaps also more of a 'high-technology, high-skill-level' nature and thus more sensitive to the location of a suburbanizing labor force. There is rather less manufacturing employment (as a proportion of total employment) in the Washington DC region, and in the data available it was grouped together with transportation, communication, and public utilities. This group in Washington DC, and the transportation, communication, and public utilities employment in Minneapolis–St Paul showed rather low λ values suggesting a locationally static employment group most likely to be found in zones where it was previously concentrated. This was also the case for government employment in Washington DC.

A subsequent set of calibrations was done for the Kansas City metropolitan area. Equation (6.26) was again the model calibrated. The employment-type definitions were:

1. agriculture and related industry, mining, construction, wholesale trade, transportation, communication, and public utilities;

Table 6.8. EMPAL calibration results and sectoral definitions: Minneapolis–St Paul data.

Employment type [a]	Parameter						R^2
	a	b	α	β	λ	$1-\lambda$	
1	0·598	0·891	0·578	−0·156	0·429	0·571	0·857
2	0·420	0·681	0·569	−0·095	0·156	0·844	0·856
3	0·492	0·584	0·501	−0·183	0·502	0·498	0·863
4	0·544	0·630	0·570	−0·100	0·260	0·740	0·950
5	0·399	0·497	0·570	−0·131	0·520	0·480	0·719

[a] 1 manufacturing; 2 transportation, communication, utilities; 3 retail trade; 4 services; 5 miscellaneous.

Table 6.9. EMPAL calibration results: Washington DC data.

Employment type [a]	Parameter						R^2
	a	b	α	β	λ	$1-\lambda$	
1	0·448	0·366	0·735	−0·205	0·210	0·790	0·950
2	0·533	0·436	0·475	−0·609	0·412	0·588	0·884
3	0·324	0·485	0·438	−0·757	0·154	0·846	0·946
4	0·736	0·596	0·574	−0·346	0·157	0·843	0·957

[a] See text.

2. manufacturing;
3. services, finance, insurance, real estate, government;
4. retail trade.

The time points were 1975 for t and 1970 for $t-1$; calibration results are given in table 6.10. Again, the results are consistent with the previous results from Minneapolis–St Paul and Washington DC. In summary, after what may still be said to be preliminary but promising tests on three cities, this seems to be an employment-model formulation with great potential for future use and development. In addition, there have been recent EMPAL calibrations for the Seattle metropolitan area and for St Louis, as well as for T'ai-pei and Kao-hsiung in Taiwan. All of these additional calibrations have yielded consistent results, and thus bring to seven the cities for which EMPAL demonstrates good descriptive validity.

One last set of results is of interest here, and has to do with the question of how worthwhile the use of a model such as EMPAL may be for forecasting purposes. For the Minneapolis–St Paul data base, a correlation was run between each employment sector, by zone, at time t, and that same employment sector, by zone, at time $t-1$. These results are shown in the first column of table 6.11. First, it must be admitted that the simple extrapolation for each sector accounts for rather substantial shares of the 'variation'. In fact, the EMPAL results are not all that much better. Yet, considered in terms of the additional percentage of unexplained residual variation which can be explained with EMPAL, there are notable gains to be made. Furthermore, a simple extrapolation leaves rather little scope for policy testing. When an extrapolation procedure is used, there are no policy variables to manipulate. In this connection it should be mentioned that the use of total zonal area as an attractiveness measure also leaves limited scope for policy manipulation. In future versions of this model it is intended to examine the possible use of other land variables in place of total zonal land.

The same type of comparison was also possible for the Washington DC data set, and is shown in table 6.12. Again the variation 'explained' by a simple extrapolation from time $t-1$ (1970) to time t (1976) is quite high. This, of course, suggests a very considerable degree of locational stability of employment in the Washington DC metropolitan area, at least for the

Table 6.10. EMPAL calibration results: Kansas City data.

Employment type[a]	Parameter						R^2
	a	b	α	β	λ	$1-\lambda$	
1	0·44	0·07	1·02	−0·27	0·18	0·72	0·96
2	−0·10	0·35	0·52	−1·55	0·07	0·93	0·89
3	0·56	0·14	0·97	−0·56	0·19	0·81	0·95
4	0·73	0·11	0·94	−0·49	0·55	0·45	0·84

[a] See text.

time period of the data. Indeed, in the sober light of ten or twenty years hence, this author may have occasion to regret not having 'left well enough alone'. Nonetheless, the EMPAL calibrations do no worse in any sector, and actually show an improvement of 9% of the residual variance (from extrapolation) 'explained' by EMPAL. All in all, these results are certainly better than some previous urban employment-location models. (Careful scrutiny of the results of some of them will reward the reader with the knowledge that a straight line extrapolation would have been somewhat more reliable than the forecasts of the model.) However, this does raise the question of the statistical significance of the results.

Finally, it should be noted that a problem with the type of calibration procedure used for EMPAL is that the statistical properties of the parameters estimated are, for the most part, unknown. Yet another similar question is whether either the spatial interaction component or the lag component of EMPAL makes a statistically significant improvement in EMPAL's performance. It is possible to construct a test to answer this question. After the calibrations were completed the parameters and data were used to calculate two variables, as below:

$$X_{1,j}^{k} = \sum_{i} P_{i,t-1} A_{i,t-1}^{k} W_{j,t-1}^{k} c_{i,j,t}^{\alpha} \exp(\beta c_{i,j,t}) , \tag{6.32}$$

Table 6.11. EMPAL calibration results compared with simple extrapolation: Minneapolis-St Paul data.

Employment type [a]	R^2 (extrapolation)	Residual	R^2 (EMPAL)	Residual reduction (%) [b]
1	0·77	0·23	0·86	39
2	0·86	0·14	0·86	0
3	0·74	0·26	0·85	42
4	0·93	0·07	0·94	14
5	0·62	0·38	0·72	26

[a] 1 manufacturing; 2 transportation, communication, utilities; 3 retail trade; 4 services; 5 miscellaneous

[b] $[R^2(\text{EMPAL}) - R^2(\text{extrapolation})]/\text{residual}$.

Table 6.12. EMPAL calibration results compared to simple extrapolation: Washington DC data.

Employment type [a]	R^2 (extrapolation)	Residual	R^2 (EMPAL)	Residual reduction (%) [b]
1	0·97	0·03	0·97	0
2	0·89	0·11	0·90	9
3	0·96	0·04	0·96	0
4	0·96	0·04	0·96	0

[a] See text. [b] See table 6.11.

where, from equation (6.28),

$$W_{j,t-1}^{k} = (E_{j,t-1}^{k})^{a} L_{i}^{b} , \tag{6.33}$$

for type k employment and, from equation (6.29),

$$A_{i,t-1} = \left[\sum_{i'} (E_{i',t-1}^{k})^{a} L_{i}^{b} c_{i,j,t}^{\alpha} \exp(\beta c_{i,j,t})\right]^{-1} . \tag{6.34}$$

Then, the second variable was simply

$$X_{2,j}^{k} = E_{j,t-1}^{k} . \tag{6.35}$$

These variables were then used in a linear multiple regression:

$$Y_{j}^{k} = s + uX_{1,j}^{k} + vX_{2,j}^{k} . \tag{6.36}$$

Of course, the values of R^2 were high for all four regressions, and both variables entered in all cases. The important question was of the statistical significance. In all sectors, and both for the Washington DC and for the Minneapolis–St Paul data sets, the coefficients of both variables (u, v) were statistically significant at the 5% level. Thus even though the contribution to explained variance of the spatial interaction term of EMPAL was sometimes small, it was in all cases statistically significant.

Taken all together, the results of these investigations yield a practical and apparently reliable model for intraurban employment-location forecasts. Clearly, further work remains to be done in disaggregating the terms of this model in attempts to understand urban employment-location phenomena better. Yet, for current practice the model derived here is superior to the somewhat spotty previous work in the field[(10)].

6.6 Calibration equations for the employment model

Although the discussion of the calibration procedure developed for this type of model is given in the next chapter, the necessary equations are given here. The calibration procedure uses a gradient-search algorithm. Thus it is necessary to calculate the partial derivatives of some criterion function with respect to each of the parameters of the model. For reasons to be discussed in the next chapter the criterion used is a likelihood measure, and the partial derivatives are calculated analytically.

The criterion function is

$$\mathsf{L} = \sum_{j} \hat{E}_j \ln \hat{E}_j - \sum_{j} E_j \ln E_j , \tag{6.37}$$

where $\hat{E}$ is the estimated variable. The superscript for employment sector is omitted here for simplification purposes but, as is evident from the results described above, calibrations are done for each employment sector.

[(10)] A more detailed review of this work may be found in the author's somewhat dated but still relevant article (Putman, 1972).

The equation for $\hat{E}_j$ is equation (6.27). The general form of the derivatives which must be taken is illustrated for the α parameter. This is

$$\frac{\partial \hat{E}_{j,t}}{\partial \alpha} = \lambda \sum_i P_{i,t-1} \left[\left(\frac{\partial \Phi_{ij}}{\partial \alpha} \right) \Theta_j + \left(\frac{\partial \Theta_j}{\partial \alpha} \right) \Phi_{ij} \right] , \tag{6.38}$$

where, for ease of notation, the following definitions are used

$$\Phi_{ij} = E^a_{j,t-1} L^b_j c^\alpha_{ij} \exp(\beta c_{ij}) , \tag{6.39}$$

and

$$\Theta_j = \left[\sum_m E^a_{m,t-1} L^b_m c^\alpha_{i,m} \exp(\beta c_{im}) \right]^{-1} ; \tag{6.40}$$

thus, of course,

$$\Theta_j = \left(\sum_m \Phi_{im} \right)^{-1} . \tag{6.41}$$

In these terms, equation (6.27) becomes

$$\hat{E}_{j,t} = \lambda \sum_i P_{i,t-1} \Phi_{ij} \Theta_j + (1-\lambda) E_{j,t-1} . \tag{6.42}$$

To evaluate equation (6.38), it is necessary to take the derivatives of Φ_{ij}, equation (6.39), with respect to each parameter:

$$\frac{\partial \Phi_{ij}}{\partial a} = \ln E_{j,t-1} \Phi_{ij} , \tag{6.43}$$

$$\frac{\partial \Phi_{ij}}{\partial b} = \ln L_j \Phi_{ij} , \tag{6.44}$$

$$\frac{\partial \Phi_{ij}}{\partial \alpha} = \ln c_{ij} \Phi_{ij} , \tag{6.45}$$

$$\frac{\partial \Phi_{ij}}{\partial \beta} = c_{ij} \Phi_{ij} . \tag{6.46}$$

Next, the partial derivative of Θ_j must be taken. From equation (6.41), we obtain

$$\frac{\partial \Theta_j}{\partial a} = -\left(\sum_m \Phi_{im} \right)^{-2} \left[\frac{\partial \left(\sum_m \Phi_{im} \right)}{\partial a} \right] = -\Theta_j^2 \sum_m \left(\frac{\partial \Phi_{im}}{\partial a} \right) . \tag{6.47}$$

Substitution of equation (6.43) into equation (6.47) yields

$$\frac{\partial \Theta_j}{\partial a} = -\Theta_j^2 \sum_m \ln E_{m,t-1} \Phi_{im} ; \tag{6.48}$$

and, for the b parameter,

$$\frac{\partial \Theta_j}{\partial b} = -\left(\sum_m \Phi_{im} \right)^{-2} \left[\frac{\partial \left(\sum_m \Phi_{im} \right)}{\partial b} \right] = -\Theta_j^2 \sum_m \left(\frac{\partial \Phi_{im}}{\partial b} \right) = -\Theta_j^2 \sum_m \ln L_m \Phi_{im} . \tag{6.49}$$

And, by the same process, we obtain

$$\frac{\partial \Theta_j}{\partial \alpha} = -\Theta_j^2 \sum_m \ln c_{im} \Phi_{im} , \tag{6.50}$$

$$\frac{\partial \Theta_j}{\partial \beta} = -\Theta_j^2 \sum_m c_{im} \Phi_{im} . \tag{6.51}$$

To get the full expression for each derivative, beginning with parameter a, we substitute equations (6.43) and (6.48) into equation (6.38), to yield

$$\begin{aligned}\frac{\partial \hat{E}_{j,t}}{\partial a} &= \lambda \sum_i P_{i,t-1} \left[\ln E_{j,t-1} \Phi_{ij} \Theta_j - \Theta_j^2 \left(\sum_m \ln E_{m,t-1} \Phi_{im} \right) \Phi_{ij} \right] \\ &= \lambda \sum_i P_{i,t-1} \Phi_{ij} \Theta_j \left[\ln E_{j,t-1} - \Theta_j \left(\sum_m \ln E_{m,t-1} \Phi_{im} \right) \right] .\end{aligned} \tag{6.52}$$

For b, we substitute equations (6.44) and (6.49) into equation (6.38):

$$\begin{aligned}\frac{\partial \hat{E}_{j,t}}{\partial b} &= \lambda \sum_i P_{i,t-1} \left[\ln L_j \Phi_{ij} \Theta_j - \Theta_j^2 \left(\sum_m \ln L_m \Phi_{im} \right) \Phi_{ij} \right] \\ &= \lambda \sum_i P_{i,t-1} \Phi_{ij} \Theta_j \left[\ln L_j - \Theta_j \left(\sum_m \ln L_m \Phi_{im} \right) \right] .\end{aligned} \tag{6.53}$$

Substituting equations (6.45) and (6.50) into equation (6.38), we find

$$\begin{aligned}\frac{\partial \hat{E}_{j,t}}{\partial \alpha} &= \lambda \sum_i P_{i,t-1} \left[\ln c_{ij} \Phi_{ij} \Theta_j - \Theta_j^2 \left(\sum_m \ln c_{im} \Phi_{im} \right) \Phi_{ij} \right] \\ &= \lambda \sum_i P_{i,t-1} \Phi_{ij} \Theta_j \left[\ln c_{ij} - \Theta_j \left(\sum_m \ln c_{im} \Phi_{im} \right) \right] ;\end{aligned} \tag{6.54}$$

and by substituting equations (6.46) and (6.51) into equation (6.38),

$$\begin{aligned}\frac{\partial \hat{E}_{j,t}}{\partial \beta} &= \lambda \sum_i P_{i,t-1} \left[c_{ij} \Phi_{ij} \Theta_j - \Theta_j^2 \left(\sum_m c_{im} \Phi_{im} \right) \Phi_{ij} \right] \\ &= \lambda \sum_i P_{i,t-1} \Phi_{ij} \Theta_j \left[c_{ij} - \Theta_j \left(\sum_m c_{im} \Phi_{im} \right) \right] .\end{aligned} \tag{6.55}$$

Finally, the derivative with respect to λ is quite simple, from equation (6.42):

$$\frac{\partial \hat{E}_{j,t}}{\partial \lambda} = \sum_i P_{i,t-1} \Phi_{ij} \Theta_j^{-1} - E_{j,t-1} . \tag{6.56}$$

The use of these derivatives, along with the reason for the selection of the criterion function in equation (6.37), will be discussed in the next chapter.

7

New residence-location models: development and calibration

7.1 Introduction

The policy analyses described in chapter 4 were accomplished with a version of the integrated transportation and land-use package which used a modified form of the IPLUM model to forecast residence location (as well as nonbasic employment location). The use of IPLUM in that project was justified by the facts that (a) it was operational, and (b) it did represent the state of practice in the USA at the time.

Several of the problems inherent in the original version of IPLUM were resolved during the work on the prototype of ITLUP, described in chapter 3. Other problems remained, among which were questions as to what variables ought to be included in a residential attractiveness measure, and how, in general, models such as this should be calibrated. A research project undertaken after the integrated model package project involved the comparison of then operational (1973–1975) urban residence-location models. During work on this project it was discovered that no model of the IPLUM type (that is, Lowry derivative) had ever been calibrated in practice in the USA (Putman, 1976). Work done in the course of that project eventually led to the development of the disaggregated residential allocation model (DRAM), and to its successful calibration on data sets from cities all around the world. Eventually the DRAM model replaced IPLUM in the integrated package, and an entire new series of policy analysis tests was performed. These are described in chapter 9. The present chapter is devoted to describing the employment and testing of DRAM.

7.2 Background

The residence-location components of virtually all Lowry derivative models had as their residential allocation function some form of the following expression

$$N_i = g \sum_j E_j p_{ij} \,, \qquad (7.1)$$

with

N_i the number of residential locators locating in area i;
p_{ij} the probability of living in area i and working in area j;
E_j the number of employees employed in area j;
g a scaling factor such that the sum of the N_i over all i equals an exogenous control total for the region.

These models often contained other scaling or multiplier factors to convert from employees to households and to assure internal consistencies of various types.

The p_{ij} term is the most important component of equation (7.1). In the original Lowry model, the function used was:

$$p_{ij} = d_{ij}^{-1\cdot 33K} , \tag{7.2}$$

where d_{ij} is the airline distance between the centroid of area i and the centroid of area j, and K the number of zones in an annulus c_{ij} miles from the origin.

In various of the Lowry derivative models, p_{ij} was modified to include measures of the attractiveness of area i. In particular in the ITLUP form of IPLUM, as described in chapter 3,

$$p_{ij} = U_i \mathrm{f}(c_{ij}) , \tag{7.3}$$

where

$$\mathrm{f}(c_{ij}) = \left(\frac{\beta}{c_{ij}^2}\right) \exp\left(\alpha - \frac{\beta}{c_{ij}}\right) , \tag{7.4}$$

and U_i is a measure of residential 'opportunities' in zone i, c_{ij} is the travel time between centroids of zones i and j, and α, β are empirically derived parameters.

The measure of opportunities was an adjusted measure of residential holding capacity (previous level of residential density multiplied by amount of available land). An adjustment factor G_i was devised, which was a logistic function of the proportion of the developable land in zone i that had been developed by the end of the base time period. This gave the form

$$U_i = L_i^{\mathrm{v}} \frac{H_i}{L_i^{\mathrm{r}}} G_i , \tag{7.5}$$

where L_i^{v} is the vacant (nonindustrial) land acreage in zone i, H_i the number of housing units, L_i^{r} the residential acreage, and G_i the 'developability' (development-level factor) of the zone; G_i has the form

$$G_i = 1 - \frac{\gamma}{(1-\gamma)\exp(\delta x_i^2)} , \tag{7.6}$$

where γ, δ are variable parameters, and x_i is the proportion of developable land area in zone i which has been developed [see equation (3.10)].

The parameters of the trip function, equation (7.4), had been estimated by trial-and-error fitting to observed work-trip distributions from the San Francisco Bay Area. The parameters of the development-level factor, equation (7.6) had never been statistically estimated, nor had the complete p_{ij} function, equation (7.3), been fitted to any actual data. It was precisely this fitting of all the parameters of the complete p_{ij} function which was necessary, but which had never been done in work with Lowry model derivatives in the USA. After consideration of numerous possible means of resolving this problem it was determined that the complete reformulation of the residence-location submodel of ITLUP would, in the long run, be the best strategy.

7.3 Formulation of the residence-location model

The essence of the residence-location process as represented in these models centers on the work trip (home-to-work or work-to-home) or a combination of the work trip with measures of attractiveness of the potential residence locations. Implicit, therefore, in any estimates of residence locations made by these models, is a set of work-trip estimates. Very little use has been made of this fact in the USA in practice. Yet, it is precisely these implicit trip matrices that eventually led to a more satisfactory method of estimating the parameters of these models. The use of IPLUM in the ITLUP package as described in chapter 3 was a particular exception to the usual neglect of these implicit trips. In that case, those implicit work trips were made explicit by extraction of the trips from the model directly, and were later used to load the transportation network. It was probably that view of the residence-location process which led to consideration of the entropy-maximizing approach to analysis of these models.

On the basis of this approach the Lowry model may be rewritten as

$$T_{ij} = E_j \, \mathrm{f}(c_{ij}) \, , \tag{7.7}$$

where T_{ij} is the number of persons (tripmakers) working in zone j and residing in zone i, E_j the number of persons working in zone j, and c_{ij} the impedance (usually travel time or travel cost) between centroids of zone i and zone j.

An important problem of the original Lowry formulation was that there was no constraint on the sums of trips. Without such a constraint there was no reason to expect that

$$\sum_i T_{ij} = E_j \, . \tag{7.8}$$

This implies that the number of employees in zone j would not necessarily equal the sum of the employees residing in all zones i who 'claim' to work in zone j.

A simple residence-location model may be derived, following entropy-maximizing principles, by letting

$$T_{ij} = A_i B_j O_i E_j \, \mathrm{f}(c_{ij}) \, , \tag{7.9}$$

where

T_{ij} is the number of trips between zones i and j (*or* the number of persons living in zone i and working in zone j);

O_i is the number of trip origins (*or* employed persons resident) in zone i;

E_j is the number of trip destinations (*or* employees employed) in zone j;

A_i is a balancing factor for trip origins;

B_j is a balancing factor for trip destinations;

$\mathrm{f}(c_{ij})$ a trip (impedance) function.

It is possible to replace the trip origins O_i in equation (7.9) by a measure of attractiveness of the origin zone, W_i. This eliminates the need for the origins balancing factor A_i, thus

$$T_{ij} = B_j W_i E_j \mathrm{f}(c_{ij}) . \tag{7.10}$$

For the constraint on the sums of trip destination, equation (7.8), to be met, we must have

$$B_j = \left[\sum_i W_i \mathrm{f}(c_{ij}) \right]^{-1} . \tag{7.11}$$

It is informative to substitute this expression back into the original equation (7.10):

$$T_{ij} = E_j \left[\frac{W_i \mathrm{f}(c_{ij})}{\sum_i W_i \mathrm{f}(c_{ij})} \right] . \tag{7.12}$$

If the term $W_i \mathrm{f}(c_{ij})$ is called an 'accessibility attractiveness' measure, then the fraction in equation (7.12) is a relative measure of the accessibility attractiveness of zone i to zone j compared with all other zones i. Furthermore, it is clear that the total number of employed residents in zone i is given by

$$N_i = \sum_j T_{ij} ; \tag{7.13}$$

and, substituting equation (7.12) into equation (7.13), we obtain

$$N_i = \sum_j E_j \left[\frac{W_i \mathrm{f}(c_{ij})}{\sum_i W_i \mathrm{f}(c_{ij})} \right] . \tag{7.14}$$

If it is asserted that an acceptable definition of p_{ij} is

$$p_{ij} = \frac{W_i \mathrm{f}(c_{ij})}{\sum_i W_i \mathrm{f}(c_{ij})} ,$$

then equation (7.14) is equivalent to

$$N_i = \sum_j E_j p_{ij} , \tag{7.15}$$

which is the same function as the Lowry model described by equation (7.1), without the now unnecessary scaling factor g.

The formulation given by equation (7.12) can be disaggregated by type of locator. For that matter, it could as well be disaggregated by type of locator and by type of housing, to yield a variable T_{ij}^{nw} on the left-hand side of the equation, where T_{ij}^{nw} might be the number of employees of type n who work in zone j and who live in type w housing in zone i. In most cases the data requirements for estimation of the parameters of such models precludes their actual application. Furthermore, if housing type were an endogenous component of the residence-location forecasting procedure, then of course it would be necessary to develop a procedure for housing forecasts too. This was definitely not in line with the planned research effort, and thus it was necessary to consider only residence-location formulations which did not contain an endogenous housing component.

Equation (7.12) was thus disaggregated only by locator type, as follows

$$T_{ij}^n = B_j^n E_j W_i^n \mathrm{f}_n(c_{ij}) \, , \tag{7.16}$$

with

$$B_j^n = \left[\sum_i W_i^n \mathrm{f}_n(c_{ij})\right]^{-1} ; \tag{7.17}$$

this yields

$$N_i^n = \sum_j E_j \left[\frac{W_i^n \mathrm{f}_n(c_{ij})}{\sum_i W_i^n \mathrm{f}_n(c_{ij})}\right] . \tag{7.18}$$

Now, if the employment E_j is also to be considered by type, then it will be necessary to have a conversion matrix which will convert employees by type into employed residents by type. Thus equation (7.18) becomes

$$N_i^n = \sum_j \left(\sum_k a_{kn} E_j^k\right) \left[\frac{W_i^n \mathrm{f}_n(c_{ij})}{\sum_i W_i^n \mathrm{f}_n(c_{ij})}\right] , \tag{7.19}$$

where E_j^k is the number of employees of type k employed in zone j, N_i^n the employed residents of type n residing in zone i, and a_{kn} a region-wide ratio of the number of type n residents per type k employee (note: normally $0{\cdot}0 < a_{kn} < 1{\cdot}0$).

What remains are the crucial questions of the definitions of W_i^n and $\mathrm{f}_n(c_{ij})$.

The great majority of models of this type have used one-parameter trip functions either of declining exponential form or of negative power form: that is,

$$\mathrm{f}(c_{ij}) = \exp(-\beta c_{ij}) \, , \tag{7.20}$$

or

$$\mathrm{f}(c_{ij}) = c_{ij}^{-\beta} \, . \tag{7.21}$$

Both of these functions are of the general shape shown in figure 7.1(a). A two-parameter trip function known as the Tanner function (March, 1971) had been used successfully elsewhere (Cripps and Foot, 1969). This function is a simplified gamma function, of the form

$$\mathrm{f}(c_{ij}) = c_{ij}^{\alpha} \exp(\beta c_{ij}) \, , \tag{7.22}$$

where it is usually assumed that $\beta < 0$ for all relevant cases. This function has the desirable property that when $\alpha < 0$ it has the same shape as do the functions given in equations (7.20) and (7.21). However, if $\alpha > 0$ then equation (7.22) yields the shape shown in figure 7.1(b), which is in accord with empirical evidence on trip distributions for many urban areas. Thus the two-parameter Tanner function was used in equation (7.19). The subscript n for the functional form simply implies the need for different values of α and β for each locator type; thus

$$f_n(c_{ij}) = c_{ij}^{\alpha^n} \exp(\beta^n c_{ij}) \tag{7.23}$$

was the trip-function formulation actually used in the final form of the model.

Next, there is the matter of defining W_i^n for actual use in the model. The residential attractiveness per se of a zone is not a directly measurable attribute. In some of the earliest work in the USA on residence-location models long lists of variables were proposed as components of residential attractiveness[11]. The problems arose first from not having a general model structure or framework in which such variables could be included, and second, from the implicit difficulties of updating these variables for use in forecasting procedures.

Most of the work done in the UK with residence-location models developed from entropy-maximizing procedures included no attractiveness variables at all. A few of these models attempted the use of variables such as number of dwellings (again introducing the problem of endogenous procedures for updating the variable during forecasting) or total population in each zone (Cripps and Foot, 1969). Another study used zonal usable land area as the surrogate measure of residential attractiveness (Barras et al, 1971). These efforts were reasonably successful, but fell far short of including an adequately rich and flexible attractiveness measure that would allow for interesting and reliable policy analyses with the model.

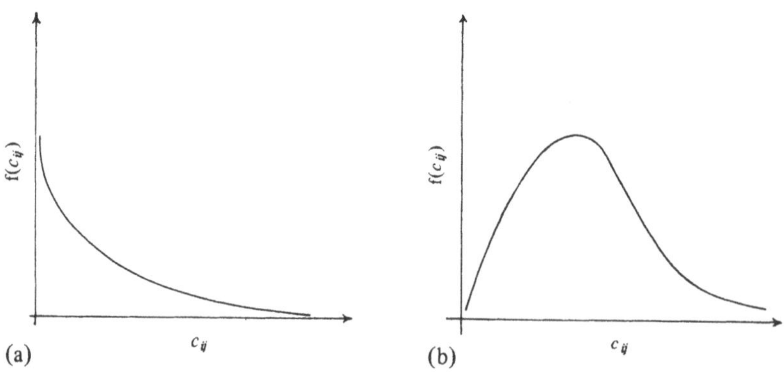

Figure 7.1. (a) Declining exponential or negative power trip function; (b) Tanner function with $\alpha > 0$, $\beta < 0$.

[11] See Putman (1979) for a review of these.

An obvious direction for investigation was towards the combination of the multivariate attractiveness measures of the early US models with the British entropy-maximizing spatial interaction models. In doing this one would also wish to keep in mind the problem of having to update all endogenous variables during the forecasting process. In this vein, it should not be too difficult to conceive of an attractiveness measure consisting of two principal components. The first of these would be some measure of the inherent attractiveness of a zone, whereas the second component might represent the 'size' or 'capacity' of the zone for additional residential locators.

Many measures have been suggested to represent the inherent attractiveness of an area. Researchers have sometimes suggested the use of measures of levels of public services such as educational facilities or, say, police and fire protection. Other measures suggested have included average slope, housing composition or house prices, and various indices of the socioeconomic composition and/or status of the population of an area. A consistent, although of course not perfect, correlate of most of these measures is the income distribution of the resident population. Furthermore, if the income distribution of the resident population of the zone were used as part of the attractiveness measure, and if the locating population were defined in terms of income class, then no intermediate steps would be necessary to update this variable in the forecasting process. This would provide both simplicity and reliability.

The advantage of eliminating intermediate variables may be easily shown. Suppose, for example, that the prior income distribution of a zone could explain 85% of the zonal variation in residence location by income class. Suppose further that knowledge of some other attribute of the zone would explain 90% or 95% of the variation, but that this other attribute had, itself, to be forecast in order to be used in a residence-location forecast. To produce a net increase, over 85%, in forecast reliability, the attribute would have to be capable of being forecasted with (roughly) over 90% reliability. As most such attributes would best be forecasted with the income distribution of residents as an independent variable, it is clear that it would be unlikely that the use of such attributes as endogenous independent variables in forecasting could result in an improvement in forecast reliability.

The second component of the attractiveness measure is intended to represent size and/or capacity of the zone for residence location. An obvious first ingredient here is the amount of vacant developable land. As discussed in chapter 3, this variable alone may sometimes misstate the actual situation in the zone, as some of the developable land may lack the necessary infrastructure for easy development. Thus it might be well to include a variable such as proportion of developable land already developed, to represent the availability of infrastructure. Moreover, as neither of these first two variables provide information on the extent to which the zone is already a residential zone, the inclusion of a residential land variable may also be of some usefulness. It is true, of course, that these variables will themselves have to be forecasted to be used as independent variables in the residence-location forecasts. Yet the land-use variables are a required output from most urban simulation models, and in that sense at least represent no addition in complexity beyond what is already required. However, as these variables will, undoubtedly, be forecasted as part of some land-use accounting procedure which will use the residence-location forecasts as input, some care will be necessary to prevent their degrading the forecast reliability of the model.

After some preliminary statistical examinations it was determined that the most effective functional form for the attractiveness measure was

$$W_i^n = (L_i^{\mathrm{v}})^{q^n} (x_i)^{r^n} (L_i^{\mathrm{r}})^{s^n} \prod_{n'} \left[\left(\frac{N_i^{n'}}{\sum_n N_i^n} \right)^{b_n^{n'}} \right] ; \tag{7.24}$$

here

W_i^n is the attractiveness of zone i to residential locators of type n;
L_i^{v} is the vacant developable land in zone i;
x_i is the proportion of developable land in zone i which has already been developed;
L_i^{r} is the residential land in zone i;
N_i^n is the number of type n residents residing in zone i;
q^n, r^n, s^n, b^n are empirically derived parameters.

In preliminary tests with the forecasting form of the model, it was found that the use of the proportion variables, which ranged in value from zero to one, led to instabilities under certain circumstances. To eliminate this problem two minor changes were made. First, x_i, was redefined such that

$$\tilde{x}_i = 1 + x_i \ .$$

Second, the product term in equation (7.24) was redefined so that

$$W_i^n = (L_i^{\mathrm{v}})^{q^n} (\tilde{x}_i)^{r^n} (L_i^{\mathrm{r}})^{s^n} \prod_{n'} \left[\left(1 + \frac{N_i^{n'}}{\sum_n N_i^n} \right)^{b_n^{n'}} \right] . \tag{7.25}$$

This attractiveness measure was substituted into equation (7.19), along with the two-parameter trip function of equation (7.23), to get the final form of the residential allocation model. Thus the final form of DRAM is

$$N_i^n = \sum_j \left(\sum_k a_{kn} E_j^k \right) \times \frac{(L_i^{\mathrm{v}})^{q^n} (\tilde{x}_i)^{r^n} (L_i^{\mathrm{r}})^{s^n} \prod_{n'} \left\{ \left[1 + \left(N_i^{n'} \Big/ \sum_n N_i^n \right) \right]^{b_n^{n'}} \right\} c_{ij}^{\alpha^n} \exp(\beta^n c_{ij})}{\sum_i (L_i^{\mathrm{v}})^{q^n} (\tilde{x}_i)^{r^n} (L_i^{\mathrm{r}})^{s^n} \prod_{n'} \left\{ \left[1 + \left(N_i^{n'} \Big/ \sum_n N_i^n \right) \right]^{b_n^{n'}} \right\} c_{ij}^{\alpha^n} \exp(\beta^n c_{ij})} . \tag{7.26}$$

Thus the final formulation of this new residence-location model combined a multivariate attractiveness measure and a two-parameter trip function within the framework of a singly constrained spatial interaction model.

As mentioned above, the idea of a multivariate attractiveness measure is not new. However, virtually all the applications of singly constrained spatial interaction models relied on one (rarely two) parameters for calibration of the equation structure to the data base. The combination of the multivariate, multiparameter attractiveness measure and the two-parameter trip function in the spatial interaction model framework is new, and provides unique flexibility for calibration of the equation structure to the data bases for which it is to be used. Furthermore, it should be noted that these parameters are to be estimated simultaneously. In many early model applications attempts were made to estimate model parameters independently. However, the need for simultaneous parameter estimation may easily be seen. Consider first that the attractiveness of an area results from the distribution of activities within that area. That distribution depends, at least in part, on the accessibility of the area. Similarly, the trip distributions of a region reflect the travel behavior of its population. Clearly the travel behavior of the population depends, at least in part, on the attractiveness of the zones to which they might have travelled. Thus the independent estimation either of attractiveness parameters or of trip-function parameters will, of necessity, include effects from both. If these independently estimated parameters are then combined in one equation, there will obviously be some degree of double counting of the effects. A discussion of the problems of parameter estimation will begin in the next section of this chapter. Before that there are a few more general points to be raised about the model structure.

Although the overall problem is not at all near being resolved, it is appropriate here at least to raise the question of the 'dynamics' of models such as the one described above. For example, the original version of IPLUM was intended to be of the general form (Putman, 1977a):

$$\Delta N_i = g\sum_j \Delta E_j p_{ij} \,. \tag{7.27}$$

Thus, increments of employment were to result in increments of residents according to a rather simple formulation of p_{ij} such as equation (7.3). Yet, it is not strictly true that the only changes in spatial location of the population of a region are those due to changes in employment location. At a minimum, new locators in such a scheme would have to be classified as in-migrants (to the region) as a result of new employment, intrametropolitan movers relocating owing to the relocation of employment, and intrametropolitan movers relocating for other reasons. This would further necessitate the inclusion of a procedure for determining where (from what zones) intrametropolitan movers have moved from, in order that the proper accounting of population in the region may be done.

As can well be imagined, an attempt to introduce these incremental variables will result in a very substantial increase in the complexity of the model's equation structure. Moreover, the data requirements for estimating the parameters of such an equation structure would be quite substantial, and perhaps unattainable.

If an incremental-variable formulation were used, then, in forecasting, these changes would be added to the numbers of residents for the base year to get new forecast spatial distributions of residents. If, on the other hand, a 'cross-section' type of formulation such as that of equation (7.26) is used, it is less clear as to what the dynamics of the system actually represent. Suppose, for example, that the parameters of equation (7.16) were to be estimated with data all from the same time period (that is, cross-sectional estimation). The question then arises as to how this equation, with the cross-sectionally estimated parameters, is to be used in forecasting. One possibility is to substitute ΔE_j^k for E_j^k in equation (7.26). This would imply (1) that the N_i^n being forecasted by equation (7.26) are actually changes in residential location by zone and type, ΔN_i^n; and (2) that one must be willing to have the model respond to both cases, $\Delta E_j^k > 0$ and $\Delta E_j^k < 0$, in a similar fashion. Further, as cross-sectional parameters are being used, it is assumed that the ΔN_i^n are behaving spatially in exactly the same manner as the whole population in the time period (in fact, time point) in which the calibration data were collected.

Another possibility is simply to use equation (7.26) as it stands, and forecast the entire spatial distribution of residences for each forecast period. If the calibrations of the equations have yielded reasonably good fits of the model to the base year data, then this should give equally good forecasts. This procedure was in fact followed in several other modelling efforts, with the addition of a type of stability constraint. This took the form of a restriction on the extent of allowable change in a given type of population of a zone. The TOMM model (Crecine, 1964) and the Leeds model (Wilson et al, 1977, page 211) both use this notion. However, this procedure does seem rather arbitrary, and an equally likely possibility would be to use a lagged population distribution in the attractiveness portion of equation (7.26). On further reflection, this seemed the most sensible path to follow. If the time periods involved were not too long, say five years or so, the desired stability would be more or less ensured by the presence of these lagged variables, yet the need for arbitrary stability constraints would have been obviated. It is unfortunate, as will be seen in the discussion of empirical results from calibration efforts later in this chapter, that the data necessary for proper calibration of a model in this form have not been available. There is some hope though that in the not too distant future this data constraint will be eliminated, and the results of calibrating this form of model may then be examined.

7.4 Calibration procedures for residence-location models

The term model calibration has been used in various contexts. Here it has but one meaning: the adjustment of the parameters in the equations of a model so as to maximize the goodness of fit between the estimate of a particular variable of a model and the observed value of that variable, both for the same 'base' time period. In early work with residence-location models, those formulated in terms of linear equations were calibrated by means of multiple regression procedures. Nonlinear models were calibrated by various curve-fitting procedures or, in the case of gravity models, by elaborate trial-and-error processes of one type or another. With the development of spatial interaction models via entropy-maximizing principles, the calibration problem was more clearly defined. Yet, virtually all of the residence-location forms of those models were one-parameter models. The development of a residence-location model involving a multivariate, multi-parameter attractiveness measure plus a two-parameter trip function required the development of new calibration procedures as well. In this section, these procedures are developed and used to examine the response surfaces of several simple numerical examples of such models.

There are many strategies and procedures[12] for calibrating these models, and it is not always clear how they relate to each other. 'Spatial interaction model' in this context means the singly constrained type of model used as a location model. This is but one of a set of types of such models. As a quick review of what has been presented in earlier chapters, at one extreme is the unconstrained model of the form

$$T_{ij} = kO_iD_j\,\mathrm{f}(c_{ij}) \; ; \tag{7.28}$$

here

T_{ij} is the number of trips from zone i to zone j;
O_i is the number of trips originating at zone i;
D_j is the number of trips terminating at zone j;
c_{ij} is the 'cost' of a trip between zones i and j,
k is an empirically determined constant.

This form of model is intrinsically linear, and for usual forms of $\mathrm{f}(c_{ij})$ can be linearized.

At the other extreme is the doubly constrained model of the form

$$T_{ij} = A_iB_jO_iD_j\,\mathrm{f}(c_{ij}) \; , \tag{7.29}$$

where there are trip-end constraints of the form

$$\sum_j T_{ij} = O_i \; , \tag{7.30}$$

[12] A more detailed survey of many of these may be found in Batty and Mackie (1972) and an even more complete discussion is given in Batty (1976, chapters 6, 7). Other recent discussions may be found in Baxter and Williams (1975), Openshaw (1976), and Williams (1976).

and

$$\sum_i T_{ij} = D_j \; ; \tag{7.31}$$

and where

$$\sum_i \sum_j T_{ij} = \sum_i O_i = \sum_j D_j = \text{const} \; . \tag{7.32}$$

The A_i and B_j of equation (7.29) which replace the k of equation (7.28) have the form

$$A_i = \left[\sum_j B_j D_j \, \mathrm{f}(c_{ij})\right]^{-1} , \tag{7.33}$$

$$B_j = \left[\sum_i A_i O_i \, \mathrm{f}(c_{ij})\right]^{-1} . \tag{7.34}$$

Between the two extremes are two singly constrained models. The production-constrained model has the form

$$T_{ij} = A_i O_i D_j \, \mathrm{f}(c_{ij}) \; , \tag{7.35}$$

subject to

$$\sum_j T_{ij} = O_i \; ; \tag{7.36}$$

thus

$$A_i = \left[\sum_j D_j \, \mathrm{f}(c_{ij})\right]^{-1} . \tag{7.37}$$

The attraction-constrained model is

$$T_{ij} = B_j O_i D_j \, \mathrm{f}(c_{ij}) \; , \tag{7.38}$$

subject to

$$\sum_i T_{ij} = D_j \; ; \tag{7.39}$$

thus

$$B_j = \left[\sum_i O_i \, \mathrm{f}(c_{ij})\right]^{-1} . \tag{7.40}$$

None of these models has any parameters to estimate other than those which may appear in the $\mathrm{f}(c_{ij})$ term. One of the simplest functions in common usage is

$$\mathrm{f}(c_{ij}) = \exp(-\beta c_{ij}) \; , \tag{7.41}$$

involving, therefore, only the estimation of β.

Before continuing, there is a notational matter to be settled. The customary use of a spatial interaction model for forecasting residence location defines the resident population in zone i, N_i, as the sum over all zones j, of T_{ij}. This trips variable T_{ij} is customarily defined as the number of trips from place of residence i to place of work j. Yet, in chapter 3, it was mentioned that it was the work-to-home trips—that is, the 'afternoon peak'—that were most appropriate for use in the integrated model package. If, to be consistent with the development of the equations presented in previous chapters, the place of residence is taken to be zone i and the place of work is taken to be zone j, then a work-to-home trip would be defined as a T_{ji} trip.

Consequently, in the remainder of this discussion of calibration, the subscript i will refer to the zone of residence, and the subscript j to the zone of work. In the spatial 'cost' variable, c, the first subscript will refer to the origin (place-of-work) zone and the second to the destination (place-of-residence) zone. Thus in the following notation c_{ji} is used in the model description. This usage is different from that found in the bulk of the current literature. However, unless it is explicitly assumed that the cost matrix is symmetric, the use of c_{ij} to denote the cost of a work-to-home trip would be incorrect.

The specific concern here is with calibrating residence-location models in the absence of interaction data (the T_{ji} matrix). Thus, the following simple location (production-constrained) model is formulated

$$\hat{N}_i = \sum_j A_j O_j D_i \exp(-\beta c_{ji}) \ , \tag{7.42}$$

where

$$A_j = \left[\sum_i D_i \exp(-\beta c_{ji})\right]^{-1} \ . \tag{7.43}$$

If employment in zone j is used as a measure of O_j and the attractiveness of zone i is used as a measure of D_i, one then gets

$$\hat{N}_i = \sum_j E_j \left[\frac{W_i \exp(-\beta c_{ji})}{\sum_i W_i \exp(-\beta c_{ji})}\right] \ , \tag{7.44}$$

where $\hat{N}_i$ is the estimated number of employed residents of zone i, W_i the attractiveness of zone i, E_j the employment in zone j, c_{ji} the zone j–zone i interaction cost, and β an empirically derived parameter.

It is important to make note here of what is known as the 'bogus calibration' problem. This difficulty can appear when the observed attractiveness, W_i, of zone i is a linear or log-linear transform of the estimated variable, in this case, of the estimated employed residents, $\hat{N}_i$, as in equation (7.44). In situations of this sort where, for example, observed population in a zone is used as a measure of attractiveness for estimating the population of zones, or where retail employment is used as a measure of attractiveness for retail employment, it is possible that the $\hat{\beta}$ will tend to zero; this gives a false solution to the calibration problem. Thus care must be taken to ensure that any W_i selected for actual calibration purposes is not a linear or log-linear transform of the dependent (estimated) variable.

Furthermore, care must also be taken that the parameters are not allowed to move to this false or bogus solution.

The question now arises as to how the numerical value of the parameter $\hat{\beta}$ may be estimated. This requires first that a criterion be selected by which one may measure the 'goodness-of-fit' of the model formulation to the data set being used for calibration. This criterion, C, will be a function of an observed variable and a set of estimates of the same variable which are, in turn, a function of the parameter value. Thus we can write

$$\mathsf{C} = \mathrm{f}(N_i, \hat{N}_i, \beta) \; , \tag{7.45}$$

where N_i is the observed employed residents of zone i.

The optimum value of the criterion may be a maximum or a minimum, depending upon the criterion definition. Two general strategies may be followed in determining that value of β which optimizes the criterion. The first of these, often followed in British practice, involves developing a set of equations using Langrangian multipliers and then partial derivatives which describe the criterion at its optimum. These equations are then solved simultaneously (usually by some iterative technique) for the value of β at the criterion's optimum. These equations are also sometimes solved by search techniques, but the overall strategy remains unchanged.

The second strategy, which is of principal concern here, uses the derivative of the criterion with respect to the parameter β to guide a search technique for finding the optimum value of the criterion. If one is dealing with a model for which the derivatives are readily calculated, this calibration strategy seems to afford the advantages of simplicity of approach and computational efficiency, even for problems involving the simultaneous estimation of as many as nine parameters (the largest size of problem tested to date). Furthermore, it may be shown that for situations where only trip-end data (that is, activity location or in this case, the N_i) are available, the first strategy is not practical. The first strategy typically uses the following likelihood function as a criterion:

$$\mathsf{L}_1 = \prod_j \prod_i p_{ji}^{T_{ji}} \; , \tag{7.46}$$

where the probability p_{ji} is given by

$$p_{ji} = \frac{W_i c_{ji}^{\beta}}{\sum_i W_i c_{ji}^{\beta}} \, . \tag{7.47}$$

The second strategy uses the following likelihood function as a criterion:

$$\mathsf{L}_2 = \prod_i p_i^{N_i} \; , \tag{7.48}$$

where the probability p_i is given by

$$p_i = \frac{\hat{N}_i}{\sum_i \hat{N}_i} \; , \tag{7.49}$$

and where

$$\hat{N}_i = \sum_j E_j \left[\frac{W_i c_{ji}^{\beta}}{\sum_i W_i c_{ji}^{\beta}} \right] . \tag{7.50}$$

In order to make the necessary calculations feasible, in both cases the natural logarithm of the likelihood function is taken. Thus for the first strategy,

$$\ln \mathsf{L}_1 = \sum_j \sum_i T_{ji} \ln p_{ji} , \tag{7.51}$$

and, from equation (7.47),

$$\ln \mathsf{L}_1 = \sum_j \sum_i T_{ji} \left[\ln \left(\sum_i W_i c_{ji}^{\beta} \right) + \ln W_i + \beta \ln c_{ji} \right] . \tag{7.52}$$

By use of Lagrange multipliers to incorporate the necessary constraints and partial derivatives with respect to β, the equations for the maximum of $\ln \mathsf{L}_1$ may be derived. This set of equations may then be solved simultaneously for the maximizing value of β. Some exploration of the nature of the criterion functions and the shapes of response surfaces of criteria as functions of parameters has been done for the first strategy and is described in the references cited at the beginning of this section.

For the second strategy, the log-likelihood function may be written as

$$\ln \mathsf{L}_2 = \sum_i N_i \ln p_i , \tag{7.53}$$

and, from equation (7.49),

$$\ln \mathsf{L}_2 = \sum_i N_i \ln \left(\frac{\hat{N}_i}{\sum_i \hat{N}_i} \right) . \tag{7.54}$$

From the definition of $\hat{N}_i$ in equation (7.50) it is clear that the form of this last equation makes it difficult (perhaps impossible) to derive the necessary equations for maximizing $\ln \mathsf{L}_2$ in a fashion similar to that of the first strategy. Consequently, in situations where only the trip-end (activity-location) data are available, the second is the one which must be followed.

7.5 The R^2 response surface: one-parameter case

One of the most commonly used criteria of goodness of fit is the coefficient of determination, R^2. Thus it seemed sensible to begin the investigation of the properties of the response surface to be encountered in calibration of these models with an exploration of the response surface for R^2 using a small numerical example. First, R^2 is calculated as per the familiar equation:

$$R^2 = 1 - \frac{\sum_i (\hat{N}_i - N_i)^2}{\sum_i (N_i - \bar{N})^2} , \tag{7.55}$$

where $\bar{N}$ is the mean of N_i.

We begin the exploration of the characteristics of the R^2 and gradient of R^2 functions by using the three zone test problem given in table 7.1. From these data, with an assumed value of $-2{\cdot}00$ for β, estimates of the number of employed residents in each zone were calculated by means of equation (7.44):

$$\hat{N}_1 = 110{\cdot}8718\ ; \qquad \hat{N}_2 = 129{\cdot}2546\ ; \qquad \hat{N}_3 = 209{\cdot}8734\ .$$

The gradient of a function, when evaluated at a specific point on the function surface, indicates the direction of the steepest slope of the surface from the point and the steepness of that slope. Gradients may be calculated numerically by evaluating the function at points small distances from the particular point at which the gradient is desired, or analytically by substituting numerical values into the derived expression for the gradient of the specific function involved. By knowing the gradient it is possible, as will be shown shortly, to move efficiently to maxima or minima on the function surface. An important qualification here is that there be no local maxima or minima on which the gradient-search procedure can become stalled and thus prevented from finding the true, or global, maximum or minimum of the function. It is to confirm the absence of these local optima that these explorations are undertaken. In general, gradient search is more efficient when analytical, rather than numerical, derivatives of the function can be calculated. Thus for these spatial interaction models, analytical derivatives are used, as the relationship between R^2 and β, although rather complex, does yield differentiable functions.

Note that the minus sign associated with β in equation (7.44) is dropped in the following equations; thus the value of β, henceforth, may be either positive or negative, depending upon the actual value used in any particular calculation. The equation for the gradient, ∇R^2, is developed as follows,

$$\nabla R^2 = \frac{\partial R^2}{\partial \beta} = \frac{\partial R^2}{\partial \hat{N}_i}\frac{\partial \hat{N}_i}{\partial \beta} \tag{7.56}$$

$$= -2\left[\sum_i (\hat{N}_i - N_i)\frac{\partial \hat{N}_i}{\partial \beta}\right] \bigg/ \sum_i (N_i - \bar{N})^2\ . \tag{7.57}$$

Table 7.1. Data for three zone test case.

Employment	Attractiveness	Cost matrix		
$E_1 = 100$	$W_1 = 3$	$c_{11} = 1{\cdot}5$	$c_{12} = 2{\cdot}5$	$c_{13} = 4{\cdot}0$
$E_2 = 150$	$W_2 = 4$	$c_{21} = 2{\cdot}5$	$c_{22} = 2{\cdot}0$	$c_{23} = 3{\cdot}5$
$E_3 = 200$	$W_3 = 5$	$c_{31} = 3{\cdot}0$	$c_{32} = 3{\cdot}5$	$c_{33} = 1{\cdot}5$

After a number of intermediate steps, and by use of equation (7.44), the resulting equation is

$$\nabla R^2 = -2\left\{\sum_i (\hat{N}_i - N_i)\left[\sum_j \left(\frac{E_j J_{ji}}{J_{j*}}\right)\left(c_{ji} - \frac{\sum_{i'} c_{ji'} J_{ji'}}{J_{j*}}\right)\right]\right\} \Big/ \sum_i (N_i - \bar{N})^2 , \tag{7.58}$$

where

$$J_{ji} = W_i \exp(\beta c_{ji}) , \tag{7.59}$$

$$J_{j*} = \sum_i J_{ji} = \sum_i W_i \exp(\beta c_{ji}) . \tag{7.60}$$

Thus it is possible to calculate R^2, ∇R^2, and the $\hat{N}_i$ for various values of β. The results for a few selected values of β are given in table 7.2; these show that there is a perfect R^2 of 1·000 at $\beta = -2\cdot00$, which of course is to be expected as the 'observed' values of N_i were obtained by use of a value for β of $-2\cdot00$. Furthermore, it can be seen that the gradient ∇R^2 is negative for $\beta > -2\cdot00$, and positive for $\beta < -2\cdot00$; this suggests how one may choose to change (increase or decrease) β, at any point, to improve the R^2 value. A graph of R^2 and ∇R^2 versus β is given in figure 7.2 (see over).

An immediately obvious and possibly serious problem appears in figure 7.2. Had the gradient-search procedure been initiated with an assumed value, $\beta \gtrsim 0\cdot4$, the value of ∇R^2 would have been positive and the search routine would have yielded, at a local optimum, a parameter estimate of $\sim 1\cdot6$ for β. The discovery of the existence of a local optimum was a disturbing result ameliorated only partly by the suggestion that perhaps all search procedures could use starting values, β, less than zero. The key question was whether the presence of local optima on the response surface was inherent to the functional form, or a chance artifact of the data set. The desire to know more about these functions led to a series of further experiments with three-zone test problems.

Table 7.2. Illustrative values of R^2, ∇R^2, and the $\hat{N}_i$ as a function of β, from the test data in table 7.1.

β	R^2	∇R^2	$\hat{N}_1$	$\hat{N}_2$	$\hat{N}_3$
0·00	0·83167	−0·40718	112·500	150·000	187·500
−0·40	0·95214	−0·15049	103·589	142·536	203·875
−0·80	0·97252	−0·00502	100·791	134·328	214·881
−1·00	0·97455	−0·01882	101·560	131·559	216·881
−1·40	0·98785	−0·03818	105·254	128·898	215·848
−1·80	0·99874	−0·01319	109·279	128·788	211·932
−2·00	1·00000	−0·00000	110·872	129·255	209·873
−2·40	0·99665	+0·01405	112·913	130·718	206·369
−2·80	0·99031	+0·01655	113·567	132·537	203·897
−3·00	0·98701	+0·01635	113·492	133·500	203·008
−4·00	0·97053	+0·01761	111·013	138·224	200·763

Several variations of the three zone problem were tried. The original cost matrix was asymmetric. New runs were made with a symmetric cost matrix. New runs were also made with equal values of attractiveness for all zones. Note that in all these tests, the 'observed' numbers of residents in each zone is the exact value which is yielded by equation (7.44) with a value of $-2 \cdot 00$ for β. By so doing, it was always known where the optimum of the criterion function should occur, and thus it was possible to compare results from the various tests.

Figure 7.3(a) shows a graph of R^2 and ∇R^2 versus β for the three-zone test problem with equal W_i and the symmetric c_{ji} matrix. Again, there was a local optimum of R^2, in this case, at $\beta \approx 1 \cdot 0$. Interestingly enough, the local optimum was much less pronounced in this case than in figure 7.2, where the asymmetric c_{ji} matrix had been used. Finally, figure 7.3(b) shows a graph of R^2 and ∇R^2 versus β for the three zone problem with unequal W_i and the symmetric c_{ji} matrix. In this case the local optimum does not occur, there being just the single optimum value of R^2 at $\beta = -2 \cdot 0$. It is, however, important to notice in this case how flat the R^2 curve is in the region of the optimum value of β. This property will return to complicate the calibration problem throughout all the discussion which follows.

The three zone tests have shown that: (a) at this problem scale, changes in the data set can have dramatic effects on the shape of the $R^2(\beta)$ function, even to the extent of producing local optima; (b) though the R^2 function is sometimes bimodal, the highest optimum occurs at the proper value of β, and no local optima occur for $\beta < 0$; (c) the R^2 function tends to be rather flat in the region of the desired optimum.

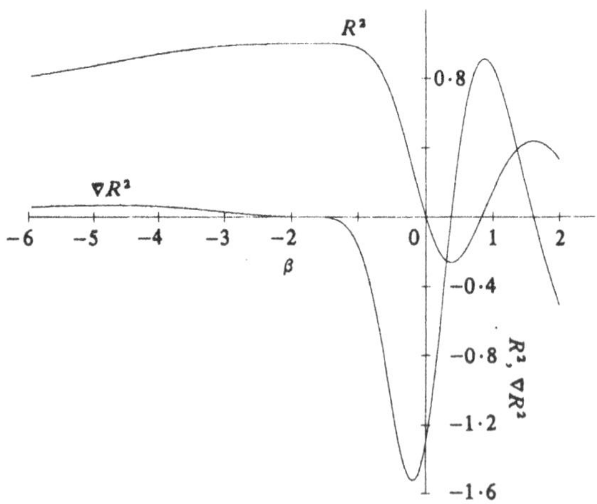

Figure 7.2. Three zone test problem: plots of R^2 and ∇R^2 obtained from data in tables 7.1, 7.2.

The next step then was to repeat these same tests for a slightly larger problem.

To examine how the above results change with an increase in the number of zones, data for a nine zone problem were constructed, as shown in table 7.3 (over). Again, the R^2 criterion and the gradient ∇R^2 were calculated and graphed. Note also that the nine zone data, too, were used with an assumed value $\beta = -2{\cdot}00$ to generate the 'observed' data. Thus, a perfect fit, where $R^2 = 1{\cdot}0000$ and $\nabla R^2 = 0{\cdot}0000$, will be found for $\beta = -2{\cdot}00$ as in the previous examples for the three zone problem. For this nine zone case, several test runs again were made. The graphs resulting from two of these runs are shown in figures 7.4(a) and 7.4(b). Both runs use the symmetric cost matrix, but in the second case, figure 7.4(b), all the W_i were set equal to $4{\cdot}0$, to illustrate their effect on the shapes of the curves.

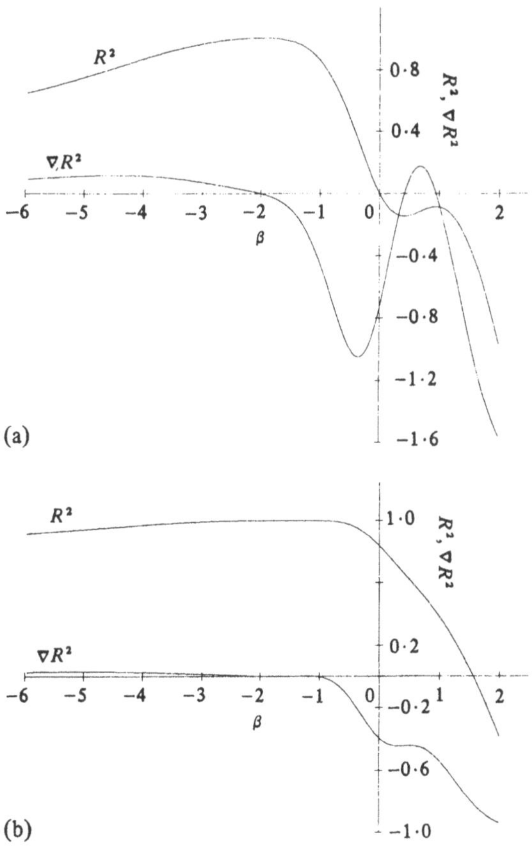

Figure 7.3. Three zone test problem: plots of R^2 and ∇R^2 versus β from a symmetric c_{ij} matrix. (a) All W_i equal; (b) unequal W_i.

Table 7.3. Data for nine zone test problem.

Employment	Attractiveness	Cost matrix								
$E_1 = 100$	$W_1 = 3$	1·5	2·5	3·5	4·5	4·0	4·0	4·0	2·0	2·0
$E_2 = 50$	$W_2 = 4$	2·5	1·0	1·5	2·0	1·5	3·0	3·5	3·0	2·5
$E_3 = 25$	$W_3 = 5$	3·5	1·5	1·0	3·0	1·0	3·5	4·0	4·0	3·0
$E_4 = 0$	$W_4 = 8$	4·5	2·0	3·0	1·0	1·0	1·5	3·5	3·5	2·0
$E_5 = 25$	$W_5 = 6$	4·0	1·5	1·0	1·0	1·0	1·5	3·5	3·5	2·0
$E_6 = 75$	$W_6 = 4$	4·0	3·0	3·5	1·5	1·5	1·5	2·0	2·5	2·5
$E_7 = 200$	$W_7 = 2$	4·0	3·5	4·0	3·5	3·5	2·0	2·0	2·0	2·5
$E_8 = 250$	$W_8 = 1$	2·0	3·0	4·0	3·5	3·5	2·5	2·0	2·0	2·5
$E_9 = 500$	$W_9 = 2$	2·0	2·5	3·0	2·0	2·0	2·5	2·5	2·5	2·5

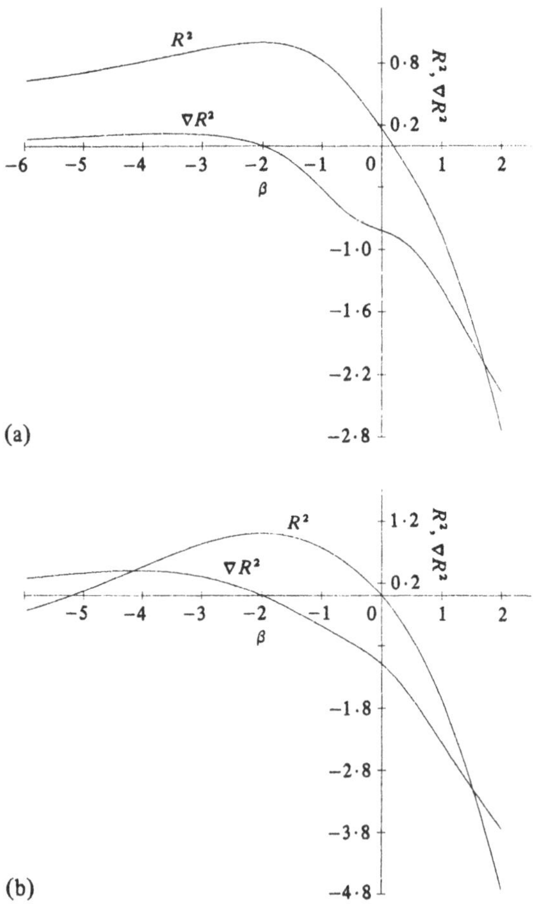

Figure 7.4. Nine zone test problem: plots of R^2 and ∇R^2 versus β (a) from the data given in table 7.3; (b) as (a), but with all $W_i = 4$.

A welcome result here was that the R^2 function now seems to be unimodal. It was also encouraging to note that the curvature $R^2(\beta)$ in the region of the optimum has increased. The peak remained sharper to the right of the optimum than to the left. Again, the different data sets did

Table 7.4. Employment and attractiveness data for twenty-five zone test problem.

$E_1 = 25$	$W_1 = 5$	$E_{10} = 50$	$W_{10} = 4$	$E_{19} = 100$	$W_{19} = 4$
$E_2 = 25$	$W_2 = 5$	$E_{11} = 75$	$W_{11} = 2$	$E_{20} = 50$	$W_{20} = 4$
$E_3 = 50$	$W_3 = 5$	$E_{12} = 250$	$W_{12} = 2$	$E_{21} = 150$	$W_{21} = 3$
$E_4 = 25$	$W_4 = 5$	$E_{13} = 400$	$W_{13} = 1$	$E_{22} = 150$	$W_{22} = 3$
$E_5 = 25$	$W_5 = 5$	$E_{14} = 150$	$W_{14} = 1$	$E_{23} = 75$	$W_{23} = 4$
$E_6 = 75$	$W_6 = 4$	$E_{15} = 75$	$W_{15} = 3$	$E_{24} = 75$	$W_{24} = 4$
$E_7 = 50$	$W_7 = 4$	$E_{16} = 100$	$W_{16} = 3$	$E_{25} = 50$	$W_{25} = 4$
$E_8 = 100$	$W_8 = 4$	$E_{17} = 200$	$W_{17} = 2$		
$E_9 = 75$	$W_9 = 3$	$E_{18} = 300$	$W_{18} = 1$		

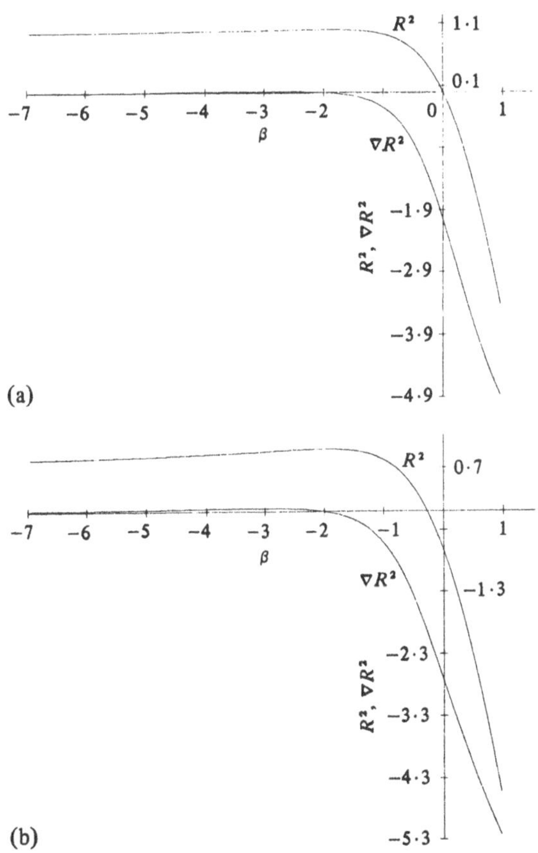

Figure 7.5. Twenty-five zone test problem: plots of R^2 and ∇R^2 versus β, (a) from the data given in table 7.4; (b) as (a), but with all $W_i = 4{\cdot}0$.

Table 7.5. Cost matrix for twenty-five zone test problem.

	1	2	3	4	5	6	7	8	9	10	11	12	13	14	15	16	17	18	19	20	21	22	23	24	25
1	1·5	1·0	2·0	3·0	4·0	1·0	1·4	2·2	3·2	4·1	2·0	2·2	2·8	3·6	4·5	3·0	3·2	3·6	4·2	5·0	4·0	4·1	4·5	5·0	5·7
2	1·0	1·5	1·0	2·0	3·0	1·4	1·0	1·4	2·2	3·2	2·2	2·0	2·2	2·8	3·6	3·2	3·0	3·2	3·6	4·2	4·1	4·0	4·1	4·5	5·0
3	2·0	1·0	1·5	1·0	2·0	2·2	1·4	1·0	1·4	2·2	2·8	2·2	2·0	2·2	2·8	3·6	3·2	3·0	3·2	3·6	4·5	4·1	4·0	4·1	4·5
4	3·0	2·0	1·0	1·5	1·0	3·2	2·2	1·4	1·0	1·4	3·6	2·8	2·2	2·0	2·2	4·2	3·6	3·2	3·0	3·2	5·0	4·5	4·1	4·0	4·1
5	4·0	3·0	2·0	1·0	1·5	4·1	3·2	2·2	1·4	1·0	4·5	3·6	2·8	2·2	2·0	5·0	4·2	3·6	3·2	3·0	5·7	5·0	4·5	4·1	4·0
6	1·0	1·4	2·2	3·2	4·1	1·5	1·0	2·0	3·0	4·0	1·0	1·4	2·2	3·2	4·1	2·0	2·2	2·8	3·6	4·5	3·0	3·2	3·6	4·2	5·0
7	1·4	1·0	1·4	2·2	3·2	1·0	1·5	1·0	2·0	3·0	1·4	1·0	1·4	2·2	3·2	2·2	2·0	2·2	2·8	3·6	3·2	3·0	3·2	3·6	4·2
8	2·2	1·4	1·0	1·4	2·2	2·0	1·0	1·5	1·0	2·0	2·2	1·4	1·0	1·4	2·2	2·8	2·2	2·0	2·2	2·8	3·6	3·2	3·0	3·2	3·6
9	3·2	2·2	1·4	1·0	1·4	3·0	2·0	1·0	1·5	1·0	3·2	2·2	1·4	1·0	1·4	3·6	2·8	2·2	2·0	2·2	4·2	3·6	3·2	3·0	3·2
10	4·1	3·2	2·2	1·4	1·0	4·0	3·0	2·0	1·0	1·5	4·1	3·2	2·2	1·4	1·0	4·5	3·6	2·8	2·2	2·0	5·0	4·2	3·6	3·2	3·0
11	2·0	2·2	2·8	3·6	4·5	1·0	1·4	2·2	3·2	4·1	1·5	1·0	2·0	3·0	4·0	1·0	1·4	2·2	3·2	4·1	2·0	2·2	2·8	3·6	4·5
12	2·2	2·0	2·2	2·8	3·6	1·4	1·0	1·4	2·2	3·2	1·0	1·5	1·0	2·0	3·0	1·4	1·0	1·4	2·2	3·2	2·2	2·0	2·2	2·8	3·6
13	2·8	2·2	2·0	2·2	2·8	2·2	1·4	1·0	1·4	2·2	2·0	1·0	1·5	1·0	2·0	2·2	1·4	1·0	1·4	2·2	2·8	2·2	2·0	2·2	2·8
14	3·6	2·8	2·2	2·0	2·2	3·2	2·2	1·4	1·0	1·4	3·0	2·0	1·0	1·5	1·0	3·2	2·2	1·4	1·0	1·4	3·6	2·8	2·2	2·0	2·2
15	4·5	3·6	2·8	2·2	2·0	4·1	3·2	2·2	1·4	1·0	4·0	3·0	2·0	1·0	1·5	4·1	3·2	2·2	1·4	1·0	4·5	3·6	2·8	2·2	2·0
16	3·0	3·2	3·6	4·2	5·0	2·0	2·2	2·8	3·6	4·5	1·0	1·4	2·2	3·2	4·1	1·5	1·0	2·0	3·0	4·0	1·0	1·4	2·2	3·2	4·1
17	3·2	3·0	3·2	3·6	4·2	2·2	2·0	2·2	2·8	3·6	1·4	1·0	1·4	2·2	3·2	1·0	1·5	1·0	2·0	3·0	1·4	1·0	1·4	2·2	3·2
18	3·6	3·2	3·0	3·2	3·6	2·8	2·2	2·0	2·2	2·8	2·2	1·4	1·0	1·4	2·2	2·0	1·0	1·5	1·0	2·0	2·2	1·4	1·0	1·4	2·2
19	4·2	3·6	3·2	3·0	3·2	3·6	2·8	2·2	2·0	2·2	3·2	2·2	1·4	1·0	1·4	3·0	2·0	1·0	1·5	1·0	3·2	2·2	1·4	1·0	1·4
20	5·0	4·2	3·6	3·2	3·0	4·5	3·6	2·8	2·2	2·0	4·1	3·2	2·2	1·4	1·0	4·0	3·0	2·0	1·0	1·5	4·1	3·2	2·2	1·4	1·0
21	4·0	4·1	4·5	5·0	5·7	3·0	3·2	3·6	4·2	5·0	2·0	2·2	2·8	3·6	4·5	1·0	1·4	2·2	3·2	4·1	1·5	1·0	2·0	3·0	4·0
22	4·1	4·0	4·1	4·5	5·0	3·2	3·0	3·2	3·6	4·2	2·2	2·0	2·2	2·8	3·6	1·4	1·0	1·4	2·2	3·2	1·0	1·5	1·0	2·0	3·0
23	4·5	4·1	4·0	4·1	4·5	3·6	3·2	3·0	3·2	3·6	2·8	2·2	2·0	2·2	2·8	2·2	1·4	1·0	1·4	2·2	2·0	1·0	1·5	1·0	2·0
24	5·0	4·5	4·1	4·0	4·1	4·2	3·6	3·2	3·0	3·2	3·6	2·8	2·2	2·0	2·2	3·2	2·2	1·4	1·0	1·4	3·0	2·0	1·0	1·5	1·0
25	5·7	5·0	4·5	4·1	4·0	5·0	4·2	3·6	3·2	3·0	4·5	3·6	2·8	2·2	2·0	4·1	3·2	2·2	1·4	1·0	4·0	3·0	2·0	1·0	1·5

show slightly different shaped curves, although they were much closer in shape than were the various three zone tests. This suggested, as had been hoped, that as the number of zones is increased the general shapes of the R^2 and ∇R^2 curves become more stable in character. The obvious next step, in the face of these encouraging results, was a still larger data set.

The data for the twenty-five zone test problem are given in table 7.4. The cost matrix for the twenty-five zone problem is shown in table 7.5. Again, R^2 and ∇R^2 were plotted against β. Also, as before, the optimum value of β is found to be $-2{\cdot}00$, the value used to generate the 'observed' data. The graphs of two test runs are shown in figure 7.5. Here the R^2 curve shows signs of levelling off, an unwelcome result that again suggests that calibration of such a model by use of the R^2 criterion might become rather tedious as a search for an optimum along such a flat surface would proceed rather slowly. A positive result here was that the differences between the $R^2(\beta)$ curves for the cases of equal W_i and unequal W_i are rather minimal. This suggested that when data sets have twenty-five or so zones, data-set differences should not seriously affect the calibration procedure, though they will of course give different optimal values of β.

The results of these investigations of the R^2 criterion for various data sets were, in general, supportive of the calibration strategy as originally conceived. However, the flatness of the $R^2(\beta)$ function in the region of the optimum did not augur well for calibration of such models in practice. Thus it was deemed desirable to investigate the possibility of using a different criterion function.

7.6 The likelihood response surface

7.6.1 One-parameter case

An alternative (actually in most cases a preferable one) to the R^2 criterion is the likelihood function. A likelihood function was defined above in equation (7.48). If the actual estimate of population $\hat{N}_i$ is substituted in equation (7.48) for the probability p_i, then this function takes the form

$$\mathsf{L}_2 = \prod_i \hat{N}_i^{N_i} , \tag{7.61}$$

where, again, $\hat{N}_i$ is the estimated number of households locating in zone i, and N_i the observed number of households in zone i. Owing to the algebraic difficulty which would be encountered in dealing with the product function $\prod_i$, it is convenient to proceed in terms of the logarithm of the likelihood function:

$$\ln \mathsf{L}_2 = \sum_i N_i \ln \hat{N}_i . \tag{7.62}$$

It is now possible to construct a likelihood criterion function such that for a perfect fit between observed and estimated values of N_i, the function yields a value of L equal to zero. We redefine the log-likelihood function such that

$$\mathsf{L} = \sum_i N_i \ln \hat{N}_i - \sum_i N_i \ln N_i . \tag{7.63}$$

The problem then is to locate the maximum value of L, which will, of course, have L = 0. As was the case for the R^2 criterion, this involves taking the derivative of the criterion function with respect to β, to calculate the gradient:

$$\nabla \mathsf{L} = \frac{\partial \mathsf{L}}{\partial \beta} = \frac{\partial \mathsf{L}}{\partial \hat{N}_i} \frac{\partial \hat{N}_i}{\partial \beta} \tag{7.64}$$

$$= \sum_i \frac{N_i}{\hat{N}_i} \frac{\partial \hat{N}_i}{\partial \beta} . \tag{7.65}$$

In equation (7.58), the partial derivative of N_i with respect to β has already been worked out. Thus, the equation for the gradient of the log-likelihood criterion is

$$\nabla \mathsf{L} = \sum_i \frac{N_i}{\hat{N}_i} \left[\sum_j \left(\frac{E_j J_{ji}}{J_{j\bullet}} \right) \left(c_{ji} - \frac{\sum_{i'} c_{ji'} J_{ji'}}{J_{j\bullet}} \right) \right] , \tag{7.66}$$

where J_{ji} and $J_{j\bullet}$ are as defined in equations (7.59) and (7.60).

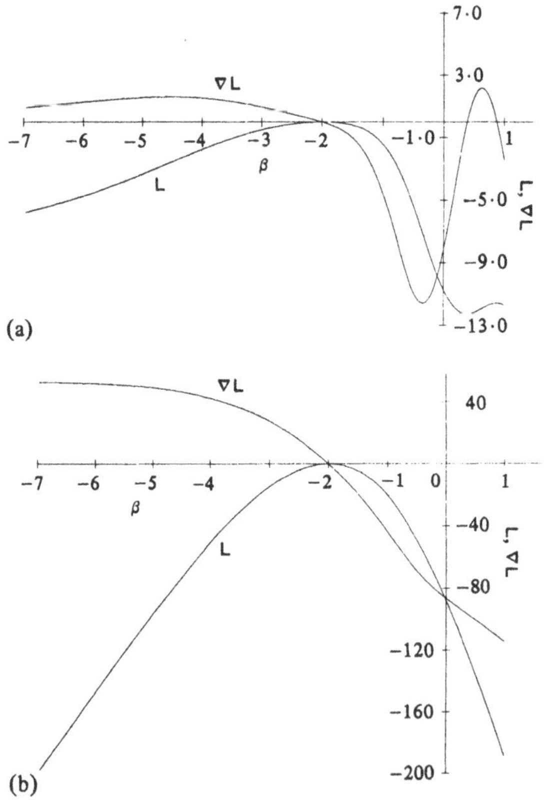

Figure 7.6. Plots of L and ∇L versus β (test data with all W_i equal). (a) Three zone problem; (b) nine zone problem.

Again, the three-zone and nine-zone test problems were examined first. Of the many test runs done, only two are shown here, in figure 7.6. The three zone problem [figure 7.6(a)] again shows the presence of a local optimum which vanishes in the nine zone problem [figure 7.6(b)]. Also, note that the nine zone problem shows a more sharply peaked curve for L than any of the R^2 curves shown earlier. Figure 7.7 gives plots of L(β) for the three, nine, and twenty-five zone problems, all with the same vertical axis scale. The obvious contrast between the rather flat peak of the three zone problem and the much sharper peaks of the nine zone and twenty-five zone problems is quite remarkable. Note that the latter two peaks overlapped, rather than the twenty-five zone peak being entirely within the nine-zone peak as might have been expected. This was due to the differences in the data bases for the two problems. The result, as had been hoped, did clearly confirm that the peak of the likelihood function became very sharp for the nine zone problem, but that it did not get much sharper as the number of zones increased beyond that up to twenty-five.

It is possible to draw the following conclusions about the use of a likelihood criterion function. First, different data sets did yield different shapes of criterion and gradient functions (as was also true with the R^2 criterion), but these differences became less significant as the number of zones increased and are probably of no consequence at the numbers of zones to be used for actual applications of these models. Second, as hoped, the sharpness of the peak of the function near its optimum is very much greater for the likelihood criterion than for the R^2 criterion; this confirms the appropriateness of the use of the likelihood function for calibration applications. From these encouraging results, the next step was to examine a slightly more complex model formulation.

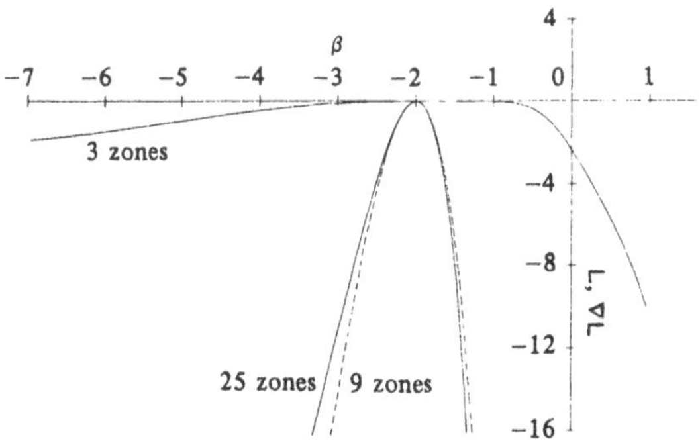

Figure 7.7. Plots of L(β) for three test problems with different numbers of zones (all to same scale).

7.6.2 Two-parameter case

The trip function used throughout the examples given above was the simple exponential function,

$$f(c_{ji}) = \exp(\beta c_{ji}) , \qquad (7.67)$$

where β was expected to be negative in all cases. As described in section 7.3, a function which has been found to yield better data fits for residence-location models is the two-parameter Tanner function,

$$f(c_{ji}) = c_{ji}^{\alpha} \exp(\beta c_{ji}) , \tag{7.68}$$

where β is expected to be negative, and α to be positive. Substitution of this function into equation (7.44) yields

$$\hat{N}_i = \sum_j E_j \left[\frac{W_i c_{ji}^{\alpha} \exp(\beta c_{ji})}{\sum_i W_i c_{ji}^{\alpha} \exp(\beta c_{ji})} \right] . \tag{7.69}$$

The gradient is in fact a vector of partial derivatives. This, of course, means that equation (7.66) for ∇L will be changed, as we now require partial derivatives of L both with respect to α and with respect to β. To simplify what follows, it will be convenient to let

$$B_j = \sum_{i'} c_{ji'}^{\alpha} \exp(\beta c_{ji'}) . \tag{7.70}$$

Then, noting that this B_j is not the balancing factor of, say, equation (7.17), we obtain

$$\hat{N}_i = \sum_j E_j [c_{ji}^{\alpha} \exp(\beta c_{ji}) B_j^{-1}] . \tag{7.71}$$

Now, we begin by taking derivatives, and first recall that

$$\frac{\partial(c_{ji}^{\alpha})}{\partial\alpha} = c_{ji}^{\alpha} \ln c_{ji} , \tag{7.72}$$

and that

$$\frac{\partial[\exp(\beta c_{ji})]}{\partial\beta} = c_{ji} \exp(\beta c_{ji}) . \tag{7.73}$$

Now, by the chain rule for derivatives, we obtain

$$\frac{\partial\hat{N}_i}{\partial\alpha} = \sum_j E_j \left\{ c_{ji}^{\alpha} \exp(\beta c_{ji}) \frac{\partial B^{-1}}{\partial\alpha} + B_j^{-1} \frac{\partial[c_{ji}^{\alpha} \exp(\beta c_{ji})]}{\partial\alpha} \right\} . \tag{7.74}$$

We continue to expand and to take derivatives

$$\frac{\partial\hat{N}_i}{\partial\alpha} = \sum_j E_j \left\{ -B_j^{-2} c_{ji}^{\alpha} \exp(\beta c_{ji}) \sum_{i'} [c_{ji'}^{\alpha} \exp(\beta c_{ji'}) \ln c_{ji'}] \right.$$
$$\left. + [c_{ji}^{\alpha} \exp(\beta c_{ji}) \ln c_{ji}] B_j^{-1} \right\} \tag{7.75}$$

$$= \sum_j E_j c_{ji}^{\alpha} \exp(\beta c_{ji}) B_j^{-1} \left\{ \ln c_{ji} - \sum_{i'} [c_{ji'}^{\alpha} \exp(\beta c_{ji'}) \ln c_{ji'}] B_j^{-1} \right\} \tag{7.76}$$

$$= \sum_j T_{ji} \left\{ \ln c_{ji} - \sum_{i'} [c_{ji'}^{\alpha} \exp(\beta c_{ji'}) \ln c_{ji'}] B_j^{-1} \right\} . \tag{7.77}$$

Next, the partial derivative with respect to β may be derived:

$$\frac{\partial \hat{N}_i}{\partial \beta} = \sum_j T_{ji}\left[c_{ji} - \sum_{i'} c_{ji'}^{\alpha} \exp(\beta c_{ji'}) c_{ji'} B_j^{-1}\right] \tag{7.78}$$

$$= \sum_j T_{ji}\left[c_{ji} - \sum_{i'} c_{ji'}^{\alpha+1} \exp(\beta c_{ji'}) B_j^{-1}\right] . \tag{7.79}$$

Finally, it is possible to write the partial derivatives of the criterion with respect to each parameter. We recall that

$$\frac{\partial \mathsf{L}}{\partial \alpha} = \frac{\partial \mathsf{L}}{\partial \hat{N}_j} \frac{\partial \hat{N}_j}{\partial \alpha} , \tag{7.80}$$

and that

$$\frac{\partial \mathsf{L}}{\partial \beta} = \frac{\partial \mathsf{L}}{\partial \hat{N}_i} \frac{\partial \hat{N}_i}{\partial \beta} . \tag{7.81}$$

Then, by substitution from equations (7.65), (7.77), and (7.80), we obtain

$$\frac{\partial \mathsf{L}}{\partial \alpha} = \sum_i \frac{N_i}{\hat{N}_i} \frac{\partial \hat{N}_i}{\partial \alpha} , \tag{7.82}$$

$$= \sum_i \frac{N_i}{\hat{N}_i} \left[\left[\sum_j T_{ji}\left\{\ln c_{ij} - \sum_{i'} [c_{ji'}^{\alpha} \exp(\beta c_{ji'}) \ln c_{ji'}] B_j^{-1}\right\}\right]\right] ; \tag{7.83}$$

and, by substitution from equations (7.65), (7.79), and (7.81),

$$\frac{\partial \mathsf{L}}{\partial \beta} = \sum_i \frac{N_i}{\hat{N}_i} \frac{\partial \hat{N}_i}{\partial \beta} \tag{7.84}$$

$$= \sum_i \frac{N_i}{\hat{N}_i} \left[\left[\sum_j T_{ji}\left\{c_{ji} - \sum_{i'} [c_{ji'}^{\alpha+1} \exp(\beta c_{ji'})] B_j^{-1}\right\}\right]\right] . \tag{7.85}$$

Before discussing the use of these partial derivatives in gradient search, it will be informative to do some explorations of the shape of the likelihood function surface, with the Tanner function in place of the simple declining exponential function. First, using the same nine zone problem from before, we assign the values $\alpha = 2{\cdot}0$, $\beta = -1{\cdot}5$ and calculate a new set of N_i which will fit the model perfectly at the given parameter values. Next, it will be possible to plot the values of the log-likelihood function against values of the parameters to get an impression of the shape of the surface. The process is complicated by the need now to attempt the portrayal of a three-dimensional surface.

Figure 7.8(a) is a plot of L versus β for specified values of α. Here, a series of relatively sharp peaks may be seen whose optima are clearly shifted with respect to β by changes in the value of α. That each curve is rather like the next suggests that the general shape of the L surface is a relatively smooth ridge. Figure 7.8(b) is a plot of L versus α for specified values of β. This plot is somewhat less clear than the previous one.

Again, there is a series of peaks, this time much less sharp and greatly shifted with respect to α by changes in β. Thus the graph shows only the left-hand portion of the peaks where $\beta = -3{\cdot}0$, $-2{\cdot}5$, $-2{\cdot}0$, and $-1{\cdot}5$, whereas only the right-hand portions are shown where $\beta = -1{\cdot}0$, $-0{\cdot}5$, and 0. While this confuses the picture a bit, it is quite clear that the values of α and β are interrelated. Additionally, it seems that within a moderate range of the perfect-fit parameters, although variation either in α or β will result in a variation in the other with respect to the function's optimum, the actual value of the optimum (in terms of L) will not be very different.

All this may be made more understandable by reference to figure 7.9, where a contour plot of the $L(\alpha, \beta)$ surface is given. Here the curves shown in figure 7.8(a) are cross-sectional slices through the surface at fixed values of β. Thus, for example, the curve labelled $\alpha = 2{\cdot}5$ in figure 7.8(a) corresponds to a slice through the figure 7.9 surface parallel

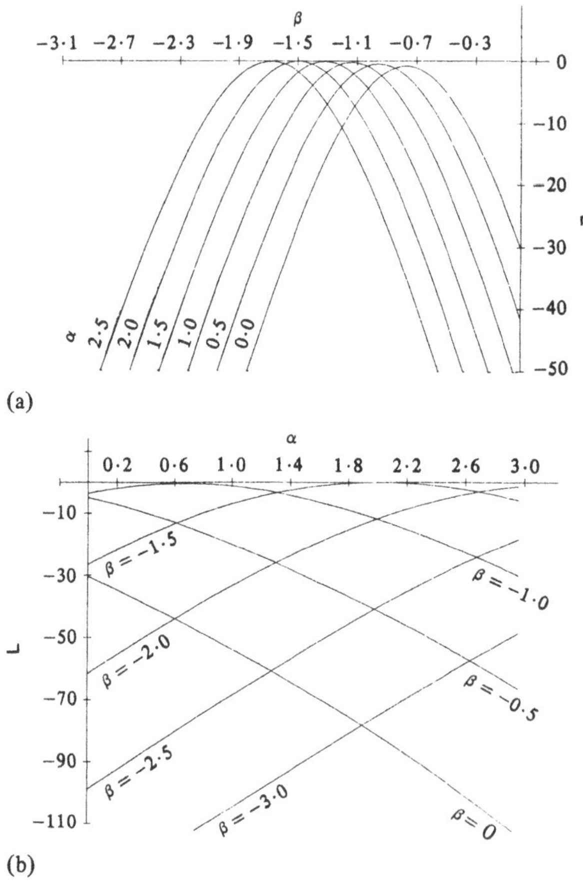

Figure 7.8. Nine zone test problem: plots of L versus α and β. (a) $L(\beta)$ for various values of α; (b) $L(\alpha)$ for various values of β.

to the β axis; the curve labelled $\beta = -1 \cdot 0$ in figure 7.8(b) corresponds to a slice through the figure 7.9 surface parallel to the α axis. Thus it is readily seen that the response surface, in the case of the Tanner function, forms a rather long and narrow ridge which slopes very gently along its long axis, and rather sharply across its short axis. From the shape of this surface, the considerable relationship between the two parameters of the Tanner function is clearly demonstrated. This represents, of course, a form of collinearity between these two parameters. From one given value of α to another, the likelihood maximizing value of β changes substantially. Although the change is less in α from one given value of β to another, it is still significant. From this, two conclusions should be noted before proceeding. First, any calibration of a model using the Tanner function must involve simultaneous estimation of α and β. Second, a simultaneous search technique such as gradient search will be more efficient than a sequential search procedure in finding the optimal values of α and β.

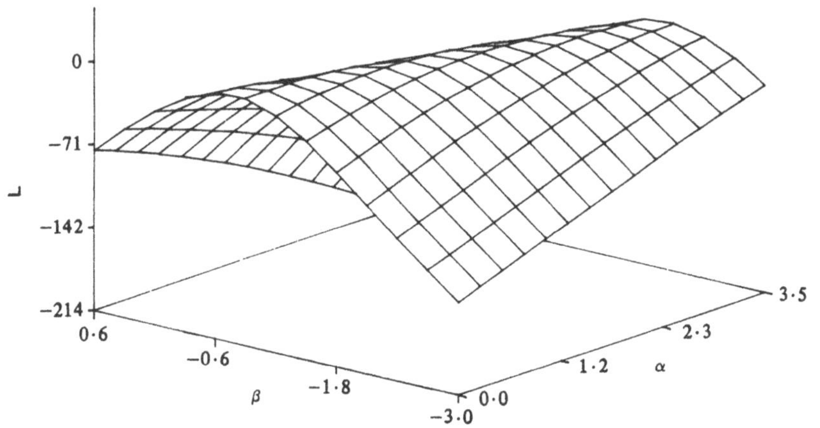

Figure 7.9. Likelihood response surface: nine zone, two parameter sample problem.

7.6.3 Three parameter case

In the prior discussion, the attractiveness measure W_i has been used with $1 \cdot 0$ as an implicit exponent. The development of the residence-location model earlier in this chapter shows that it may be necessary to consider the exponent of W_i as a parameter to be estimated empirically. Furthermore, W_i itself will, in any actual application (data permitting), be a multivariate function, with several parameters to be estimated [see, for example, equation (7.24)]. The question to be asked here is to what extent will the introduction of additional parameters alter the shape of the criterion surface.

To investigate this question, the equation for estimating zonal numbers of residents was altered by the addition of a parametric exponent to the

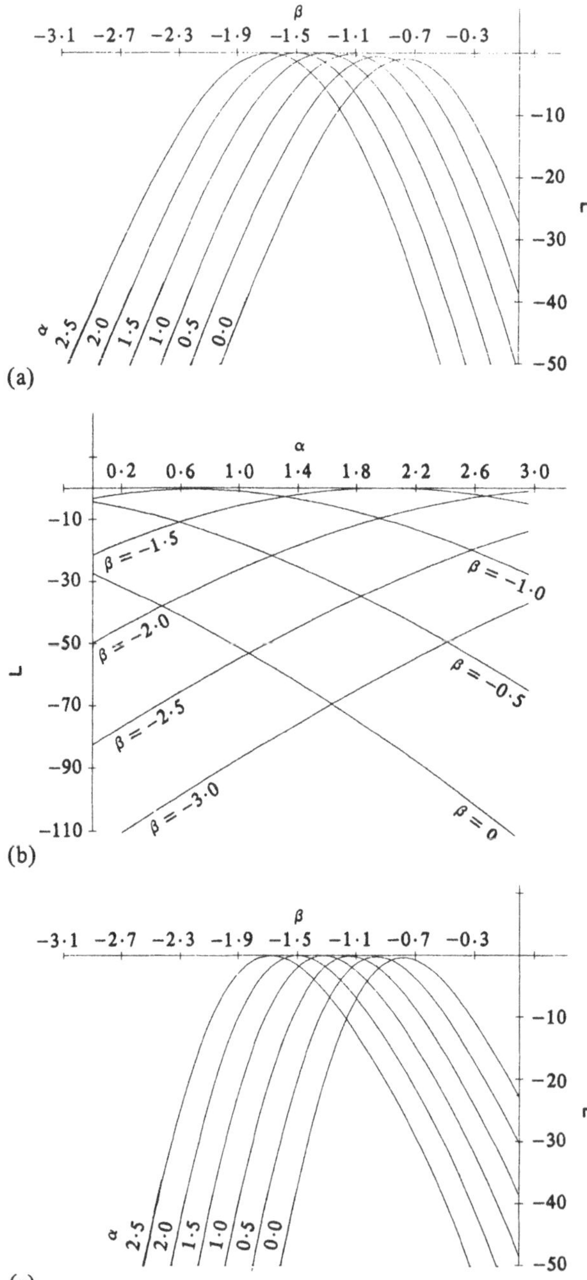

Figure 7.10. Likelihood plots $L(\alpha, \beta)$ for the nine zone, three parameter test problem. (a) $L(\beta)$ for various α at $\delta = 0{\cdot}5$; (b) as (a), but $L(\alpha)$ for various β; (c), (d), as (a), (b), for $\delta = 3{\cdot}0$; (e), (f) as (a), (b), for $\delta = -3{\cdot}0$.

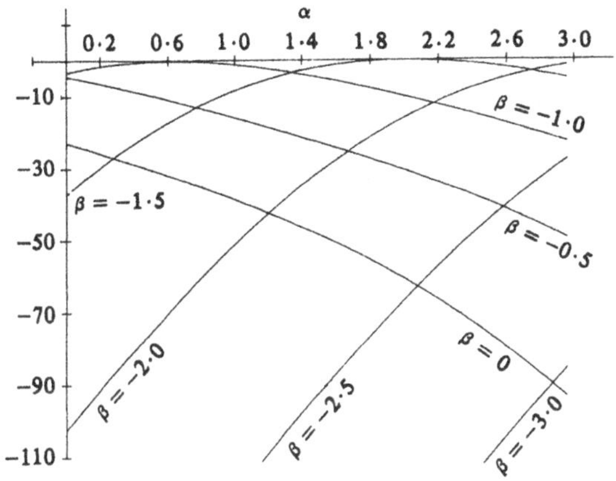

(d)

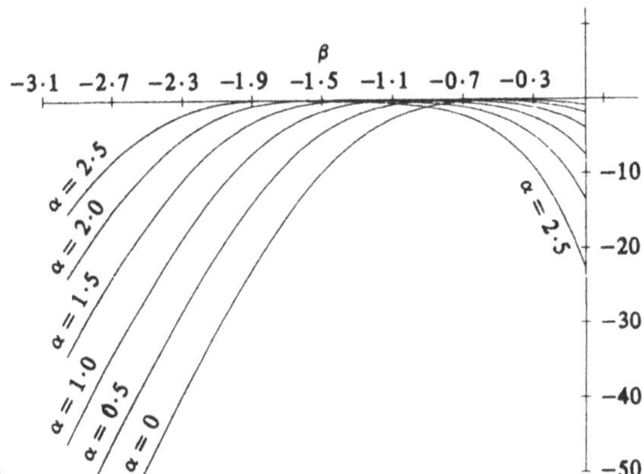

(e)

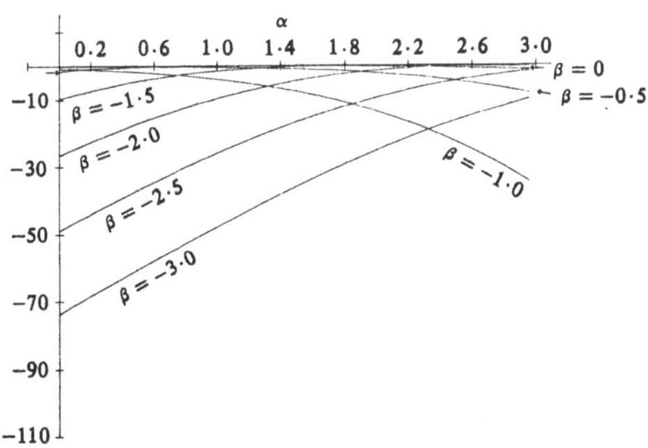

(f)

Figure 7.10 (continued)

attractiveness measure. Thus equation (7.69) becomes

$$\hat{N}_i = \sum_j E_j \left[\frac{W_i^\delta c_{ji}^\alpha \exp(\beta c_{ji})}{\sum_i W_i^\delta c_{ji}^\alpha \exp(\beta c_{ji})} \right] . \tag{7.86}$$

In adding the additional parameter δ to the model it becomes necessary to add an additional derivative calculation, to get the partial derivative of L with respect to δ. Without going through all the steps, it is clear that this derivative will be of the same form as the one for L with respect to α in equation (7.78):

$$\frac{\partial \hat{N}_i}{\partial \delta} = \sum_j T_{ji} \left\{ \ln W_j - B_j^{-1} \sum_{i'} [W_{i'}^\delta c_{ji'}^\alpha \exp(\beta c_{ji'}) \ln W_j] \right\} , \tag{7.87}$$

and noting that in this case,

$$B_j^{-1} = \sum_{i'} W_{i'}^\delta c_{ji'}^\alpha \exp(\beta c_{ji'}) . \tag{7.88}$$

Again, we recall that

$$\frac{\partial \mathrm{L}}{\partial \delta} = \frac{\partial \mathrm{L}}{\partial \hat{N}_i} \frac{\partial \hat{N}_i}{\partial \delta} , \tag{7.89}$$

and, substituting from equations (7.65), (7.87), and (7.89), we obtain

$$\frac{\partial \mathrm{L}}{\partial \delta} = \sum_i \frac{N_i}{\hat{N}_i} \left\{ \sum_j T_{ji} \left[\ln W_j - B_j^{-1} \sum_{i'} W_{i'}^\delta c_{ji'}^\alpha \exp(\beta c_{ji'}) \ln W_j \right] \right\} . \tag{7.90}$$

The equations for $\partial \mathrm{L}/\partial\alpha$ and $\partial \mathrm{L}/\partial\beta$ remain the same as for the two-parameter case. It should be clear that if the model were extended to include a multivariate attractiveness measure with more than one parameter, further partial derivative calculations would be necessary, but these would be of the same general form as equation (7.90).

The 'observed' numbers of residents in zone i are then calculated from equation (7.86) with $\delta = 0{\cdot}5$, $3{\cdot}0$, and $-3{\cdot}0$ and with $\alpha = 2{\cdot}0$ and $\beta = -1{\cdot}5$ in all cases. Then a series of plots may be drawn, analogous to figures 7.8(a) and 7.8(b). In figure 7.10(a) the $\mathrm{L}(\beta)$ curves for $\delta = 0{\cdot}5$ and specified values of α is shown. The shapes of the curves in this figure do not differ markedly from those in figure 7.8(a). Similarly, figure 7.10(b) shows $\mathrm{L}(\alpha)$ curves for $\delta = 0{\cdot}5$ and specified values of β. This figure is similar to figure 7.8(b). Similar plots for δ values of $3{\cdot}0$ and $-3{\cdot}0$ are given in figures 7.10(c)–(f). From all these it may be seen that addition of the δ parameter to the model does shift the optimal values of $\hat{\alpha}$ and $\hat{\beta}$, but does not alter the general shapes of the $\mathrm{L}(\alpha)$ and $\mathrm{L}(\beta)$ response surfaces.

The next question is: what shape of $\mathrm{L}(\delta)$ response surface exists for various values of α and β? In figure 7.11(a) the plots are given for L versus δ for a fixed value of $\beta = -1{\cdot}5$ and specified values of α. Here again, the surface appears to be another long sharp ridge. In figure 7.11(b) the plots are given for $\mathrm{L}(\delta)$ for $\alpha = 2{\cdot}0$ and specified values of β. In figure 7.12(a) a contour plot of the L(β, δ) surface is given, and figure 7.12(b) shows the contour plot of the L(α, δ) surface. In both cases the above-mentioned ridges may be clearly seen.

To help clarify these figures, the following may be of use. The line labelled $\alpha = 2\cdot0$ in figure 7.10(a) corresponds to a slice taken through the surface along the dotted line shown in figure 7.12(a). The line labelled $\beta = -1\cdot5$ in figure 7.10(b) corresponds to a slice taken through the surface along the dotted line shown in figure 7.12(b). The line labelled $\alpha = 2\cdot0$ in figure 7.11(a) corresponds to a slice taken through the surface along the dotted line in figure 7.12(a). Finally, the line labelled $\beta = -1\cdot5$ in figure 7.11(b) corresponds to a slice taken through the surface along the dotted line in figure 7.12(b). The arrows by the dashed and dotted lines in figure 7.12 indicate the point of view of the slices as shown in their respective figures. Having done these explorations of the shape of

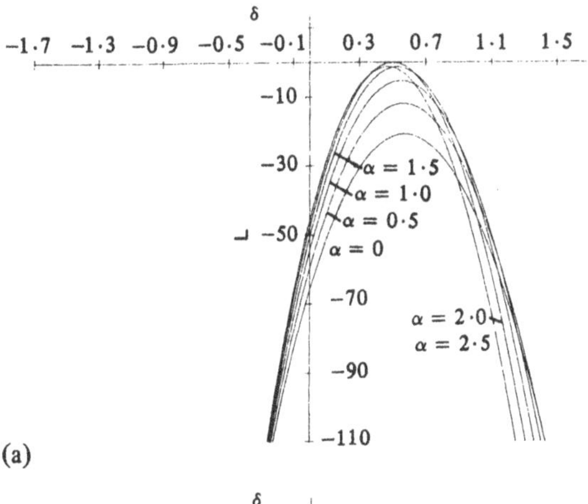

(a)

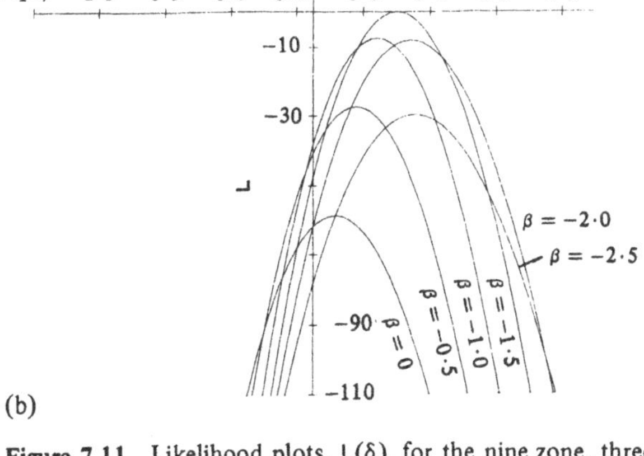

(b)

Figure 7.11. Likelihood plots, L(δ), for the nine zone, three parameter test problem. (a) $\beta = -1\cdot5$, various values of α; (b) $\alpha = 2\cdot0$, various values of β.

the log-likelihood surface, it is possible to return to the question of gradient-search determination of best-fit parameters.

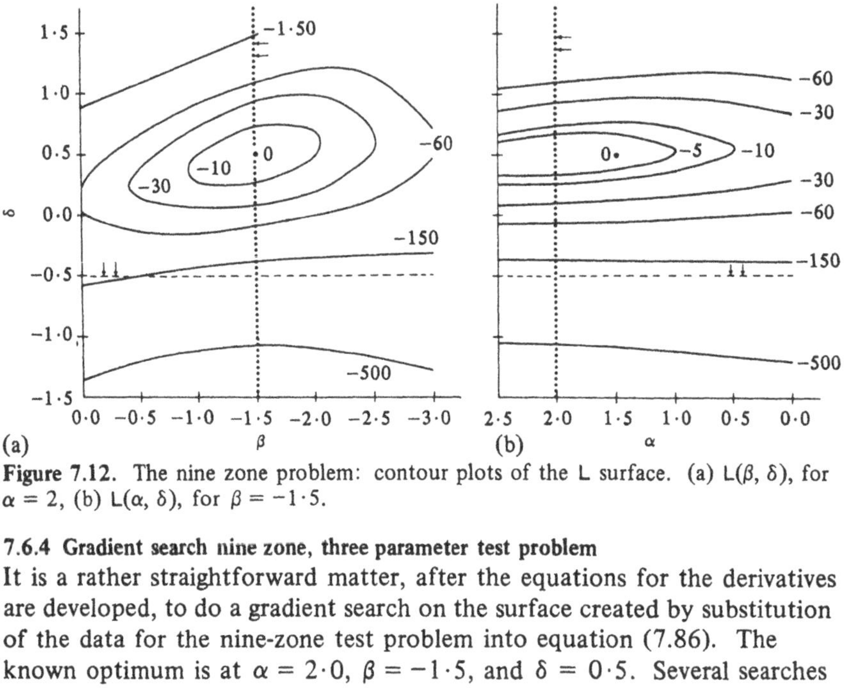

Figure 7.12. The nine zone problem: contour plots of the L surface. (a) $L(\beta, \delta)$, for $\alpha = 2$, (b) $L(\alpha, \delta)$, for $\beta = -1{\cdot}5$.

7.6.4 Gradient search nine zone, three parameter test problem

It is a rather straightforward matter, after the equations for the derivatives are developed, to do a gradient search on the surface created by substitution of the data for the nine-zone test problem into equation (7.86). The known optimum is at $\alpha = 2{\cdot}0$, $\beta = -1{\cdot}5$, and $\delta = 0{\cdot}5$. Several searches were run, each from a different starting point. Searches initiated at different starting points took different numbers of steps to reach the optimum. All searches reached the same optimum. One sample search is shown in table 7.6 for a starting point $\alpha = 2{\cdot}5$, $\beta = -3{\cdot}0$, and $\delta = -1{\cdot}5$.

Table 7.6. Nine zone, three parameter test problem: sample values in gradient-search process.

Iteration	L	α	β	δ
1	−671·4975	2·5000	−3·0000	−1·5000
2	−10·4296	2·8776	−2·0269	0·3264
3	−0·4113	2·9639	−1·8154	0·4836
4	−0·1367	2·9351	−1·8688	0·5242
5	−0·1330	2·9220	−1·8604	0·5248
6	−0·1308	2·9116	−1·8595	0·5225
7	−0·1289	2·9094	−1·8583	0·5236
8	−0·1269	2·8975	−1·8542	0·5213
9	−0·1042	2·8159	−1·8210	0·5217
10	−0·0009	2·0125	−1·5025	0·4993

The parameters could have been pushed closer to their known optimal values by further iterations of the search procedure, but the ten iterations shown in the table suffice for illustration.

Here it is again important to comment on the bogus calibration problem. In this three-parameter form of the model, the calibration procedure could conceivably lead to bogus calibrations. This could occur if the W_i in equation (7.86) was a linear or logarithmic transform of N_i, the observed number of employed residents. In practice this has never been the case in any of the work with the models described here. Furthermore, when using the Tanner function, it would be required that both α and β become 0·0 to have a bogus calibration. This is, of course, a most unlikely result given the shape of the $L(\alpha, \beta)$ response surface as described above. A suggested research strategy might be to calculate the simple correlation between N_i and any proposed form of W_i (or its component factors). If this correlation should turn out to be very high then the researcher will be warned as to the possibility of a bogus calibration result.

7.6.5 Summary

The following conclusions should again be noted. First, in the absence of spatial interaction data it is difficult (if not impossible) to calibrate singly constrained spatial interaction models by methods other than those described here. Second, the procedure suggested for use in such cases involves selecting a criterion function and then using the equations for the partial derivative(s) of the criterion with respect to the parameter(s) to guide a search for the best-fit parameter values. With regard to the model form itself, the investigation was begun in section 7.5 by using R^2 as the criterion and a three zone, one parameter sample problem. It was found that in this case differences in the data set produced substantial differences in the shape of the criterion and gradient functions. The investigation then shifted to a nine zone, one parameter sample problem. It was found that the differences in data sets were reduced. The R^2 criterion was found to be very flat in the region of its optimum; this suggested the desirability of finding a different criterion. The procedure was repeated for a twenty-five zone, one parameter problem, and similar results were obtained. Subsequently, the three, nine, and twenty-five zone, one parameter problems were rerun with the use of a likelihood criterion. At the nine zone and twenty-five zone levels, the likelihood criterion showed a much sharper peak near the optimum than did the R^2 criterion. The likelihood criterion was, therefore, adopted for further investigation, and as both the nine zone and twenty-five zone problems had yielded similar results, for the sake of economy the remainder of the investigation was performed at the nine zone level. A nine zone, two parameter sample problem was developed, in which strong correlation between the two parameters was found. Nonetheless, the surface was plotted in several ways and found to be unimodal.

Finally, a nine zone, three parameter sample problem was developed. Correlations were found to exist between all three parameters, and the surface was again plotted in several ways. It appeared again that the likelihood surface was unimodal, although searches over this surface would probably be complicated by the existence of ridges in the surface. By way of illustration, a sample 'trace' of a gradient search over the nine zone, three parameter likelihood surface was presented. It was found that the number of search iterations could vary considerably, depending upon the choice of starting points. Moreover, given the interrelatedness of the parameters, it became quite obvious that it would always be necessary to estimate their values simultaneously rather than (as was often the case in past practice with land-use models) individually.

Having developed the equation forms for the partial derivatives and thus the gradient, the possibility of gradient-search calibration of residence-location models becomes an actuality. There are several remaining problems of calibration which will be discussed in the final section of this chapter. In chapter 8, the results from extensive calibration tests of residence-location models of this type will be presented, along with the results of sensitivity tests of model response to changes in variable inputs.

7.7 Further issues in calibration

As mentioned above, the next chapter will present the results of calibration attempts for residence-location models for more than two dozen cities. One problem which arises is that of comparing the goodness of fit of these various results. The use of R^2 to compare results is helpful, but as R^2 is not the calibration criterion used, its use for comparison is inadequate. However, the likelihood measure used as the calibration criterion, equation (7.63), is not appropriate either, as its value is a function of the data set used. Thus any comparison of the likelihood criterion per se on different data sets is meaningless. Thus it becomes necessary to develop some goodness-of-fit measure relating to the likelihood criterion used in calibration.

Following the general meaning of the R^2 measure as a normalized sum of squared deviations, it is possible to develop a likelihood-related best/worst achievement ratio. The notion of this ratio is to measure the improvement that the best-fit parameters, according to a likelihood criterion, provide above some base level 'worst-fit' parameters. Although there are numerous concepts of worst fit, one which is computationally attractive is the uniform distribution or equiprobable location fit. In such a case, all zones are equally probable locations for residents and thus the residents of a region are distributed equally to all zones. The best/worst achievement ratio is merely a measure of the extent to which the best-fit parameters are an improvement over a uniform distribution.

In a sense this measure is analogous to R^2. First, recall that the expression for R^2 is

$$R^2 = 1\cdot 0 - \left[\frac{\sum_i (\hat{X}_i - X_i)^2}{\sum_i (X_i - \bar{X})^2}\right] , \tag{7.91}$$

where X_i is the observed value of some attribute X in zone i, $\hat{X}_i$ the estimated value of X in zone i, and $\bar{X}$ the mean of X_i. If $\bar{X}$ were used for all values of $\hat{X}_i$, then the fractional part of equation (7.91) becomes $1\cdot 0$, and $R^2 = 0$. Thus the use of $\bar{X}$ as the uniform or equiprobable estimate of $\hat{X}_i$ represents a worst reasonable case, a form of a Bernoulli equiprobable-outcomes philosophy for estimating X_i.

Consider then the measure of likelihood used to calibrate these models, as given in equation (7.63) and repeated here for convenience:

$$\mathsf{L} = \sum_i X_i \ln[\mathrm{p}(\hat{X}_i)] , \tag{7.92}$$

where $\mathrm{p}(\hat{X}_i)$ is the probability of observing $\hat{X}_i$ in zone i.

If the equiprobable assumption is made, then $\bar{X}$ is the estimated value of X_i for each zone, and $\mathrm{p}(\bar{X}_i)$ the probability of observing $\bar{X}$ in zone i, where

$$\bar{X} = \sum_i X_i/N . \tag{7.93}$$

Next, consider this worst reasonable estimate to yield a 'worst' likelihood value of L_w:

$$\mathsf{L}_\mathrm{w} = \sum_i X_i \ln[\mathrm{p}(\bar{X})] . \tag{7.94}$$

The best possible estimate of X_i is X_i, so the 'best' likelihood value L_b may be defined as

$$\mathsf{L}_\mathrm{b} = \sum_i X_i \ln[\mathrm{p}(X_i)] . \tag{7.95}$$

If we take these quantities together in a form analogous to R^2, it is possible to define the best/worst achievement ratio, φ, as follows:

$$\varphi = 1\cdot 0 - \frac{\mathsf{L} - \mathsf{L}_\mathrm{b}}{\mathsf{L}_\mathrm{w}} . \tag{7.96}$$

This ratio has the property that for a perfect-fit model, $\mathsf{L} = \mathsf{L}_\mathrm{b}$, and therefore $\varphi = 1$. For something less than a perfect fit, as all $\mathsf{L} \leqslant 0$, the absolute value of L will be greater than the absolute value of L_b, but less than the absolute value of L_w. Thus the value of the fractional term will be less than one, but positive. Consequently the value of φ will fall between zero and unity, as do R^2 values. A slight difficulty is that the value of φ at the point $\mathsf{L} = \mathsf{L}_\mathrm{w}$ will be near but not equal to zero.

However, experience has shown that $|L_w| \gg |L_b|$ in most cases so that for all intents and purposes the range of the φ ratio is indeed from zero to one. Now, as a ratio measure, φ is independent of the data base, and thus comparisons may be made from one calibration result to another.

The greatest problems encountered in application of the calibration procedure all relate to the lack of knowledge of the statistical distributions and properties of the parameters estimated by these procedures. This prevents a proper statistical comparison of the empirically estimated parameters; it also prevents the evaluation of statistical significance of parameters. Furthermore, it makes difficult the decision as to whether the search procedure should be continued for additional iterations as the optimum is approached. All of these questions are, however, the subject of ongoing research.

Finally, despite the fact that actual calibration and use of a model will always require artful use of the procedures, they are relatively straightforward in application. This means that more than two decades after the initial attempts at developing residence-location models, there is, as shall be amply demonstrated in the next chapter, a reasonable, robust, and reliable model available to do the job.

8

Verification tests of residence-location models

8.1 Introduction

The advancement of knowledge in the sciences typically follows a continuing cycle of theory development and theory verification. Frequently these cycles operate within a general sequence of phenomenon description and phenomenon explanation. Each new discovery, whether descriptive or explanatory, suggests the need for more of both. A new development of theory suggests the need for empirical verification, while pointing out the directions for further theoretical endeavors. Similarly, new empirical discoveries suggest needs for further theoretical extensions, while also suggesting further empirical examinations. This cyclical behavior can be seen, to varying degrees, in all branches of science.

One of the principal tools of science is the controlled experiment used for purposes of theory verification. This technique is, however, not available to many branches of the social sciences in general, nor to investigators of urban transportation and location phenomena in particular. Thus, rather than having, say, half a dozen identical Londons, in each of which certain variables might be adjusted in order to observe the consequences, verification of hypothesized interactions between land use and transport depends on passive observation and analysis of data. Furthermore, these phenomena, involving the behaviors of hundreds of thousands or millions of individuals, inevitably raise conflicts between data availability, data acquisition and management, and the availability of the time and resources necessary for data processing and interpretation. Thus there often exists a noticeable gap between transportation and location theory, and the empirical evidence necessary to its support[13].

The results described in this chapter are a move towards empirical verification of the residence-location model developed in chapter 7. Two forms of verification are attempted. The first type of verification is in terms of fitting the equation structure of the model to various cross-sectional data sets. Although, as discussed in the previous chapter, this is not as satisfactory as if there were multiple-time-point data available for fitting dynamic model formulations, the results are nonetheless quite positive. The second type of verification is in terms of sensitivity tests—that is, tests of the response of the model to variations in input data. This is done for the model as calibrated for one of the data sets which will be described in the discussions of cross-sectional calibration.

8.2 Verification from model calibrations: complete data sets

The models which are referred to generically as spatial interaction models have been extensively discussed in chapter 7. Briefly, the type of model

(13) The situation is not peculiar to this field alone. See, for example, Leontief's (1971) exhortation to economists and econometricians.

whose calibration results are discussed here is a singly constrained model used for estimating the location of residence, and has the form given in equation (7.26). The model is (as in earlier chapters) referred to as the disaggregated residential allocation model (DRAM). If we set aside, for the moment, the exact functional forms for the trip function of the model and for its attractiveness measure, the general assertion of equation (7.26) is quite clear. The location of residents in a particular zone (subarea) of an urban area is directly proportional to the residential attractiveness of the zone and to its accessibility to places of employment. This assertion has been empirically verified time and time again. It may be considered a valid macrodescriptive theory of urban residential spatial patterns. The difficulty with this descriptive theory is that it does not provide an explanation of why urban residential spatial patterns can be so described. Even with the use of multivariate attractiveness measures and multi-parameter trip functions, there remains a certain lack of connection between macrodescriptive theory and microbehavioral (explanatory) theory. Nearly ten years ago, Senior (1974, page 389) wrote:

> "The problem of establishing *direct* links, and hence consistency, between microlevel theory and the intended operational representation of it in a macrolevel model has remained remarkably resistant to any general analytical solution".

Since then there has been progress: some of this is described by Los (1979), alternative hypotheses are presented by Smith (1977), and some empirical work has been done by Anas (1979b) and by others, but the problems are far from being resolved.

In the face of this general situation, the extensive calibration experiments whose results are presented here were undertaken with one principal goal in mind. This was to demonstrate the descriptive validity of this class of model for urban areas at widely separated points in time and space. The results presented here clearly accomplish this aim. There are two principal implications of this accomplishment. First, from a practical point of view, models of this type can now be put to practical use for a wide range of metropolitan policy analyses. Second, these results provide a firm empirical foundation for further research into the causes and explanations of these seemingly universally true urban spatial patterns.

By way of review, note first that DRAM uses a modified gamma function for the trip function [equation (7.23)]:

$$f(c_{ji}) = c_{ji}^{\alpha^n} \exp(\beta^n c_{ji}) \ , \tag{8.1}$$

where α^n and β^n are empirically derived parameters for each class of tripmaker, n [14]. Second, in DRAM (data permitting) a multivariate

[14] Note that in most of the DRAM analyses (where appropriate data were available) the afternoon travel peak was used for the cost data. As this is the work-to-home trip, the model is usually formulated with a c_{ji} term rather than the more standard c_{ij}.

attractiveness measure is used [from equation (7.25)]:

$$W_i^n = (L_i^{\mathrm{v}})^{q^n}(\tilde{x}_i)^{r^n}(L_i^{\mathrm{r}})^{s^n}(\tilde{N}_i^1)^{b_1^n}(\tilde{N}_i^2)^{b_2^n}(\tilde{N}_i^3)^{b_3^n}(\tilde{N}_i^4)^{b_4^n} ; \qquad (8.2)$$

here,

L_i^{v} is the vacant, buildable, land in zone i;
$\tilde{x}_i$ is one plus the proportion of buildable, land in zone i which has already been built upon;
L_i^{r} is the residential land in zone i;
$\tilde{N}_i^1$ is one plus the proportion of employed residents of zone i who are in the lowest income group, 1;
$\tilde{N}_i^2$ is one plus the proportion of employed residents of zone i who are in the lower-middle income group, 2;
$\tilde{N}_i^3$, $\tilde{N}_i^4$ are similar quantities for the upper-middle and upper income groups, 3 and 4;
q, r, s, b are empirically derived parameters superscripted for each locating group n ($n = 1$–4).

Thus in estimating the parameters of DRAM it is necessary to estimate as many as nine parameters. As the model directly forecasts the location of employed residents in each of the four income groups, it is therefore necessary to estimate four such sets of parameters. The procedures used in producing these estimates were discussed in chapter 7, and involve using a gradient-search technique with a maximum likelihood criterion.

It is appropriate next to discuss expectations as to the signs of these parameters. The trip function, equation (8.1), may be considered as the product of two factors. The first factor, c_{ji}^{α}, is a measure of residential location opportunities encountered with increasing distance travelled from the workplace in zone j. The second factor, $\exp(\beta c_{ji})$, is a measure of people's willingness to make increasingly 'long' (in time or cost) trips between work and home. Thus it is to be expected that in most cities α will be positive and β will be negative. In prose, the greater the distance travelled from place of work, the more residential location opportunities encountered, and the longer the work trip, the less willing will people be to make it. Furthermore, one might expect higher income groups to be more willing (and/or able to afford) to make long trips than are lower income groups. Although it is expected in most cases to observe $\alpha > 0$, and $\beta < 0$, it is possible to obtain other combinations. In the empirical work described in this chapter, the overwhelming number of cases (urban areas) yielded the expected $\alpha > 0$, $\beta < 0$. However, a few cases (to be described shortly) emerged where $\alpha < 0$, and $\beta < 0$, and where $\alpha < 0$ and $\beta > 0$ but $|\alpha| \gg |\beta|$. It is worth discussing the effects of these different combinations of signs on the shape of the trip function.

As mentioned in chapter 7, this trip function, also known as the Tanner function (if it is assumed that $\beta < 0$) was originally selected because various combinations of the parameters of the function would yield usefully different shapes of trip frequency distributions. The function, for purposes of these types of model, will only have nonnegative values of c_{ji}, and the following discussion makes that assumption throughout. If we examine the power term of the function first, when $\alpha > 0$, the c_{ji}^{α} portion of the function increases monotonically with increasing values of c_{ji}. When $\alpha < 0$, this portion of the function decreases monotonically with increasing values of c_{ji}, but is asymptotic to the c_{ji} axis. The exponential term may be considered in the same way. When $\beta > 0$, the $\exp(\beta c_{ji})$ portion of the function increases monotonically (from a minimum value of $1{\cdot}0$ when $c_{ji} = 0$), with increasing values of c_{ji}. When $\beta < 0$, this portion of the function decreases monotonically (with a maximum value of $1{\cdot}0$ when $c_{ji} = 0$) and asymptotic to the c_{ji} axis. Thus there are four possible general situations for the function, corresponding to the possible combinations of signs of α and β. These are shown in figure 8.1. Figures 8.1(a) and 8.1(b) show the curves of $\exp(-1{\cdot}5c_{ji})$ in both instances, with those of $c_{ji}^{1{\cdot}5}$ in figure 8.1(a), and of $c_{ji}^{-1{\cdot}5}$ in figure 8.1(b). These are the two most likely sets of signs of α and β. It can now be readily seen that the product $c_{ji}^{\alpha}\exp(\beta c_{ji})$ in figure 8.1(a) will be a somewhat skewed-peaked curve, whereas in figure 8.1(b) the product of the two curves shown will itself be a declining curve similar to the two components.

The situation shown in figure 8.1(c), where $\alpha > 0$ and $\beta > 0$, is most unlikely to be observed and need not be of concern here. Its implication in terms of the model construct is that all tripmakers would make the longest possible work trips; the longer the trip, the greater the benefits to be derived therefrom. Finally, figure 8.1(d) shows the case $\alpha < 0$ and $\beta > 0$. Although this pattern of coefficients is possible, it is likely that the values of β would be quite small. In that case the behavioral implications would be that tripmakers have a slight preference for longer rather than shorter work trips, but that opportunities decline steeply at trip ends. With β near zero, the product of c_{ji}^{α} and $\exp(\beta c_{ji})$ would have very much the same shape as c_{ji}^{α}, but shifted somewhat above the c_{ji} axis on the graph.

To continue the discussion of expected values for the parameters of the model, we note that the first half of the attractiveness measure contains the three 'land' variables. Each of these three variables would be expected to have a positive exponent parameter. As the value of any one of the land variables increased, the attractiveness of the zone to residential locators would also increase. The residential land variable and the vacant land variable are, however, expected to operate together to describe the development potential of the zones and to adjust for variations in size of zone. The variable describing the proportion of buildable land already built on is expected to amplify or attenuate the first two, by virtue of its being a surrogate measure of infrastructure development. The magnitude of the exponent parameters may be expected to range between $0{\cdot}0$ and $2{\cdot}0$, averaging somewhere near $1{\cdot}0$. This would give an upward-sloping curve, but with a parameter value near $1{\cdot}0$, the curve would be almost a straight line.

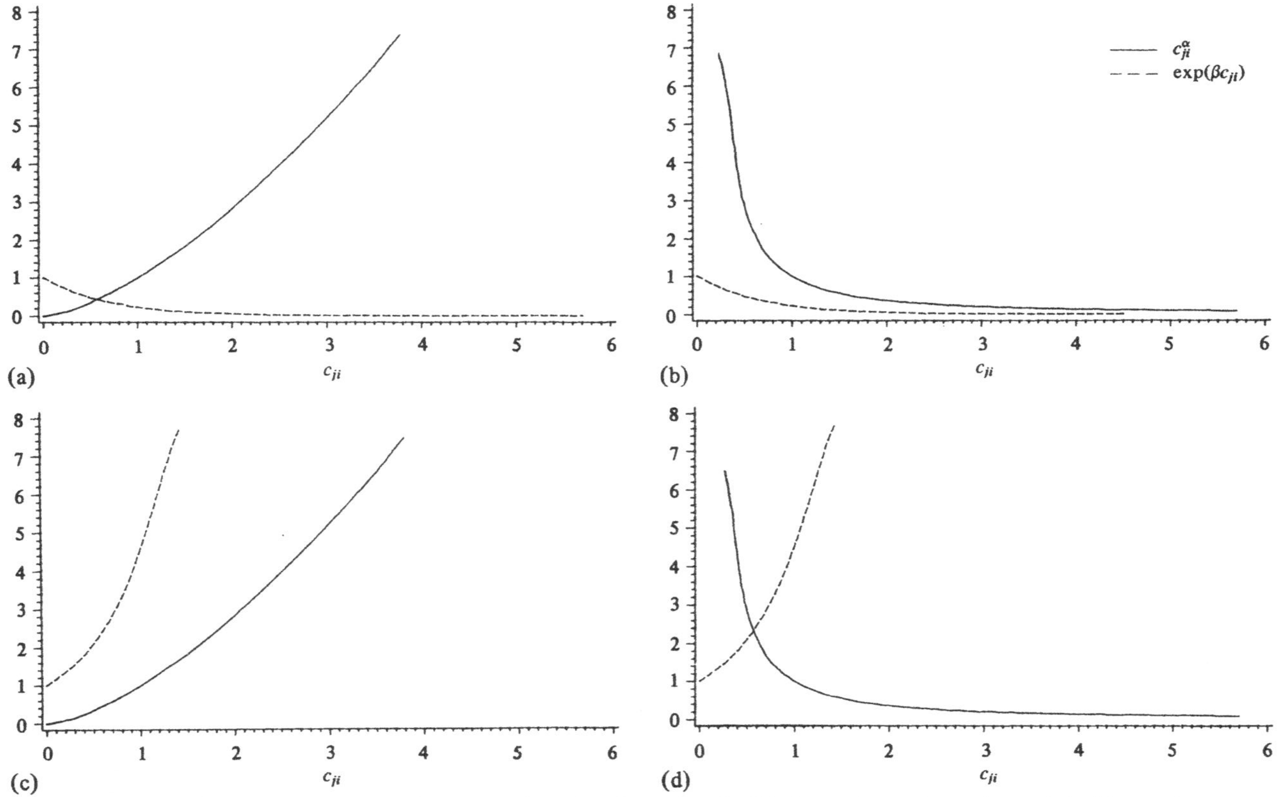

Figure 8.1. Plots of the components of the Tanner function [equation (8.1)], for various values of α and β: (a) $\alpha = 1{\cdot}5$, $\beta = -1{\cdot}5$; (b) $\alpha = -1{\cdot}5$, $\beta = -1{\cdot}5$; (c) $\alpha = 1{\cdot}5$, $\beta = 1{\cdot}5$; (d) $\alpha = -1{\cdot}5$, $\beta = 1{\cdot}5$.

The second half of the attractiveness measure contains the four household-composition variables. These variables describe the income distribution of the households residing in the zone and as such are also surrogate measures of the attractiveness of the zone. Here the parameters for 'like household-to-like household' should be significantly greater than those for each household to any other. The coefficients for similar (that is, next highest or next lowest) household quartiles may be expected to be greater than those for more dissimilar quartiles. A high value of one of these parameters—3·0, say—implies that a residential locator of a particular income group would be most likely to locate in zones having higher than average proportions of that income group. For example, for high-income residential locators, the parameter associated with the percentage of low-income residents may be very small, or even negative. If it was, say, −2·0, then high-income residential locators would be repelled from zones with higher than average proportions of low-income residents.

The issue is in all cases clouded by the interactions of the seven variables in the attractiveness term. Reasonable assertions about the effects of any individual variable on the residence-location behavior of a resident of a particular income group for a given zone may be given; yet, when all the variables operate together, then individual effects may be obfuscated by their interaction effects. This problem did not exist in the earlier simple spatial interaction models where there were no attractiveness variables and, of course, no associated parameters. Similarly this problem does not appear in the early models which had multivariate attractiveness measures with parameters that were not simultaneously estimated.

As the title to this section indicates, the concern here is with data sets which were sufficiently complete to allow calibration of the full DRAM formulation as given in equation (7.26). In such cases all parameters were estimated for each of four employed resident types which, in most cases, were classified directly from income quartiles. All parameters for a given resident type were estimated simultaneously. Thus the results described here, being for each of the four income groups (except in the case of Santiago, Chile, which had three occupation classes in lieu of income classes) for each of fourteen metropolitan areas, represent fifty-five fittings of DRAM to different data sets.

The values for α and β obtained for each income group of employed residents in each metropolitan area are given in table 8.1. The first result to notice is that in all cases, the parameter estimations yield $\alpha > 0$ and $\beta < 0$, the expected result. There are rather substantial differences between these fourteen urban areas, along numerous different dimensions. There are undoubtedly cultural differences between Hong Kong and Phoenix or, say, between Tehran and Atlanta. There are clearly differences in ages of cities, and the extent to which they are pre-automobile or post-automobile in their major development period. There are also differences in their economic bases and their relationships to the regions (and/or nations) they serve. Finally, there are substantial differences in their spatial representations, ranging from 24 zones at the least detailed, to 291 zones for the most detailed. Yet, despite these differences the signs of α and β are, in fifty-four out of fifty-five instances, exactly what were expected. This suggests a rather general descriptive validity for at least this portion of the model formulation. Intercity differences in the α values follow no apparent pattern, nor do the intracity variations in α

Table 8.1. Trip-function parameters for fourteen urban areas with complete data sets.

Area	Number of zones	Year	Parameter	Income group [a]			
				LI	LMI	UMI	UI
Atlanta, GA	183	1970	α	1·89	1·92	1·66	1·62
			β	−0·73	−0·81	−0·74	−0·72
Denver, CO	183	1970	α	2·56	2·37	2·90	2·88
			β	−0·46	−0·39	−0·18	−0·12
Minneapolis, MN	108	1970	α	1·45	2·29	2·99	2·21
			β	−1·09	−1·10	−1·27	−1·10
Philadelphia, PA	173	1970	α	1·47	1·72	1·87	1·81
			β	−1·62	−0·79	−0·47	−1·38
Phoenix, AZ	229	1970	α	1·30	1·42	0·94	1·02
			β	−0·51	−0·51	−0·35	−0·42
San Diego, CA	85	1970	α	2·06	2·10	2·08	2·09
			β	−0·15	−0·13	−0·15	−0·12
San Francisco, CA	291	1965	α	1·05	0·30	0·33	0·87
			β	−0·45	−0·26	−0·25	−0·31
Seattle, WA	227	1970	α	2·24	2·23	2·23	2·20
			β	−0·41	−0·50	−0·53	−0·56
Tulsa, OK	118	1970	α	0·46	0·30	0·36	0·18
			β	−0·41	−0·23	−0·25	−0·70
Vancouver, BC	163	1970	α	0·69	1·10	1·20	1·20
			β	−2·42	−0·99	−0·80	−0·72
Washington DC	110	1970	α	2·55	2·63	3·90	3·18
			β	−1·00	−0·96	−1·41	−1·22
Hong Kong	111	1976	α	0·66	0·41	0·12	−0·11
			β	−0·85	−0·57	−0·40	−0·04
Tehran	24	1971	α	2·70	2·49	2·37	2·55
			β	−0·03	−0·03	−0·02	−0·03
Santiago, Chile [b]	40	1967	α	2·52	2·50	2·49	
			β	−0·10	−0·09	−0·08	

[a] LI, lower income; LMI, lower-middle income; UMI, upper-middle income; UI, upper income.

[b] Occupation groups.

from income group to income group. It was hoped at first that it would be possible to associate variations of these parameters with differences in city sizes, ages, or general patterns of location of the different income groups. The intracity, income-group-to-income-group variations in β had no readily discernable *inter*city pattern either. The intercity variations did, however, yield one interesting result. The absolute values of β decrease, generally, for US cities from east to west across the continent. One inference to be drawn from this result is that people in western cities are more willing to make work trips of a given length than are people in eastern cities: westerners are likely to do more daily travelling than easterners. This result is certainly consistent with commonly-held opinion. Yet another possible inference to be drawn is that this variation in β relates to the ages of these cities and to their pre-automobile and post-automobile development. Either of these possibilities—and others—would represent an interesting finding if confirmed, and this reinforces the thought that there may be much to be learned by further, more rigorous and detailed investigation of intercity and income-group variation in these parameters, and the relation of this variation to other measures of the characteristics of urban areas.

In table 8.2, the values obtained for q^n, r^n, and s^n (n = 1–4) of equation (8.2)—that is the exponents for vacant buildable land, one plus the percentage of buildable (developable) land built upon (developed), and residential land variables, for each income group—are listed. This table shows somewhat less intercity uniformity of parameters than was found for the trip-function parameters given in table 8.1. The residential land parameter, s, behaved quite consistently for all twelve cities. It was positive in all cases, with all values falling between 0 and 1, and many falling relatively close to 1, as expected. The parameters (r) for the proportion of developable land developed were, with only a few exceptions, also within the anticipated range 0–2, with most of the values falling between 0·5 and 1·5, as expected. The results for the vacant buildable land parameter, q, did not turn out at all according to expectations. Although it was consistently close to zero for all the data sets, it showed both positive and negative signs in different urban areas, and even for different income groups in a given urban area. There are several possible explanations of these results. The most likely explanation is in terms of variable-to-variable interactions and possible multicollinearity effects. In linear multiple regression, such effects would be both more obvious and testable in terms of the statistical significance of coefficients. In these parameter-fitting experiments, such statistics are not so readily computed. Thus it is, for the present, only possible to infer such problems. The point here is that any of these three land-related variables taken alone would probably yield parameters of the proper signs and magnitudes. Indeed, two of the three variables do just that, even when taken all together in the model equation. Thus one possibility is that with two other land variables in the attractiveness measure, the vacant land variable

Table 8.2. Land-variable parameters for twelve urban areas[a] with complete data sets.

Area	Income group[b]	Parameter[c]		
		q	r	s
Atlanta, GA	LI	−0·068	0·093	0·817
	LMI	−0·095	−0·146	0·898
	UMI	−0·049	0·300	0·940
	UI	−0·011	0·421	0·874
Denver, CO	LI	0·056	0·278	0·838
	LMI	0·062	0·301	0·867
	UMI	−0·039	0·217	0·865
	UI	−0·020	0·350	0·778
Minneapolis, MN	LI	−0·068	0·716	0·693
	LMI	−0·096	0·717	0·841
	UMI	−0·130	0·531	0·904
	UI	−0·075	1·000	0·762
Philadelphia, PA	LI	0·022	1·390	0·592
	LMI	0·030	1·432	0·503
	UMI	−0·020	1·533	0·465
	UI	0·079	1·484	0·678
Phoenix, AZ	LI	−0·099	−0·032	0·625
	LMI	−0·089	0·016	0·646
	UMI	−0·069	0·154	0·578
	UI	−0·068	0·167	0·613
San Diego, CA	LI	−0·108	0·383	0·840
	LMI	−0·099	−0·242	0·940
	UMI	−0·065	0·247	0·779
	UI	−0·022	0·228	0·794
San Francisco, CA	LI	0·001	1·240	0·774
	LMI	0·030	1·028	0·829
	UMI	0·041	1·041	0·668
	UI	0·027	1·109	0·630
Seattle, WA	LI	−0·209	0·410	0·785
	LMI	−0·166	0·434	0·921
	UMI	−0·147	0·436	0·941
	UI	−0·175	0·412	0·959
Tulsa, OK	LI	−0·110	0·166	0·567
	LMI	−0·137	0·165	0·533
	UMI	−0·106	0·337	0·627
	UI	−0·039	0·337	0·647
Vancouver, BC	LI	−0·030	0·401	0·568
	LMI	−0·024	0·780	0·908
	UMI	−0·039	0·837	0·827
	UI	−0·078	0·748	0·717
Washington DC	LI	−0·088	data not available	0·495
	LMI	−0·066		0·444
	UMI	−0·123		0·468
	UI	−0·015		0·311
Hong Kong	LI	0·014	1·605	0·743
	LMI	0·098	2·180	0·614
	UMI	0·102	2·009	0·614
	UI	0·057	1·906	0·579

[a] No data available for Tehran or Santiago. [b] See table 8.1. [c] See equation (8.2).

does not contribute significantly to the explanation of variance and thus is not statistically significant.

Another possibility is that the great variation in zone size in these data sets forces the parameters of these land variables to act as adjusters for zone size, as well as indicators of the extent to which each of these variables contributes to zonal attractiveness. One last possibility to consider here is that the model structure, originally intended to be dynamic in form, is misspecified for cross-sectional data. In this specific case, if the dependent variable were defined as 'change in the region's low-income households living in zone *i*', then one would expect a positive relationship to the amount of vacant developable land in the zone in the base year. If, however, the dependent variable was defined as 'proportion of the region's low-income households living in zone *i* in the base year', then there might well be a negative relationship to vacant developable land in the base year. Zones with large quantities of vacant buildable land might well be zones with rather low population levels. These are the sorts of questions which remain to be resolved by future investigations.

In table 8.3, the values obtained for b_1, b_2, b_3, and b_4 in equation (8.2)—that is, the exponents of the four household (employed residents) income groups—are tabulated. From equation (8.2) we recall that the residential attractiveness of a specific zone has two general components, one depending upon existing household composition and the other upon several land measures. The household-composition component is represented for each

Table 8.3. Household-variable parameters for fourteen urban areas with complete data sets.

Area	Income group [a]	Parameter			
		b_1	b_2	b_3	b_4
Atlanta, GA	LI	3·17	1·23	−1·63	−2·57
	LMI	0·96	3·09	0·79	−1·89
	UMI	0·11	0·81	3·68	−0·78
	UI	−0·50	0·31	1·69	2·54
Denver, CO	LI	3·66	1·70	−1·05	−1·58
	LMI	1·12	1·93	0·63	−1·02
	UMI	−0·04	0·69	1·63	−0·06
	UI	−0·25	0·17	0·76	1·13
Minneapolis, MN	LI	3·87	0·59	−2·16	−2·46
	LMI	1·47	2·12	−1·34	−1·61
	UMI	0·03	0·38	1·19	−0·64
	UI	−0·25	−0·64	−0·65	2·74
Philadelphia, PA	LI	3·48	1·53	0·28	−1·02
	LMI	1·76	2·96	0·56	−1·02
	UMI	0·26	0·97	3·05	0·31
	UI	0·35	0·83	1·39	1·65

Table 8.3 (continued)

Area	Income group[a]	Parameter			
		b_1	b_2	b_3	b_4
Phoenix, AZ	LI	3·32	1·54	−1·49	−2·72
	LMI	1·27	3·57	0·75	−1·74
	UMI	1·27	1·05	4·16	0·44
	UI	−0·78	−0·56	0·75	1·24
San Diego, CA	LI	3·22	1·27	−0·86	−1·46
	LMI	1·77	3·04	0·57	−1·52
	UMI	−0·18	0·56	2·57	−0·10
	UI	−0·31	0·49	1·01	2·02
San Francisco, CA	LI	2·22	1·11	−0·60	−1·72
	LMI	0·83	3·00	0·23	−1·22
	UMI	−1·10	1·00	3·05	0·01
	UI	−1·77	−0·54	1·19	2·26
Seattle, WA	LI	1·65	1·13	−0·80	−1·77
	LMI	1·12	2·53	0·98	−0·73
	UMI	−0·18	0·54	2·50	1·15
	UI	−1·38	−0·51	1·48	3·02
Tulsa, OK	LI	3·46	1·60	−0·23	−3·35
	LMI	1·35	3·84	1·10	−1·40
	UMI	−0·20	0·91	3·68	0·99
	UI	−0·91	0·24	1·30	2·43
Vancouver, BC	LI	1·93	2·96	−1·42	−2·38
	LMI	0·50	3·08	−0·55	−2·51
	UMI	−1·20	0·86	2·41	−1·13
	UI	−1·62	−0·48	0·11	2·10
Washington, DC	LI	5·42	2·36	−0·79	−2·08
	LMI	1·77	4·43	0·24	−1·45
	UMI	1·73	3·79	4·85	−0·27
	UI	−1·07	0·38	1·89	2·90
Hong Kong	LI	4·45	4·02	1·59	−3·95
	LMI	3·70	5·84	3·89	−1·95
	UMI	1·61	3·61	6·44	−1·32
	UI	−0·81	2·02	5·34	1·60
Tehran	LI	2·50	1·50	−0·50	−2·00
	LMI	−1·99	9·01	0·99	−1·71
	UMI	0·24	8·00	5·75	−0·09
	UI	−0·91	5·99	3·24	2·33
Santiago, Chile[b]	LI	3·08	0·89	−1·02	
	LMI	1·21	3·21	0·49	
	UMI	−1·86	1·86	3·01	

[a] See table 8.1.

[b] Occupation groups.

of the four income groups, in the form of one plus the proportion of households of that income group in the zone. This set of variables acts as a surrogate for other zone-specific attributes. Many measures of the socioeconomic status of the population of a zone will correlate with the income distribution of that population. So too will housing quality, housing price, educational opportunities (quality of the school system), extent of various public services, and the general attractiveness of the zone both from aesthetic and from environmental viewpoints. It is worth noting here the use of an intentionally 'macro' model structure, as distinguished from, say, the microeconomic approach which would perhaps attempt to quantify and bring into the model structure other more specific zonal-attribute variables such as housing stock and house prices. From the point of view of presently practical forecasting procedures, this 'macro' approach seems to be much the more sensible method. It is more sensible not only in that it avoids the difficulties of collecting the data for these other variables, but even more so in that it avoids the problems of having to update these variables (most probably by the use of another model) as part of a recursive forecasting process.

In table 8.3, the parameters associated with the household-composition variables are arranged to show the importance of each household type as an attractor (column headings) to each household type as a locator (row headings). So in Minneapolis, as an example, for the location of lower-middle income (LMI) households, the attractiveness parameter to lower income (LI) households is 1·47, whereas for upper income (UI) households there is a negative attractiveness parameter of −1·61. In examining the overall results, note that in virtually all the metropolitan areas studied, a household in any given income class is most likely to be located with others in its same income class. There are further regularities to be found in the relationship of each income class to the other income classes. LI locators are most likely to be located, in all cities examined, with LMI households, and most unlikely to be located with UI households. In all cities but Tehran, LMI households are most likely to be located with LI households and they too, are most unlikely to be located with UI households. Upper-middle income (UMI) households are most likely, in the majority of cases, to be located with UMI households and most unlikely to be located with LI households. These regularities are quite in keeping with one's expectations and really offer no surprises, yet it is most gratifying to have this consistency show up in the empirical results from so many different urban data sets.

8.3 Verification from calibrations: limited data sets

In addition to the data sets from which the results discussed above were obtained, there were several data sets containing somewhat less disaggregated information on household incomes and/or land-use categories. None of the data sets discussed in this section contained the information upon which a household disaggregation could be based. Furthermore, for quite a number of these cities the data sets contained no information on the classes of land use within the zones. As it was still desired to explore whether these cities could be calibrated, at least in part, several simplified versions of the DRAM equation structure were used.

These simplified forms of DRAM were obtained simply by removing variables from the attractiveness function. The simplest version of the model, merely used zone size (area) as the attractiveness measure, along with the standard trip function:

$$N_i = \sum_j E_j \left[\frac{L_i^\delta c_{ji}^\alpha \exp(\beta c_{ij})}{\sum_i L_i^\delta c_{ji}^\alpha \exp(\beta c_{ji})} \right] , \tag{8.3}$$

where all variables are as previously defined except L_i, the total area of zone i, and δ, an empirically derived parameter.

The data sets for some of this new group of cities had two additional variables, residential land, and percentage of developable land developed. Thus for these data sets the form of the model used was

$$N_i = \sum_j E_j \left[\frac{(L_i^v)^q (\tilde{x}_i)^r (L^r)^s c_{ji}^\alpha \exp(\beta c_{ji})}{\sum_i (L_i^v)^q (\tilde{x}_i)^r (L^r)^s c_{ji}^\alpha \exp(\beta c_{ji})} \right] , \tag{8.4}$$

where all variables have been previously defined for equation (8.2).

The estimates of α and β obtained with these simplified forms of the model are shown in table 8.4. Here too, as was the case in table 8.1 for the cities with complete data sets, the consistency of the results is quite striking. Taken together with table 8.1, the data sets represent twenty-eight urban areas from different parts of the world. The populations of these

Table 8.4. Trip-function parameters for fourteen urban areas with incomplete data sets.

Area	Number of zones	Year	Parameter	
			α	β
Gosford-Wyong, Australia	22	1971	0·09	−0·03
Melbourne, Australia	41	1970	1·04	−0·06
Natal, Brazil	18	1976	0·93	−0·01
Rio de Janeiro, Brazil	50	1968	1·08	−0·01
Reading, England	130	1962	1·31	−0·06
Tokyo, Japan	25	1970	0·69	−0·01
Mexico City, Mexico	111	1976	−0·47	−4·89
Monclova-Frontera, Mexico	67	1977	2·73	−0·14
Ankara, Turkey	36	1970	0·64	−0·14
Gaziantep, Turkey	53	1970	1·07	−0·004
Izmit, Turkey	23	1968	0·90	−0·03
Colorado Springs, CO	81	1970	1·95	−1·90
Hazleton, PA	90	1960	−0·18	0·15
Ljubljana, Yugoslavia	98	1961	−2·10	−0·003

urban areas range from sixty thousand for Hazleton, PA, to twenty-two million for Tokyo. The cities are of widely varying ages: some of the US cities date from the early to middle nineteenth century, whereas the earliest settlement at Gaziantep can be traced back to 4000 BC. The topographical characteristics of these urban areas are also quite varied. Some of these cities are situated on rather flat plains, others are in very mountainous country. Some have no nearby body of water, others are bisected or trisected by rivers; Hong Kong is partly an island, and the San Francisco metropolitan region is almost the shape of a giant letter O, with the San Francisco Bay in its center. Yet for all this variation there appears to be consistent underlying human behavior which is rather well described by the equations being examined here.

The different data sets had various measures of the land use in their zones. Some cases were limited only to a measure of the total area of the zone. In other cases some measure of 'urbanized' area was available; yet other data sets had all the land-use variables used in the calibrations of the full data sets discussed in the previous section. The variable definitions and estimated parameters for the land variables for these limited data sets are shown in table 8.5. Because of the considerable range of variable

Table 8.5. Land-variable parameters for fourteen urban areas with incomplete data sets.

Area	Land variable(s)	Parameter	Value
Gosford-Wyong	developable	δ	−2·31
Melbourne	developable	δ	0·13
Natal	urbanized	δ	0·48
Rio de Janeiro	urbanized	δ	0·60
Reading	vacant developable	q	1·19
	proportion developable	r	8·71
Tokyo	vacant developable	q	0·04
	proportion urbanized	r	0·50
Mexico City	vacant developable	q	0·02
	proportion developable developed	r	0·90
	residential	s	0·85
Monclova-Frontera	total zonal area	δ	0·24
Ankara	total zonal area	δ	−0·31
Gaziantep	total zonal area	δ_1	0·25
	residential building area	δ_2	0·07
Izmit	total zonal area	δ	0·80
Colorado Springs	vacant developable	q	0·06
	proportion developable developed	r	1·57
	residential	s	0·68
Hazleton	vacant developable	q	−0·09
	proportion developable developed	r	1·00
	residential land	s	0·95
Ljubljana	total zonal area	δ	1·05

definitions here, it is difficult to make any general statement about these results. Some comments on these results will be given in the general discussion of the calibration results in the next section.

8.4 Verification from calibrations: general discussion

Before the implications to be drawn from these calibrations are discussed, a simple set of goodness-of-fit results is provided. As mentioned in chapter 7, the goodness-of-fit criterion used in the calibration procedure is a maximum likelihood criterion. In addition to this criterion, an R^2 value was calculated, for each calibration, between the observed and estimated household locations. This was done according to the standard definition

$$R^2 = 1 - \frac{\sum_i (\hat{N}_i - N_i)^2}{(N_i - \bar{N})^2} , \tag{8.5}$$

where $\hat{N}_i$ is the estimated number of residents of zone i, N_i the observed number of residents, and $\bar{N}$ the mean of N_i.

These goodness-of-fit results for all twenty-eight of the urban areas examined in this series of tests are shown in table 8.6. In general the results are quite good. The data sets were in all cases used 'as is' (that is, unmodified). They were supplied by government agencies and/or professional colleagues. No further processing of data was undertaken beyond conversion to a standard form for use in the calibration programs. The checking for errors and consistency that is normally done in the course of an actual model application for agency use could not be done. The use of knowledge from locally informed persons to modify these raw data files and to 'fine tune' the parameter estimates would certainly

Table 8.6. Goodness-of-fit (R^2) results for twenty-eight urban areas.

Area	R^2 by income group				Area	R^2 for total population
	LI	LMI	UMI	UI		
Atlanta	0·813	0·864	0·922	0·866	Gosford-Wyong	0·574
Denver	0·832	0·728	0·749	0·593	Melbourne	0·641
Minneapolis	0·909	0·873	0·889	0·867	Natal	0·696
Philadelphia	0·762	0·584	0·426	0·693	Rio de Janeiro	0·741
Phoenix	0·721	0·609	0·592	0·654	Reading	0·857
San Diego	0·806	0·830	0·824	0·830	Tokyo	0·844
San Francisco	0·614	0·612	0·693	0·684	Mexico City	0·776
Seattle	0·732	0·837	0·816	0·862	Monclova-Frontera	0·223
Tulsa	0·767	0·600	0·574	0·687	Ankara	0·233
Vancouver	0·795	0·712	0·887	0·847	Gaziantep	0·434
Washington DC	0·772	0·713	0·594	0·442	Izmit	0·712
Hong Kong	0·879	0·875	0·876	0·622	Colorado Springs	0·605
Tehran	0·615	0·625	0·439	0·560	Hazleton	0·868
Santiago	0·808	0·693	0·788	–	Ljubljana	0·899

improve the results. A number of the data sets, especially those for some of the non-US cities, had obvious and serious problems of inconsistency, for which there was no remedy. Under these circumstances, the results shown in table 8.6 are indeed quite good.

It is informative to consider some of these inconsistencies in the data sets, as they point out potential pitfalls for future efforts. One problem that appeared in several cities was that of obviously inconsistent and/or incomplete employment data. In the data for Monclova–Frontera, for example, the employment total was more than twice the total population. As there was very little chance that there was extensive daily in-commutation to the region for work purposes, there was clearly a considerable problem with the data. In the data for Izmit, where the population was 90000, and the estimated total employment was 36000, only about 10000 employees were included in the data set. Similarly, the Gaziantep data showed a total of 17000 employees, with a population of 194000 (which is itself at odds with a census report of 296000 population).

Inadequate data for zone-to-zone travel times and/or costs caused another area of difficulty. Several cities provided no data of this sort. To approximate the travel time/cost data, distances between zone centroids were calculated from maps of the regions. This was the case for virtually all of the non-US cities. For Mexico City, the situation was even more difficult. The population, employment, and land-use data were for a 111 zone system, but the only available travel-impedance data were zone-to-zone travel times for a 15 zone system. A 15–111 zone conversion table was available, but no 111 zone map. Thus each of the small zones (usually six to twelve in each large zone) could be identified only as being in one of the large zones. The travel time between any two small zones was set equal to the travel time within or between the large zone(s) in which the small zones were located. Needless to say this gave a very discontinuous distribution of zone-to-zone travel times for the 111 zone system. Despite all these difficulties (though it is clear that they had their detrimental effects on the results), the general model structure showed a remarkable ability to describe the spatial patterns of residences in these many urban areas. It is the general descriptive ability of this model structure which is the most significant finding. No other urban residence-location model has ever been calibrated to so many different urban data sets. There is clearly a significant gap between the macrodescriptive premises on which this model rests and the microeconomic theories of household location. Yet the descriptive ability of this model structure cannot be denied. It may be that this is a 'reduced form' of a more 'micro' model. It may also be possible that the aggregation problem which separates micromodels of residence-location decisionmaking from macrodescriptive models such as the one tested here is insurmountable. This situation has parallels in the natural sciences. The understanding of microlevel processes (subatomic particles, for example) is important in its own right, and aids in understanding observed macrolevel behavior. Yet, for many practical applications, the macrolevel descriptions of such processes (for example, the gas laws such as Boyle's) are both adequate and appropriate.

Much remains to be learned, even at the macrolevel, about the determinants of urban spatial patterns. One early objection to this class of model was that the descriptive validity would apply only to modern cities in the United States. The results shown here clearly eliminate that objection. Although it is possible that some urban areas can be found where these models will not fit the data, it seems unlikely. Another objection was that such models would be valid only for middle to late twentieth-century cities. Work with this model data set for a middle to late nineteenth-century city rebuts this assertion as well (Putman, 1977b).

Recent applications of these models have contributed new information. Both DRAM and EMPAL were recently calibrated for St Louis and Kansas City, and for T'ai-pei and Kao-hsiung in Taiwan. The results were consistent with those given above. A rather interesting phenomenon did appear in the work on Kansas City, which was later to appear in DRAM calibrations for the Californian regions of San Diego and San Joaquin. It appeared that major redefinitions of vacant land resulted in changes in the sign of α (for the trip-function portion of the model) from positive to negative. This, at first glance, would suggest that the trip distributions would change from a skewed-peaked curve to a monotonically decreasing exponential type curve. Yet, when the resulting trip distributions were plotted they retained their skewed-peaked form even with negative values of α. Apparently the variables (and their associated parameters) in the attractiveness measure combined to preserve this shape of trip distribution. This, of course, raises questions as to the interpretation of some of the calibration results. The general question here is of the interaction between the variables and parameters in the model structure.

Study of the interactions between variables requires refinement of the procedures for estimation of parameters. The criterion function appears to be very flat, with respect to some of the parameters, in the neighborhood of the optimum. Resolution of this problem is necessary to make precise comparisons of best-fit parameters. This also implies problems of multicollinearity. The resolution of the multicollinearity problem will require the examination of the few available techniques for examining the statistical distributions of the parameters of the model and of techniques for evaluating the statistical significance of these parameters. There is also some evidence of significant effects on parameter estimates from changes in zone definition. However, all of these problems exist to one degree or another in any mathematical representation of a probabilistic reality. The calibration-verification results presented here amply justify continued research in this direction.

8.5 Verification from sensitivity tests

Another means by which the validity of a model structure may be tested is by examination of its response to specific changes in its inputs. The purpose of such tests is to determine the extent to which such responses yield results which are in accord with informed expectations. A uniquely valuable facet of work with such models appears when results of tests such as these do not conform to expectations. In such instances, the tracing of the cause of the discrepancy can sometimes lead to new insights into the model structure (although on other occasions the problems turn out only to be errors in the program or data). On some occasions these discrepancies lead (not too often, but more often than the detractors of this approach would like) to useful new insights into the determinants of urban spatial patterns.

The tests described here were done with the DRAM model. The parameters were estimated for a 108 zone data set for the Minneapolis–St Paul metropolitan area. The actual values of the parameters for the area are given in table 8.3. During the normal course of developing and testing a computer model of an equation structure such as DRAM, hundreds of computer runs of the model are made. Once the usual problems of errors in the computer code and the data set are eliminated, each run provides a little more information about the response of the model to various alterations in input. This information is quite distinct from that given by the numerical values of the outputs themselves. Sometimes really useful bits of knowledge about model performances are obtained from computer runs which were, in terms of forecasting, failures. This gradually accumulated information about the way the models behave is begun by doing the kinds of tests whose results are presented here.

Several different groups of tests were done, each involving specific changes in (a) population, (b) employment, (c) land use, or (d) the transportation system (accessibility). Each of these tests involved changing input data to DRAM and observing corresponding changes in outputs. All tests were made with data from the Minneapolis–St Paul metropolitan region, with 1960 as the base year and 1970 as the forecast year. All results (outputs of each test run) were compared to a base run of DRAM. The base run was a forecast made with regionwide total population and employment forecast inputs and with no zone-specific changes of variable(s). The outputs of this run were used as the standard of comparison for all the sensitivity-test runs. Some of these tests were intended to represent the types of possible policy alternatives which could be examined with a model such as this; others are clearly manipulations of the models contrived for the purposes of these sensitivity tests.

8.5.1 Changes in population

8.5.1.1 *Reductions in low-income households*

Tests involving zonal reductions of the number of LI households were made separately for several different zones in the region. One example was for zone 57 in the Minneapolis urban core (see figure 8.2). A highly simplified and quite artificial policy involving the elimination of dilapidated, substandard housing was represented by the removal of the LI households of the zone in the base year. This amounted to 697 LI households, or 80% of the total households in zone 57 in the base year. The response of DRAM to this input change was quite reasonable. Compared with the base run, the number of LI households in zone 57 was reduced by 65% in the forecast year, and there were 42% fewer LMI households. At the same time, the number of UMI households increased by 16%, and UI households showed a 15% increase. The absolute increases in the upper two household classes were, however, rather small. Thus the zone showed a net decrease in the total number of households of 47% compared to the base run. These results are very like those of actual attempts at urban renewal or slum clearance in metropolitan areas. Renewed city-center zones often remain relatively empty for a time, with perhaps some increases (significant percentages but small absolute numbers) in UI and UMI households.

It is worth noting where the displaced households relocate. With 697 LI households removed from zone 57 in the base year, 241 returned to that zone in the forecast year. Of the remainder, 194 relocated in a ring of zones adjacent to zone 57, mostly in those zones already with large numbers of LI households. The remainder relocated throughout the

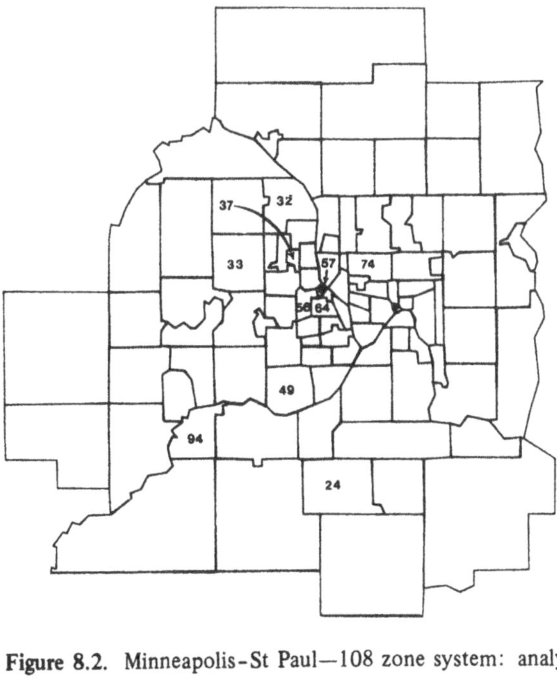

Figure 8.2. Minneapolis-St Paul—108 zone system: analysis zones.

region, with emphasis on the next adjacent ring of zones and in two zones adjacent to the St Paul urban core; in all cases, the relocation was to zones which had relatively large numbers of LI households in the base year.

A second DRAM run was made, attempting to represent the rapid rebuilding of such zones with housing of much higher cost than that which was removed: the deleted LI households were prevented from returning to zone 57. The general pattern of these results was similar to that of the previous test. In all cases there were slightly larger increases in LI households in the zones other than zone 57, to account for those prevented from returning.

Pursuing this line of experiment, a second pair of test runs was made for zone 49. In the base run this zone was a rather rapidly growing suburb of UMI and UI households. In the base year, 384 LI households, 8% of the total households in the zone, were removed for the test run. By the forecast year, there were 928 LI households, making just over 8% of the new total. In the base run the corresponding figures were 1046 and 9%. The proportions of other household types in the test run were 19% LMI, 35% UMI, and 38% UI. In the base run the corresponding figures were 21%, 34%, and 36%. Thus, the elimination of a small low-income enclave in the base year did not prevent thc return of a small proportion of LI households to the zone, but did slightly retard growth in this sector.

The test was then rerun as before, but with the return of LI households precluded (perhaps to simulate substantial increases in housing costs). In this case there was a net decrease of 10% in the population of the zone compared with the base run. There were no LI households, a 23% decline in LMI households, less than a 1% increase in UMI households, and an 11% increase in UI households. The LI households which were prevented from locating in zone 49 all located in the zones falling between suburban zone 49 and the Minneapolis urban core. Apparently the exclusion of LI households from this suburban zone had the effect of forcing those households to locate in the urbanized area immediately adjacent to the core of the city. Furthermore, the absence of LI households from the zone resulted in the zone appearing to be too 'expensive' for LMI households, and this group also declined.

After the responses in an urban core zone and in a suburban zone had been tested, one more DRAM test was made, for zone 56, midway between the core and the suburbs. In this case, 1000 LI households were removed from the base year data. This was a 10% decrease in the LI households, and a 5% decrease in the total households in zone 56. In both the test run and the base run the number of households in the zone increased substantially from the base year to the forecast year. The test run showed fewer LI and LMI households, and more UMI and UI households. From lowest to highest income, the four household types showed growths of 83%, 57%, 26%, and 2% in the test run, and of 103%, 68%, 26%, and a decline of 3% in the base run. Thus again, a decrease in LI households in a zone in the base year retards the growth of the two lower income groups, and slightly accelerates the growth of the two upper income groups. Even though the deleted LI households were not prevented from returning to

Table 8.7. Test results of deleting low-income households at base year from zone 49 of Minneapolis-St Paul system.

Sector	Households allowed to return			Return precluded		
	test run	base run	percentage difference	test run	base run	percentage difference
Households						
LI	928	1046	−11·3	0	1046	−100
LMI	2123	2377	−10·7	1831	2377	−22·9
UMI	3866	3908	−1·1	3915	3908	−0·2
UI	4267	4071	+4·8	4529	4071	+10·1
Total	11184	11402	−1·9	10275	11402	−9·9
Employment						
Basic 1 [a]	3258	3258	0·0	3258	3258	0·0
Basic 2 [b]	667	667	0·0	667	667	0·0
Nonbasic	1981	1983	−0·1	1974	1983	−0·5
Total	5906	5908	0·0	5899	5908	−0·2

[a] Manufacturing.

[b] Transportation, communication, utilities.

Table 8.8. Test results of deleting low-income households at base year from zone 56 of Minneapolis-St Paul system.

Sector	Households allowed to return			Return precluded		
	test run	control run	percentage difference	test run	control run	percentage difference
Households						
LI	8884	9836	−9·7	8121	9836	−17·4
LMI	4994	5342	−6·5	4617	5342	−13·6
UMI	3089	3080	+0·3	3040	3080	−1·3
UI	2852	2771	+2·9	2978	2771	+7·5
Total	30887	30737	+0·5	18756	21029	−10·8
Employment						
Basic 1 [a]	1822	1822	0·0	1822	1822	0·0
Basic 2 [b]	1237	1237	0·0	1237	1237	0·0
Nonbasic	28543	28564	−0·1	28528	28564	−0·1
Total	31602	31623	−0·1	31587	31623	−0·1

[a] Manufacturing.

[b] Transportation, communication, utilities.

the zone, most relocated in adjacent zones slightly closer to the urban core. An interesting question raised, but not examined here, is what the differences were between the zone and the others tested that led to no return of LI households here, but to attempted returns for the other zones. To give the reader a better notion of the numbers involved in these tests, the results for zones 49 and 56 are shown in tables 8.7 and 8.8.

8.5.1.2 *Increases in low-income households*

In the same way that a zonal decrease in LI households may be used to describe crudely and simplistically an effect of slum clearance or urban renewal, a zonal increase in low-income households may be used to describe crudely a public housing project. A series of test runs was undertaken to study this phenomenon. Zone 37, midway between the urban core and the suburbs, was selected for the first of these tests. A total of 1000 LI households were added to this zone in the base year. The DRAM results showed forecast year increases, compared with the base run, of 41% in LI households, 31% in LMI households, and 1% in UMI households. UI households showed a 5% decrease compared with the base run, and zone 37 showed a net increase in households of 11%. Thus this overall increase of 21% on the base year total number of households in the zone yields a significant shift of the household composition towards the lower income households. In the base run, the forecast composition is 18% LI, 22% LMI, 29% UMI, and 30% UI; the corresponding figures for the test run are 23%, 26%, 27%, and 24%. Finally, we note that the households contributing to the increase in zone 37 were drawn from a ring of adjacent zones and that the UI households which left zone 37 dispersed to the ring of adjacent zones.

Two further test runs of DRAM were made for two suburban zones, to explore this phenomenon further. First, zone 32 received an increment of 1000 LI households in the base year, and in the second run the increment was put in zone 33. The results in both these runs were virtually identical, showing increases in LI and LMI households and decreases in UMI and UI households when compared with the base run. Thus the introduction of additional LI households to suburban zones in the base year produced long-term changes in the household composition of the zone, always in the direction of the lower income household groups.

A final test run with an increment of LI households was made for a rural zone. Zone 94 is a rural zone which, in the base run, showed a net decline in population from the base year to the forecast year. The bulk of this decline was in LI households, which dropped 66% from the base year to the forecast year. During the same period the number of UI households in the zone increased by 44%. The addition of 1000 LI households to this zone in the base year only partly altered its household composition in the forecast year. LI households still declined, though by a somewhat smaller percentage, and LMI households increased by 26%, as compared with a decline of 12% in the base run. The number of UMI households was relatively unchanged, but UI households increased by 28% and 4% in the test run and the base run respectively. This appeared to illustrate the classic phenomenon of suburbanizing high-income households pricing low-income households living on the urban fringe out of the market.

To summarize, addition of LI households to a zone in the base year changed forecast household distribution of the zone. The shift was in the direction of increases in LI and LMI households and decreases in UI and UMI households. The extent of the shift depends on the initial total population and the population composition of the zone, and in one rapidly suburbanizing rural zone even the addition of LI households in the base year failed to stem the influx of suburbanizing high-income households.

8.5.1.3 *Increases in upper income households*

A final set of DRAM population tests was made, where UI and/or UMI households were added in the base year. In the first of these tests 1000 UI households were added to zone 74 in the base year. This is a populous zone, well within the urban area, but not in the core. These new households represented a net increase of 17%, and a 40% increase in UI households. The results of this change were that the UI households in zone 74 increased somewhat more in the test run than in the base run, and all other income classes increased somewhat less.

The second of these tests was on zone 94 (a rural zone used above for a low-income increment test), in which an increment of 1000 UI plus 1000 UMI households was added to the zone in the base year. This produced results similar to, but not quite so pronounced as those produced by the UI household increment.

8.5.1.4 *Summary*

In this whole series of tests, DRAM shows responses consistent both with theoretical and with empirical descriptions of urban phenomena. Increases in low-income households in a zone produce decreases in high-income households, decreases in low-income households produce increases in high-income households, and increases in high-income households produce decreases in low-income households. Ripple effects are often observed in the ring of zones adjacent to the zone in which the test was effected. Although such results, confirming expectations, cannot prove the validity of the model structure, they do demonstrate that it is at least consistent with current theory.

8.5.2 Changes in employment or in land availability

A further series of test runs was conducted in a similar fashion, but involving changes in basic or nonbasic employment or land available for development in the base year.

Changing employment, either basic or nonbasic, in a specific zone gave only modest responses in model outputs. This was due primarily to the fact that employment change in a particular zone affects residence location (aside from modest changes in basic or nonbasic land use possibly resulting in changes in available land) as only one element in the calculation of an employment 'potential' surface over all the zones in the region. Though this is a specific artifact of the DRAM equation structure, it is probably a rather reasonable representation of reality. DRAM did exhibit a multiplier effect here, and although the results did not seem unreasonable they were at variance with some expectations. These contrary results led eventually to a revision of the way in which unemployment rates, labor-force participation rates, and employee-per-household rates were treated in the model.

Changes in land availability (simulating a protection policy for a flood-plain and aquifer recharge area) resulted in numerous changes in the distribution of households in the metropolitan area. The diversity of the zones to which the restrictions were applied and the subsequent diversity of responses defy conventional or brief analysis. Further study is still required here, first to identify the specific responses to zonal changes in land availability, and then to reevaluate complex multizone restrictions such as those which were tested here. Subsequently it was also discovered that there was an error in the land-use-accounting section of the program.

8.5.3 Changes in accessibility

8.5.3.1 *Regionwide changes*

Throughout the literature on transportation and urban development, one finds the observation that where transport is readily available and consequently, access is great, development tends to be spread out. Similarly when access is poor, development tends to be more concentrated and, in the case of larger regions, subnucleated. Recent experience throughout the USA amply demonstrates the generality of this phenomenon: virtually every transportation improvement is closely followed by further spread of locating activities. Measures of the ease or difficulty of interaction between activities are central components of virtually all urban spatial models. This interaction phenomenon is an integral part of the DRAM formulation, and is represented in the form of the trip function. A series of model runs was made to evaluate the response of the model structure to changes in this variable.

The first tests of response to access changes dealt with regionwide changes. These tests might, for example, represent significant increases or decreases in cost and/or availability of fuel. In the first run there was an arbitrarily imposed increase in impedance (that is, an increase in travel time and/or cost and a subsequent reduction in accessibility). The available data base provided the interzonal travel times, estimated for an unloaded (uncongested) network. Although congested travel times are preferable for these analyses, they were not available for use in these experiments.

A regionwide increase or decrease can be accomplished, for example, by multiplying the zone-to-zone travel-cost matrix by 1·1 or 0·9 respectively. Parking charges can be imposed by adding to the vector of impedances terminating in the zones where those charges are to be imposed.

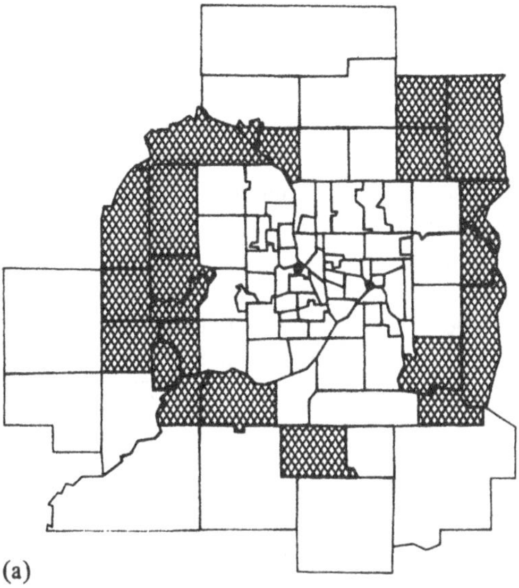

(a)

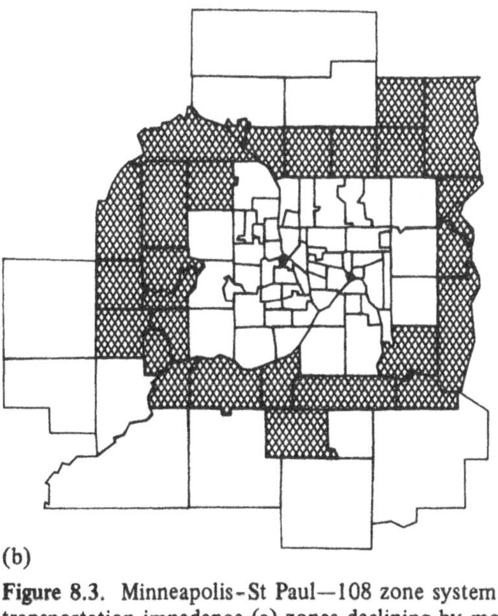

(b)

Figure 8.3. Minneapolis-St Paul—108 zone system: response to regionwide changes in transportation impedance (a) zones declining by more than 10%, owing to impedance increase of 10%; (b) zones growing more than 10%, owing to an impedance decrease of 10%.

Improvements between various zones and other zones can be imposed by multiplying the appropriate rows and/or columns of the impedance matrix.

The model outputs showed a consistent response to the first run of this set, which involved a 10% regionwide increase in impedances (highway travel times and/or costs). The response of LI households to this was mixed, with some fringe areas showing declines relative to the base run, and some showing increases. The other three income classes were, however, uniform in their relative decreases in the urban fringe zones. The implication here, which certainly needs further study, is that ceteris paribus, low-income households are the least sensitive to travel costs. It may be not so much a matter of insensitivity to changes in travel cost as a matter of inability to respond to those changes, owing to other factors such as housing discrimination or limited employment opportunities. The net effect on all households is that 15% of the zones of the region, all located at the urban–suburban fringe, showed relative declines of 10% or more relative to the base run; employment showed some degree of centralization and many small changes in and near the core. Figure 8.3(a) shows the zones with 10% or more decline in total households; the declines were absorbed (that is, matched by increases) in the urban cores.

The second run of this set involved a regionwide 10% decrease in impedances (that is, a transportation improvement). Again, the response of LI households was mixed, with equal numbers of fringe zones showing gains or declines relative to the base run. The LMI households showed a strong decentralization response to this regionwide decrease in impedance. At the same time, basic employment declined in the urban core, and showed signs of beginning suburbanization. Nonbasic employment also declined in the urban core. The net effect of these impedance decreases is a substantial decentralization of population and the beginnings of decentralization of employment, when compared with the base run. Figure 8.3(b) shows zones with a 10% or more increase in total households, the corresponding declines were all in and adjacent to the urban core. In short, relative decreases in transportation costs encouraged further urban sprawl, whereas relative increases in transportation cost discouraged sprawl and perhaps even encouraged recentralization of activities around existing major concentrations.

8.5.3.2 *Zone-specific changes*

The second group of access tests involved changes in impedance between specific zones and the rest of the region. The first in this group of tests involved a small improvement (in the form of a 5% reduction in impedance) in the accessibility of a zone on the suburban–rural fringe (zone 24), to the rest of the region. Theoretically one would expect modest increases in population and employment in the affected zone. Since the impedances (access) of adjacent zones to the region would remain unchanged, there would be little or no spillover effect to surrounding zones. It should be noted that an actual improvement of transport to one zone would simultaneously affect others, because the network would connect many zones to each other via the improved link(s). In an integrated transportation and land-use model (ITLUP, for example) this can easily be simulated, as the transportation system is described and used in link-by-link form. This was not possible in the test runs described here, as the use of DRAM in isolation from the integrated package involves the use of a matrix of impedance data in lieu of the actual networks.

The DRAM run showed increases in all household types of between 10% and 19%, with the total households in zone 24 increasing by 13%, relative to the base run. Total employment in the zone showed a slight increase. There were virtually no effects in surrounding zones, as the absolute increase in the number of households in the zone was only 221.

The second test of this pair involved a 20% improvement (impedance decrease) in accessibility of the same zone (zone 24) to the rest of the region. The expectation was that the results should be simply an amplified version of those from the previous test. The model results showed a 68% increase in the total number of households and an 8% increase in total employment. Though these changes seem large, it must be remembered that a 20% improvement in accessibility of an individual zone to the entire region is a phenomenal increase in accessibility; almost the equivalent of replacing a small unpaved local road with an expressway.

The same pair of tests was then repeated for an urban core zone (zone 64). The first of these runs was for a 20% improvement in accessibility of the zone. Typically when expressways have connected to urban core areas, these areas have experienced declines in high-income groups and increases in low-income groups. The DRAM response to this change in accessibility was a 21% decrease in UMI households, increases of 5% in LMI and 2% in UI households, and a substantial increase of 111% in LI households. A rerun of these tests with a 5% improvement in accessibility produced similar but smaller responses in all cases. It is worth noting that in both cases the LI increments were drawn from adjacent urbanized zones, whereas the UMI decline (departures) was taken up by several suburban zones.

8.5.3.3 *Summary*

These test results showed increased urban sprawl and decentralization with regionwide improvements in access, and decreased sprawl and centralization with regionwide decreases in access. When access to a specific zone is increased, DRAM shows increases in population of that zone. When the improvement was for a suburban zone, all types of population increased. When the improvement was for an urban (central) area, the principal increase was in LI households, with a decline in UMI households.

8.6 Verification from sensitivity tests: discussion

Taken all together, the results of these sensitivity tests offer additional support for the validity of the general structure of models like DRAM. In virtually all of the tests described above, the model produced sensible responses to changes in inputs. In the few cases where the results were questionable, subsequent work has found minor errors or omissions in the model structure which were subsequently remedied. Overall, these results lend credence to the proposition that models such as these can be used for urban policy analyses of many issues.

The lack of an explicit connection between the macrodescriptive equation structures of these models and the microdescriptive bases of contemporary economic theories of urban form does lead to the need for further research. Also, in terms of the use of these models, the case is made earlier in the book for the use of sets of models such as ITLUP for these policy analyses. The use of DRAM outside the context of ITLUP does, for example, lead to problems in expressing policy options in forms that are usable by the model.

To a certain extent, these sensitivity tests represented a prototype study for later model runs where similar changes in travel costs and parking charges were input. In both cases, the difficult question was one of how these changes were to be made in the existing impedance matrix. The test runs described here were made without the use of a network-simulation package, thus the transportation input to the models was in the form of a matrix of zone-to-zone travel times (the only interzonal data available for these tests). Given that this was the case, it then became a matter of some considerable interpretation to determine just how changes in, say, gasoline cost or parking charges were to be converted into equivalent changes in travel time. Time–cost equivalence is a long-standing problem in transportation and land-use analysis. Most of these models, both in theory and in practice, assume the use of a generalized travel-cost variable which is to contain some combination of both travel time and travel cost. The difficulty, of course, is the determination of just exactly how this combination may properly be calculated. The implication is that there is a way to convert travel time into equivalent dollars, or dollars into travel time. In practice, it is very difficult to determine just how these conversions will be made, as it is quite clear that the value of time to one individual will not equal the value of the same amount of time to another individual. Thus, the user of a model requiring these inputs is left having to determine what a five-minute change might mean in dollars, or a fifty-cent toll might mean in minutes, to several types of traveller.

The only sensible procedure for dealing with the time–cost conversion dilemma at present is to avoid attempts to analyze absolute increments in time or cost. Rather, whatever the impedance (generalized travel cost) matrix units may be, it is reasonable to do a series of simulation runs in which these units are varied by small percentage increments or decrements.

One may then discuss the consequences of a 5% or 10% *change* in impedance. It would then be left as a matter of interpretation to decide whether a 25% increase in toll or parking charge implied a 5% or 10% change in impedance. This is clearly an area where further research does need to be conducted, and where existing research on the value of time must be incorporated.

The results of the sensitivity-test runs and the parameter-estimation results presented earlier did, however, suggest another aspect of cities and their development which should be mentioned. Different metropolitan areas are noticeably different in their degrees of sensitivity to changes in travel times or travel costs. Another way of saying this is that some regions are either insensitive to congestion, or simply never develop enough congestion to make any difference to their tripmakers and their location decisions. This leads to the further implication that variables which describe the attractiveness of zones for residential location may in some cases be much more important determinants of residential location than travel times or travel costs. For a land-use model to be applicable in various types of metropolitan areas, including those which may not have a significant sensitivity to transportation systems, it is necessary to have a model such as DRAM where several other variables in addition to travel time or cost are ingredients of the residence-location decision.

This leads to one last note on sensitivity tests and the use of these models in policy analysis. Having demonstrated that the model structure responds more or less as it should, it is possible mathematically to develop a set of indicators of which locating activities in what particular zones are most sensitive to specific changes in inputs and/or parameters of the model. This involves the use of various partial derivatives of N_i with respect to the attractiveness or impedance variables, and to the various parameters. These will be presented in chapter 10.

In the next chapter a series of policy tests done with the modified version of ITLUP using DRAM are presented to illustrate the performance and use of the complete integrated model package.

9

Policy simulations with the improved package

9.1 Introduction

A rather complete description of the initial development of the prototype integrated transportation and land-use package was given in chapter 3. The results of a series of policy simulations done with the model package were described in chapter 4. Throughout all those runs there continued to be problems of stability of the outputs of the model package.

Briefly, the package began by generating a trial (preliminary) set of base year zone-to-zone trips. These trips were then loaded on the transportation network for the forecast year by use of a capacity-constrained assignment procedure. The zone-to-zone travel characteristics of the loaded forecast year network were then used as input to the location and land-use model. The elements of the OD trip matrix were estimated simultaneously with the location and land-use estimates. These trips were then loaded on the transportation network for the forecast year in lieu of the trial values used previously. The process was then repeated iteratively several times in the hope of achieving an equilibrium solution.

As has been recounted in chapter 3, a thorough examination of the intermediate calculations in the model package yielded some possibilities of solving the problem. The most effective of these was to use what amounted to a smoothing procedure. This involved, after a predetermined number of iterations, an averaging of zone-to-zone travel costs from one iteration to the next. This yielded substantially more stable outputs, and became standard procedure as used for the policy analyses described in chapter 4.

At a somewhat later date the problem of output stability was taken up again and a new approach was examined and adopted. The next section, 9.2, describes these examinations. Section 9.3 is a discussion of a new set of policy analyses performed with the modified version of the model package. Section 9.4 describes an extension of these analyses to evaluate the consequences of the policies on air quality. Finally, section 9.5 contains brief descriptions of additional policy simulations done with the model package.

9.2 Approximating system equilibrium

In chapter 3 it was reported that problems were encountered in reaching an equilibrium solution for activity location and traffic assignment. At the time that the work was being done there was no mathematical proof that such equilibria should exist; nor, of course, were there any algorithms which could guarantee the finding of such equilibria if they did exist. Nonetheless, long experience in the use of these models suggested that the model system should achieve an equilibrium solution under normal circumstances. As described in chapter 3, a procedure was adopted which resulted in a considerable damping of the oscillations in the solutions from the model. This procedure was invoked after the third iteration of the package, and involved averaging the OD trip matrices which resulted from successive interations and assigning these averaged trips to the network. Much more stable solutions were obtained on adoption of this procedure, and it was in this form that the model package was used for the policy simulations described in chapter 4 as well as for a number of others done in subsequent years.

However, the procedure used seemed to be rather too arbitrary, and consequently a thorough reexamination of the model package was undertaken. The configuration of ITLUP at this time has already been shown in figure 3.5. Each step of the process was examined, and various intermediate outputs were calculated and printed. The source of the oscillations could not be localized to any particular submodel. Attention then turned to the implied time relation of the ITLUP iterations. Although (just as in the case of the original Lowry model construct a decade earlier) the individual iterations in ITLUP had no explicit time reference, they were often thought of as intermediate points on a time line between the base and forecast years. Thought of in these terms, ITLUP was taking a single large step (in this case fifteen years) from the base year to the forecast year and then trying to iterate to a stable solution. A possibility that became apparent when the problem was viewed in this light was to take several smaller steps instead of the single large one.

For want of a better name this small step procedure was called the 'new process' (NP) form of ITLUP. From the very beginning of the work with ITLUP each of the submodels was designed to be (as nearly as possible) an independent module. By this means, the testing of alternative forms of models and configurations of model-to-model interconnections was facilitated. Thus it was possible, with a minimum of difficulty, to rearrange ITLUP to conform to the new process. The essence of the new process was that, rather than take a fifteen year step from base year to forecast year and then try to achieve an equilibrium solution, it took a series of three five year (approximately) steps, and then iterated to equilibrium.

Before we continue, it must be noted that the first experimentation with the new process was undertaken when residence location was still being estimated by IPLUM. This first configuration of ITLUP is shown in figure 9.1. First, it should be noted that it is not strictly necessary to have explicit intermediate time points. The fifteen-year forecast increments (regional totals of population, employment, and trips) were in this instance arbitrarily divided into three equal increments. Of course, had there been five-year forecast increments available they could have been used instead of the arbitrary increments. Now, we can follow through the steps implied in figure 9.1.

1. The base year was taken to be 1965 and the projection year was 1980, as for the initial ITLUP tests.
2. As in the 'old process', an OD matrix of trips for 1965 was generated and then assigned to the 1980 network. Subsequently, the zone-to-zone travel impedances (costs) were calculated.
3. From these zone-to-zone impedances, and the first-increment (I1) values of regional population, employment, and trip totals, IPLUM was used to estimate the first-increment population distribution and consequent trip matrices.
4. Next, the base trips for 1965 were subtracted from these first-increment trips. The difference was the actual incremental trip matrix. These incremental trips were then assigned to the 1980 network (which was already partially loaded with the 1965 trips), and new zone-to-zone impedances were calculated.
5. With these new impedances and the second-increment (I2) values of regional population, employment, and trip totals, IPLUM was used to estimate the second-increment population distribution and trip matrices.
6. The trip matrices from step 4 were subtracted from these new trip matrices and the remaining new trips were loaded on the partially loaded 1980 network. New zone-to-zone impedances were again calculated.

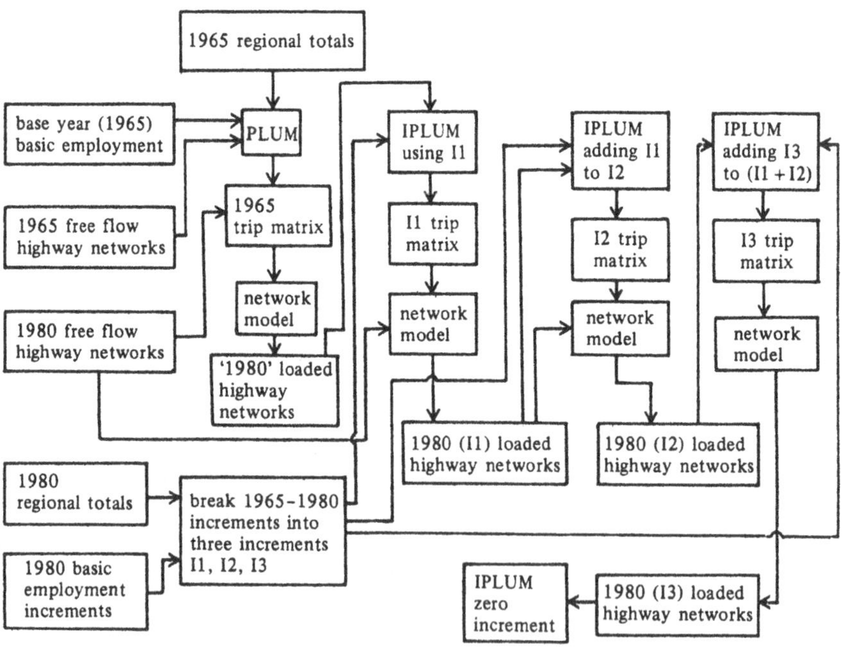

Figure 9.1. Schematic of new process for ITLUP.

7. With these impedances, the full regional control totals and the final one-third of the basic-employment increments as inputs, IPLUM was run and new population distributions and trip matrices were generated.
8. The trip matrices from step 6 were subtracted from these new trip matrices and the remaining new trips were loaded on the partially loaded 1980 network. New zone-to-zone impedances were calculated.
9. With these impedances, the full regional control totals, and increments in basic employment of zero, IPLUM was run again and should then have produced an equilibrium solution of population distribution and trip matrices.

On its first test, this new process was run through to step 8. The resulting population projections, where I1 corresponds to the values obtained at step 3, I2 to those at step 5, and I3 to those at step 7, are shown in figure 9.2. The dotted lines in figure 9.2 represent the final (sixth-iteration) projection values from run 3.AVG, as shown in figure 3.12. It can be seen that the new process came rather close after three recursions to where the old process was after six iterations.

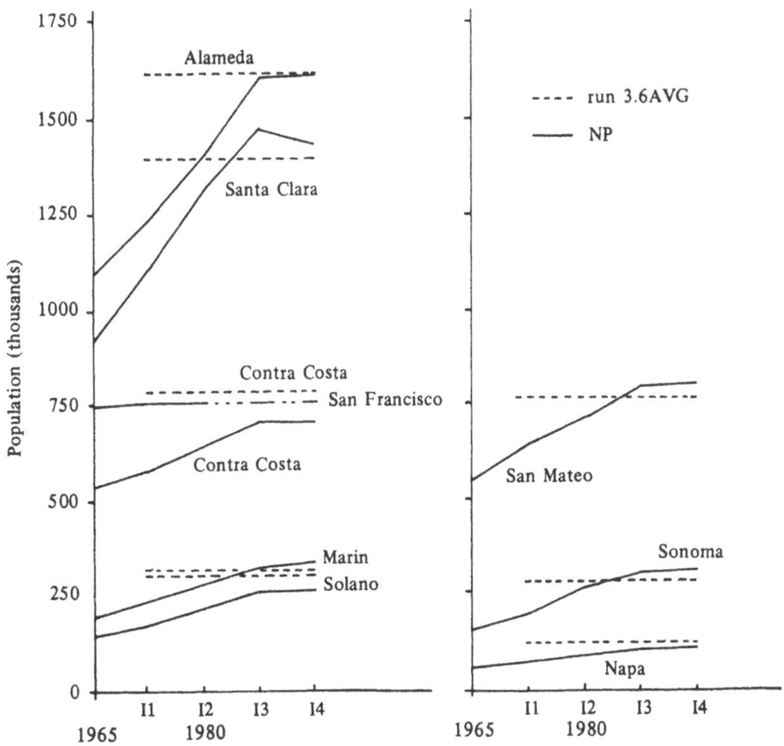

Figure 9.2. ITLUP—new process: population projections for 1980, compared with those of run 3.6AVG.

The fourth recursion of the new process (I4) was not run as described above, because of an unexpected operational deficiency in the version of IPLUM then being used. The model was theoretically capable of projecting the response, in terms of spatial redistribution of local-serving employment and population, to changes in basic employment and/or transportation facilities. However, it could not operate with changes in basic employment of zero. As there were not sufficient resources or time to correct this deficiency in IPLUM it was not possible, in these preliminary tests, to run the fourth recursion of the new process as originally intended in step 9. As an alternative test, the impedances from step 8 were used and IPLUM was rerun as in step 7. Thus the last growth increment was added to the previous two-thirds, but with the impedances derived from the *preceding* estimate of the last growth increment. The results of these two estimates of the spatial distribution of the last growth increment (one based on a two-thirds full network and the other on a completely full network) should, if the new process had found an equilibrium in the first three steps, have been nearly identical. In fact, the largest difference, at the county level, was less than 5%. With the last iteration shown as point I4 in figure 9.2, these differences were as follows:

San Francisco	+0·08%	Sonoma	+1·76%	Contra Costa	−0·12%
Marin	+4·07%	Napa	+2·29%	Santa Clara	−2·64%
San Mateo	+0·77%	Solano	+1·72%	Alameda	+0·31%

Further testing was clearly desirable, but the implication of these results was that by this new process the whole system came rather close to reaching an equilibrium point in just three iterations, as compared to the five or six iterations required by the original procedure.

A later version of ITLUP incorporated DRAM (as described in chapter 7) as a replacement for IPLUM, though the nonbasic-employment-locating component of IPLUM was retained. With this intermediate version of ITLUP an additional series of tests was run to explore further the equilibrium properties of the system. In the first of these tests the model system was run through once to step 8 of the new process. Then the location and land-use portion of the model package was rerun, with the same set of basic employment increments and the same regional control totals as those used in step 7. This produced a set of population distributions and trip matrices which were labelled 1980A. These 1980A trip matrices were then loaded on the 1980 network, in place of the trips for 1980 used in the previous step. The zone-to-zone impedances were recalculated, and the recursive process was continued to yield forecasts labelled 1980B, 1980C, 1980D, and 1980E. The county-level results for total numbers of households continued to show small to moderate oscillations, but the values tended to remain within an 'envelope' and appeared as if they might well have converged after further iterations. Thus the problem, although somewhat reduced, remained to be solved.

Finally, yet another set of runs was done; in this case the size of the increments was again reduced. Instead of three increments of approximately five years followed by several equilibrium iterations, there were two '5 year' increments, followed by two '2½ year' increments, followed by several equilibrium iterations. Again, the oscillations were reduced, this time down to just a few percent for each iteration. A further test showed that, for the most part, the oscillations that remained resulted from congestion–decongestion oscillations on the network. These, in turn, resulted from the incremental tree-by-tree loading procedure (described in chapter 5). Thus it appears that to eliminate these oscillations completely will require some modification in the capacity-constrained assignment procedure. This might take the form of a considerable reduction in the sizes of the trip increments to be assigned; or perhaps the use of a multipath assignment procedure might be examined within the operation of the model package. Finally, it is possible that current work being done with models of combined location, mode split, and assignment or equilibrium assignment will produce some suggestions that will help to resolve this problem.

Quite apart from the question of convergence to an equilibrium solution, the overall model package seemed to work rather well when the procedure of taking small incremental steps toward a final forecast year was followed. The ITLUP package was used in this form to do a series of policy simulations for the USDoT. Results of these simulations are described in the next section.

9.3 New policy-simulation results(15)

The simulations described here were performed with the intention of demonstrating the usefulness of the method, as well as the potential effects of several possible policy options. There were three sets of simulations: (a) variations in regional growth rates; (b) variations in regionwide transportation costs; and (c) variations in specific links of the transportation system. The first of these clearly illustrated the effects of transportation–location relationships and feedbacks, while setting a context within which the results of the other simulations could be evaluated. The second set of runs illustrated the effects of regionwide changes in transportation costs, which might result from energy-cost increases, or from taxes or subsidies. The last group of simulations demonstrated the effects of more specific changes in the transportation system such as parking charges and bridge tolls. All of these results taken together demonstrated considerable potential for policy alterations in the transportation–location system, and at the same time suggest the considerable complexity of the interactions and the consequent need for systematic, integrated analysis of any proposed policy actions.

(15) Portions of the material in this section appeared in a report prepared by the author for the Office of Transportation Economic Analysis, USDoT, in August 1980 (Putman, 1980b).

9.3.1 Spatial consequences of regional growth or decline

The first of this new series of policy tests was done, in part, for the purpose of establishing a range, or envelope, of forecast responses against and within which to evaluate subsequent model runs. The six simulation runs examined the consequences of changes in the rate of growth of the population of the region and nonbasic employment on the spatial distributions of these two variables. The importance of using an integrated transportation-location simulation package for policy analysis was clearly demonstrated in the results from this set of runs.

Consider first the central hypothesis of virtually all urban location models: the probability of residential location in a particular area is directly proportional to a measure of the 'attractiveness' of the area, and inversely proportional to the generalized 'cost' of travel between the area and the locator's workplace. The resulting probability estimates, for all locators in all zones, are transformed into a vector of the proportions (or shares) which each area will have of the total population (of each locating group) in the region. One direct result of this formulation is that for virtually all operational simulation models of urban location, a regionwide population increase of, say, 5%, will result in an across-the-board increase of 5% in the population of each zone (subject in some cases to exogenously imposed constraints on land use or density).

Even at first glance, this is an unreasonable result: it is readily verifiable that in periods of regional growth or decline some subareas of regions seem to grow or decline faster (or slower) than others. One reason—perhaps a main reason—for this differentiated response is the transportation system of the region. As the transportation system becomes congested at some places and not at others, areal differences in growth or decline will begin to appear. An analysis technique which represents direct and indirect relationships between the transportation network and location—such as the integrated transportation-location simulation package used for these analyses—can represent these phenomena, as is shown by the rather interesting results described below.

The analyses described in this chapter made use of the same San Francisco Bay Area data base that was used in the set of prototype analyses previously described in chapters 3 and 4. The results, however, are presented here at a 30 zone level of detail (see figure 9.3), rather than at the 9 county level used in the bulk of those previous discussions. The simulations were conducted, as before, at a 291 zone level of detail, but the quantity of simulation output at that scale was too great to describe here. Each of the thirty large zones contains 10-20 of the small zones (census tracts). The first three simulations involved increments of 5%, 10%, and 15% growth above the projections used in the 'benchmark' or base run of the models, respectively; the second three simulations involved decrements of 5%, 10%, and 15% growth relative to the base run. The principal comparisons were of the spatial distributions of households. It should be noted here that it is not a trivial matter in simulation analyses to decide which output variable(s) provide the clearest and most succinct description of the results. After considerable examination of simulation outputs, it was decided that household distributions were the most reliable and illustrative variables to use. Each set of 'policy run' results was then compared with the base run (a full run of the model package giving an unaltered forecast of the spatial distribution of households).

It is helpful to summarize the results of these runs by defining five groups of response types. In each of the six simulation test runs one-half to three-quarters of the thirty zones showed values differing from those of the base run by roughly the same percentage as did the regional totals. This first group of 'same-change' zones was interesting only in comparison with the others. The one-quarter to one-half of the zones that varied at significantly different rates were of considerable interest. Some zones showed increases much greater than regionwide increases, or decreases much greater than regionwide decreases. This group of zones might be termed 'volatile' or 'highly responsive' zones. Other zones showed increases significantly less than the regionwide increases, or decreases

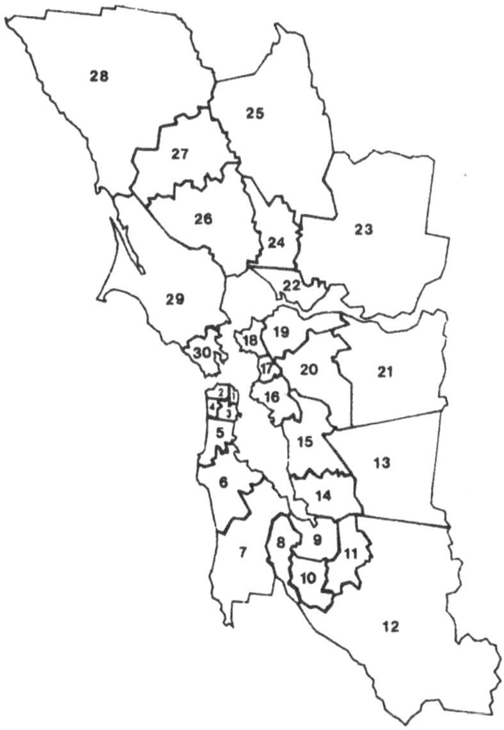

Figure 9.3. San Francisco Bay Area—30 zone system: definition of analysis zone boundaries.

significantly less than the regionwide increases. These might be termed 'stable' or 'unresponsive' zones.

A fourth group of zones differed from the regionwide average only when there was a regionwide increase in growth rates; that is, they showed increases significantly different from average when the regionwide growth rate increased, but all decreased at approximately the average rate with a regionwide decrease. Finally, a fifth group differed from the regionwide average only when there was a regionwide decrease in growth: zonal decreases were either significantly greater or significantly less than average decreases, but showed approximately average increases when the regionwide growth rate increased.

Before these results are reviewed in more detail, mention must be made of two additional output measures: the highway congestion index and the average trip-length index. It will be recalled that the integrated transportation and land-use package which produced these simulation results contains a capacity-constrained network assignment procedure. Further, trips are both generated and distributed simultaneously with the estimation of residence location. For each of the several thousand highway links in the network, information on volume of trips and link capacity are available during the assignment calculations. Thus it is possible to calculate a volume/capacity ratio for each link. The highway congestion index is calculated by calculating the mean volume/capacity ratio for a subset of all the links in each zone. This measure was discussed earlier in chapters 3 and 4.

In addition to the volume on, and capacity of, each link in the network, the trip-assignment procedure also makes use of and continuously updates the time required to travel the length of each link. Several times during the assignment procedure it is necessary to calculate the minimum-time path over the network links from each zone to each other zone. Thus at the completion of the assignment calculations there exists information describing a zone-to-zone travel-time matrix. With this information it is possible to calculate, for each zone, the average length (time) of all trips which originate in the zone. This average length (time) is called the average trip-length index and was also discussed in more detail in chapter 4. These two measures give some, albeit imprecise, indication of transportation conditions in each zone.

In the policy-simulation results mentioned above, three out of thirty zones showed rather volatile response to changes in regional growth rates. The numbers of households in these zones (zones 4, 19, and 25) grew much faster than the regional average (compared with the base, or benchmark run) for all six simulation test runs of the models. It was not clear from initial superficial comparisons why these three zones should have been the most volatile in the region. Zone 4 is an area which is relatively less developed than the other zones at the north end of the San Francisco peninsula; zone 19 is a relatively undeveloped zone separated by a mountain range from the rather heavily developed Berkeley–Oakland area; and zone 25 includes the predominantly rural Napa Valley which appears to decline in the base run. None of these zones show exceptional changes in highway congestion or average trip-length indices in conjunction with their household growth increases.

The three most stable zones, which showed less than average growth (compared with base run) with increased regional growth rates, and less than average decline with decreased regional growth rates, are zones 5, 12, and 23. Again, it was not clear on first inspection why these three zones should have been the most stable in the region, as superficially they too have nothing in common. It is worth noting that on average the three volatile zones showed a highway congestion index which was significantly less than the average for the three most stable zones. The stable zones had a congestion index 17% greater than the volatile zones. Moreover, the average trip length (for trips originating in the zones) was roughly 12% shorter (on average) for the stable zones than for the volatile zones. These results suggested that there is some transport-related cause of these behavior patterns. These volatile and stable zones are shown in figure 9.4.

Of zones which differed from the average only with regionwide growth increases, zones 2, 6, and 26 grew faster than average, whereas zones 3, 17, and 24 grew less than average. These zones, too, were superficially dissimilar. Again, however, the congestion index for the slow-growth zones was 38% greater than that for the fast-growth zones (on average),

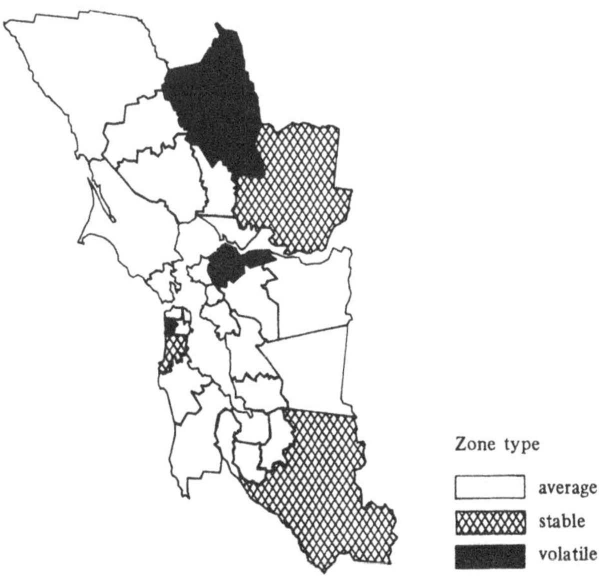

Figure 9.4. San Francisco Bay Area—30 zone system: zonal response to changes in regional growth rates.

and the average trip lengths in the slow-growth zones were 13% less than the trip lengths for the fast-growth zones.

Finally, for regionwide growth decreases, zones 18 and 16 declined by more than the average zonal decrease, and zones 8 and 29 by less than the average zonal decrease. Here again the congestion index for the fast-decline zones was found to be 54% greater than the slow-decline zones. The trip lengths (on average) were 3% less for the fast-decline zones than for the slow-decline zones. This represented a slight anomaly in the emerging pattern, which may possibly be explained by the substantial difference in congestion levels.

Additional corroborative evidence was found by observing the specific effects of the various regionwide growth-rate increases on zone 30. With a 5% increase in the regionwide growth rates, zone 30 exhibited a 19% increase in total households, almost four times the regional average (compared with the base run). A 10% increase in regionwide growth rates resulted in a 12% increase in zone 30, and the 15% regionwide increase yielded an 18% increase in zone 30. Thus there was a dramatic growth in the zone for a small change in the regionwide growth. On the other hand, increases only slightly above average were observed in the zone when there were larger regionwide increases. Examination of the congestion index shed light on this behavior. Whereas the volatile zones (zones 4, 19, and 25) showed only small increases (usually $<2\%$) in congestion for each different change in their growth, zone 30 showed virtually no change at the 5% regionwide increase, but an increase in congestion of 11% with the 10% regionwide growth increase. This increase in congestion (probably arising largely from the Golden Gate Bridge connecting zone 30 in Marin County to the San Francisco peninsula) apparently had the effect of retarding the ability of the zone to respond to high regional growth rates.

To summarize these results, zones which were superficially dissimilar were observed to have in common that the level of highway congestion (an index created by calculating the average volume/capacity ratios on the highway load links in the zone) was consistently associated with the degree of response of the zone to regional growth or decline. The greater the congestion index of the zone, the more sluggish was its response to regionwide growth and the more pronounced was its response to regionwide decline. Furthermore, a measure of trip lengths (an index created by calculating the weighted average trip length of all trips originating in a zone) showed longer trip lengths to be associated with less congested zones and thus with more vigorous responses to regionwide growth and more modest responses to regionwide decline. This observation suggests that in zones with less congestion it is easier for residents to consider employment over a wider geographic area (or employees in a less congested zone may consider residence over a wider geographic area) and thus be more responsive to growth opportunities. As a final point, it should be noted that, in general, longer trip lengths seem to result from regionwide growth increases as increased labor demand pulls in employees from more remote locations. Decreases in trip lengths seem to be associated with regionwide growth decreases, perhaps because a declining labor market makes it more difficult for those with higher personal transportation costs to compete for the available jobs. All in all, there is evidence of significant influence of the highway system on spatial patterns which result from response to changes in regional growth rates.

The discussion has, to this point, been couched in terms of differences in the spatial allocations of the total number of households. The model estimates the spatial allocations of four classes of household. These classes are defined in terms of quartiles of household income (see chapter 8): low income (LI), lower-middle income (LMI), upper-middle income (UMI), and upper income (UI). The actual monetary values of the quartile boundaries are unimportant and in fact change over time with changing employment mix in the region being analyzed. The total household estimates are the sums, in each zone, of the individual household class estimates. The class-to-class differences in response to the policy alternatives did not justify extensive analyses of all the differential responses by income class. There were, however, some noticeable differences which are worth mentioning here.

In the three regionwide-growth-increase simulation runs, all of the individual household types followed the same pattern as the total households in the zones of greatest increase relative to the base run. There were, however, some significant differences in response between particular income groups in certain other zones. With a 5% regionwide growth increase, LI households showed substantially larger increases in zone 2 in San Francisco and zone 25 in Napa. LMI households showed a significantly greater than average increase in zone 25, too, and UI households showed a larger than average increase in zone 14 in western Alameda. In contrast to these, zone 1 in San Francisco showed an increase in total households, but a relative decrease in UI households in response to the 5% increase in regionwide growth rates. The simulations of 10% and 15% regionwide growth increases both showed all four income groups following the same pattern as total households in the zones of greatest increase. The only notable exception to this pattern was, for both simulation runs, a substantial increase (relative to the base run) of UI households in zone 1 in San Francisco and zone 13 in eastern Alameda.

It is thus quite clear that there were some significant household-type-specific differences in zonal responses to changes in regional growth rates. Preliminary examination of the zones, household types, and circumstances of these differences did not suggest any consistent hypothesis to explain them. It is probable that more extensive analyses at a finer level of detail (and perhaps incorporating more local knowledge of the area) would result in a clearer understanding of the causes of these different responses.

In actual policy analysis for planning agencies, it is these sorts of 'different from average' results that suggest the need for further analysis by persons intimately familiar with the region being analyzed.

9.3.2 Spatial consequences of regionwide changes in transportation cost

It has often been suggested that with continuing increases in transportation costs, metropolitan regions will witness a concentration of activities towards the center of the region, or central business district (CBD). This notion results from an implicit assumption of monocentricity of urban areas. It is far more likely, as most metropolitan regions are polycentric, that increased (or increasing) transportation costs will result in more dense clustering of activities around existing centers to produce a subnucleated region. This second group of simulation runs was intended to examine these questions.

Four simulation runs were made in this group of tests. Two runs, involving regionwide increases in transportation cost of 5% and 10% respectively, were made. These were followed, for the sake of comparison, by two runs involving a 5% and a 10% reduction in regionwide transportation cost. The first result to be noted was that both for cost increases and for cost decreases, the 5% and 10% runs both showed similar patterns: for example, the 5% increase in transportation cost produced a similar spatial pattern of relative growth or decline to that produced by the 10% increase. In many instances the zonal growths or declines in the 5% run were roughly one-half of those in the 10% run. The many nonlinearities in the model package did, however, ensure that some of the results from the 5% runs did not simply equal one-half of the corresponding values from the 10% runs. However, for the sake of brevity, only the 10% runs are discussed below.

A 10% increase in the transportation costs between each zone and each other zone in the San Francisco Bay Area did result in a tendency towards centralization of activities around existing clusters. Figure 9.5(a) shows the zones which grow (relative to the base run) when there is such an increase in the regionwide transport cost. Of the ten zones which do show growth in this case, the six which show the greatest growth (zones 1, 2, 9, 16, 17, and 28) are clearly zones containing existing clusters of activity. It is also quite clear that the 'center' of the region—that is, the San Francisco CBD—was certainly not the only zone to benefit.

The other side of the coin is shown in figure 9.5(b) which shows zones experiencing a decline, relative to the base run, in response to the 10% cost increase. It is quite clear from this map that it was the zones which were suburban and rural in character that showed a relative decline in response to increased transportation cost. The zones which showed the greatest decline (zones 6, 14, 19, 20, 26, and 30) are clearly those which were predominantly suburban. The remaining declining zones are more rural than suburban. Thus the expected spatial distribution results, in terms both of relative growth and of relative decline, are quite apparent in the outputs of this simulation run of a regionwide transport-cost increase.

A 10% decrease in regionwide transportation costs produced results which were, for the most part, the opposite of those of the cost increase. All the zones which grew (relative to the base run) in response to a regionwide decrease in transportation cost were predominantly suburban and rural in nature. The zones which showed the greatest relative growth, zones 6, 19, 20, 25, 26, and 30 (with the exception of zone 25, which had the lowest relative growth of the top six growth zones), were predominantly suburban. All these 'growth' zones are shown in figure 9.6(a). It is worth comparing figure 9.6(a) with figure 9.5(b), from which it can easily be seen that, with few exceptions, the zones which declined with an increase in transportation cost are the same as those that grew with a decrease in transportation cost.

The few exceptions are of some interest. Zone 25, the northern half of Napa County, is rather rural in character. With the exception of zone 28, the northern end of Sonoma County, the tripmakers from zone 25 have the longest average trip lengths in the entire region. A regionwide decrease in transportation cost produces significant growth of the three highest income groups. The LI group, however, shows a decline in zone 25 in response to the cost decrease. This decline would be occasioned, no doubt, by the rise in UMI and UI households in the zone. The opposite results are seen for the regionwide cost increase. Zone 25 experiences a relative decline in UMI and UI households (also a modest decline in LMI

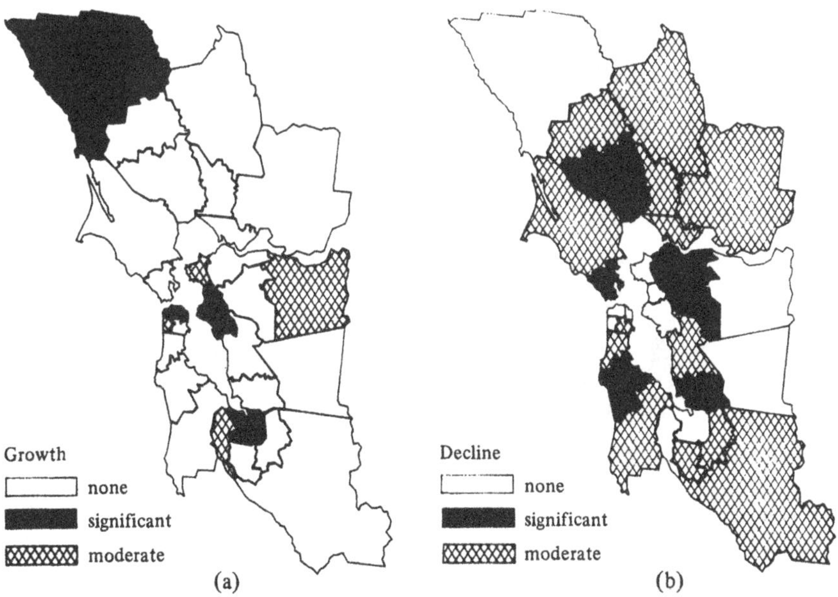

Figure 9.5. San Francisco Bay Area—30 zone system: growth or decline of zones after a regionwide increase in transportation cost of 10%. (a) Zones showing growth; (b) zones showing decline.

households) and a modest increase in LI households. It appears that these results illustrate the phenomenon of rural low-income households being displaced by high-income long-distance commuters when improved transportation facilities (that is, of lower time and/or lower cost) make the longer trips practicable.

Zone 4, which roughly includes the southeast quarter of the city of San Francisco, also gave unusual results. In response to a regionwide increase in transportation cost, there was a slight increase in LMI households and small to moderate declines in the other three household groups. In response to a regionwide decrease in transportation cost, there was a small increase in UI households and moderate decreases in the other three household groups. Overall, the zone declined both with increases and with decreases in cost. What did change in each case was the mix of household types in the zone. This rather heavily populated zone has (with the exception of zone 10) the shortest average trip lengths of all zones in the region. It is also moderately congested (within the upper quartile of all zones). The great bulk of the trips which originate in zone 4 terminate nearby in zones 3, 4, or 5. The behavior of the zone in response to regionwide changes in transportation cost is a result of the household composition and geographic location of the zone. Further analysis in conjunction with a knowledgeable local planner would probably be fruitful and would help to explain various peculiarities of individual zones.

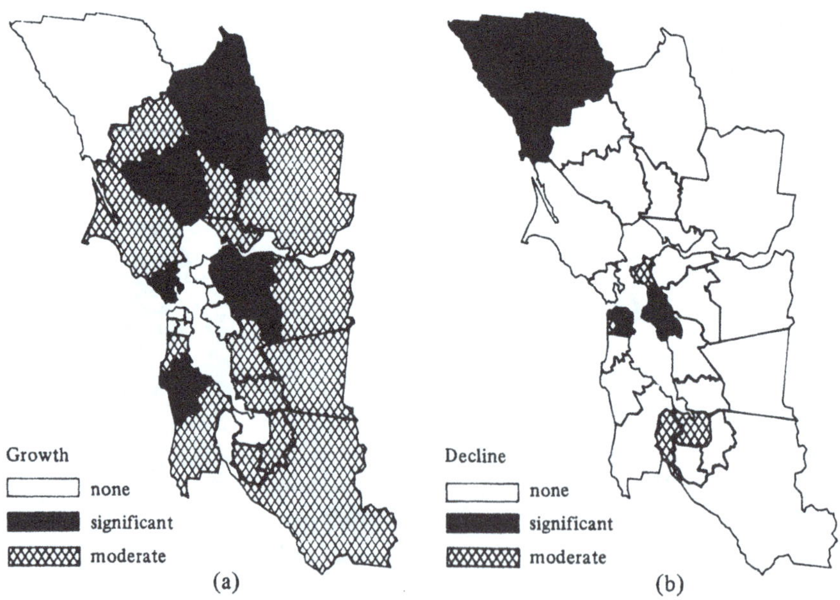

Figure 9.6. San Francisco Bay Area—30 zone system: growth or decline of zones after a regionwide decrease in transportation ost of 10%. (a) Zones showing growth; (b) zones showing decline.

Having mentioned these special cases, we shall now discuss those zones which decline with a regionwide decrease in transportation cost. Zones 1, 2, 3, 16, 17, and 28 show the greatest decline in this case. These zones, with the exception of zone 3, are exactly the same as those that grow the most (relative to the base run) as a result of a regionwide increase in transportation cost. These are the zones of greatest urbanization, or centralization, of activities. Zone 28 is of particular interest here, as it is an example of subnucleation. This zone contains the northern portion of Sonoma County, adjacent to the city of Santa Rosa. In response to the regionwide cost increase, all types of households seem to grow in this zone. This result, along with the general pattern of zones shown in figure 9.6(b), clearly demonstrate concentration around existing centers of activity.

In general, the results of these tests of spatial consequences of increases or decreases in regionwide transport cost were quite significant. Suburban areas tended to grow when transportation cost decreases and decline when cost increases. Urbanized areas did just the opposite, growing with transportation cost increases and declining with cost decreases. Within these general patterns there was some variation due to the particular characteristics of individual zones. It should also be noted that average trip lengths tended to increase with decreases in transportation cost, and to decrease with increases in transportation cost. Congestion levels showed virtually no response to cost increases and slight increases in response to cost decreases. These results confirm the hypothesis set forth at the beginning of this section.

9.3.3 Spatial consequences of specific changes in transportation cost

This last group of policy simulations was done to demonstrate and analyze the effects of changing specific links in the regional transportation network. The first set of simulations to be discussed here examined the consequences of increased auto parking charges in the San Francisco CBD on the spatial distribution of households. The second set of simulations examined the consequences of changing auto tolls on the Golden Gate Bridge and on the San Francisco–Oakland Bridge. The results of both these sets of simulation tests illustrate the possible consequences of actual implementation of these policies, as well as the use of these models for policy analysis.

Both these sets of policy simulations again raise the problem of time–cost equivalence in transportation and location models. For parking charges, the questions are:

(1) What are current parking charges?

(2) What size increments (cost changes) should be tested?

(3) How are these monetary changes converted to time equivalents in order to modify the travel-time network and zone-to-zone travel-time matrix in the models?

Similar questions arise in the case of the bridge tolls. The resolution of this difficulty would involve the use of data on the particular value-of-time relationships for the San Francisco Bay Area. Unfortunately, even when available, these data are not often broken down according to different income groups. If the intent of these policy simulations had been to produce precise estimates for an actual policy decision it would have been necessary, despite the difficulties in obtaining data, to establish a set of value-of-time relationships before proceeding with the analyses. As the intent here was less ambitious—only to illustrate the general signs and magnitudes of the effects—it was possible to sidestep the problem and do the simulations by making percentage changes in the impedance matrix (or on the network links) rather than by attempting to estimate the conversion of a change in parking fees of Y dollars into X minutes of travel time. This procedure allowed for policy analysis in the absence of time-cost conversion factors, which would have been of suspect accuracy and stability in any case.

The parking-charge tests were accomplished by adding given percentage increments to specified rows and columns of the zone-to-zone travel-time matrix. This was done immediately after the assignment phase of the network-simulation portion of the model package. The modified travel-time matrix was then used as input to the location submodel of the package. Three such simulations were run, reflecting 5%, 10%, and 20% increments in the intrazonal (or terminal) travel times for the affected zones. It was necessary to perform these changes at the 291 zone level of detail, and then to aggregate the results for smaller zones which were wholly contained by zones 1, 2, and 3 of the 30 zone system used throughout this discussion. The results of the three simulations were quite similar, differing only in degree. This suggested that although there is some degree of nonlinearity of response, there are no apparent unstable or discontinuous responses.

Figure 9.7 shows the spatial distribution of the significant population responses to an increase in parking charges (actually represented in the models by a 20% increase in intrazonal travel time) in the San Francisco CBD. Zone 1 showed a significant population decline (relative to the base run), and zones 2 and 3 showed small declines. The consequent population increases appear in zones 4, 5, 6, 16, and 30, which are the next adjacent zones (though across bridges in the case of zones 16 and 30). These apparently straightforward results nonetheless lead to some rather complex problems of interpretation. First one must consider the simple notion that if the cost (or travel time) was increased for all trips originating from or arriving in a particular zone, then the population of the zone would be expected to decline. However, declines in population and/or employment are not the intended result of a policy of parking restrictions. It is of course possible that in reality there would be substantial shift of transportation mode, with more tripmakers making use of transit or park-and-ride facilities. One might then expect the reduced congestion in the CBD to increase its attractiveness. With the then available version of DRAM–ITLUP it was not possible to test this hypothesis. All of the parking-charge simulations resulted in varying degrees of relative decline in the CBD. Thus it seems more than likely that even with the possibility of an increase in trips to and from the CBD by transit systems, an increase in parking charges in the CBD may well result in some degree of decline in population (and probably in employment).

The bridge-toll tests were conducted by altering the travel times for the links in the network which represented the Golden Gate and San Francisco–Oakland Bridges. Again, no attempt was made to give exact monetary equilivalences to these changes, owing to the lack of information on a proper cost–time conversion ratio. These tests are therefore discussed in the form of an increase in bridge tolls of *Y* percent causing an increase or decrease of population in zone *X*. The spatial distribution of results was completely in accord with expectation. A decrease in tolls on the two bridges resulted in a relative decline in population throughout the northern end of the San Francisco peninsula (zones 1–5) and increases in zones 15, 16, 18, 22, 25, 29, and 30. With the exception of the increase in zone 25 (which is anomalous), the other changes indicated a trend towards suburbanization of the population from the northern end of the San Francisco peninsula to areas north and east across the two bridges. These spatial patterns are clearly evident in figure 9.8(a).

The spatial consequences of an increase in the bridge tolls are almost the exact opposite of those of the toll decrease. The zones at the northern

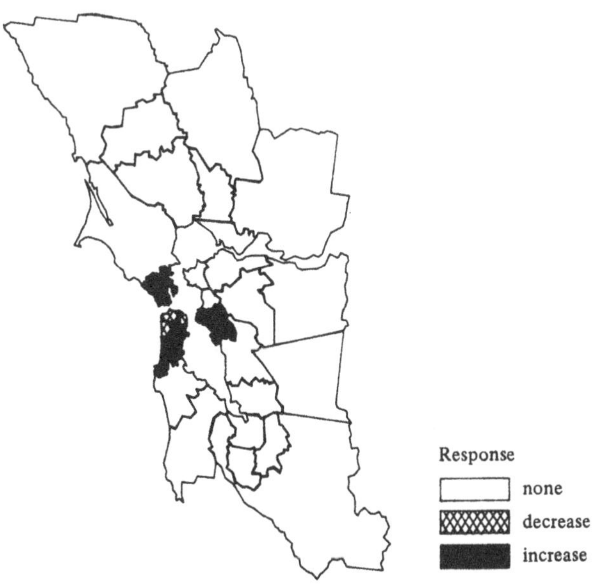

Figure 9.7. San Francisco Bay Area—30 zone system: zonal response (population) to increased CBD parking charges.

end of the San Francisco peninsula (zones 1–5) showed a relative population increase in response to the increased tolls; the zones from which travellers commute to the San Francisco peninsula (zones 15, 17, 26, 29, and 30) showed a population decrease. These results are shown in figure 9.8(b). In both the bridge-toll simulations the effects described hold true for each income group, as well as for total population.

Finally, it is interesting to compare the average trip lengths of trips originating in the growing or declining zones in these two simulation tests. In the case of bridge-toll increases, the average trip lengths for growing zones showed no significant change. However, the average trip lengths for declining zones showed a very substantial increase. Moreover, the number of trips which originate in zones 15, 17, and 30, and which terminate in zones 1–5, declined dramatically. In the case of the bridge-toll decreases, the average trip lengths for declining zones changed very little, whereas the average trip lengths for growing zones decreased substantially. The number of trips which originate in zones 15, 17, and 30, and which terminate in zones 1–5 increased dramatically. All of these results were in accord with what would be expected. As an interesting aside, it should be mentioned that the improvements to the Golden Gate Bridge (for example, additional traffic lanes) which have occasionally been suggested would, by relieving congestion, have a similar effect to that of the toll decrease. In the simulation results discussed above, zone 30 was the most significantly affected zone. The concern of Marin County (zones 29 and 30) planning

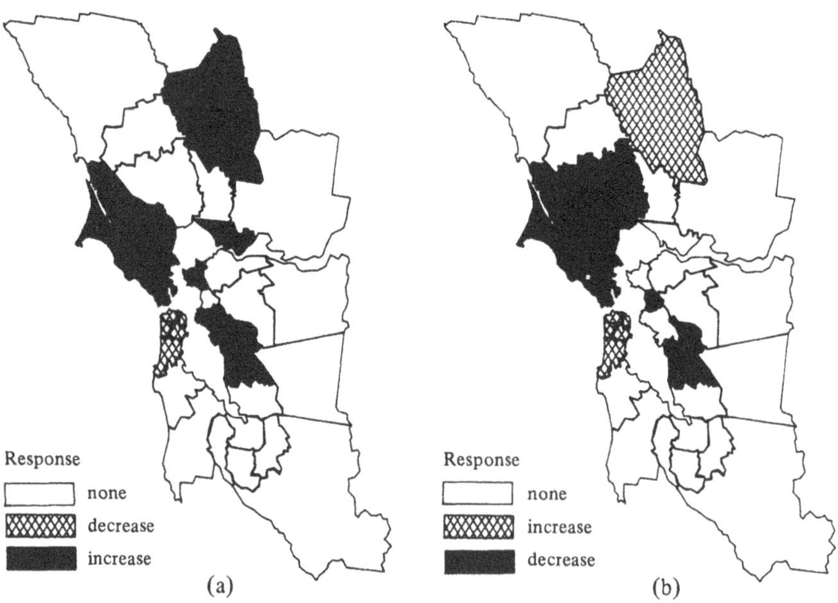

Figure 9.8. San Francisco Bay Area—30 zone system: zonal response (population) to variation in bridge tolls. (a) Decrease in tolls; (b) increase in tolls.

officials, for example, that bridge improvements would make more difficult the control of land use in the county, seems to be strongly supported by these results.

9.3.4 Conclusions from simulation results

These first simulation runs, in considering the transportation system as existing regional infrastructure, yield results not previously discussed in the literature. In classical economic comparative advantage analysis, regions are described in terms of their ability to respond to changes in demand for their products. One aspect of that responsive ability may be the presence or lack of transportation facilities or infrastructure. In the simulations described here a similar phenomenon was observed.

In these analyses, regionwide increases of different sizes in population and employment growth were tested. It was observed that some zones grew faster in response to these regionwide changes than did others. It was also found that the growth rate of the zones was related to their existing transportation facilities. The zones which grew the most in response to regionwide increases of population and employment were zones where there was an apparent excess of capacity in the transportation system. Thus the growth was accommodated without significant increases in traffic congestion. The zones which grew least in response to regionwide population increases already had congested transportation facilities and, further, these same zones also showed the greatest response (declines) in response to regionwide population decreases.

A traditional highway impact analysis considers only the impacts of the new construction on traffic flow and, perhaps, on land development. These analyses do not ordinarily consider the longer term impacts of the highways on the ability of the zones they serve to respond to changing economic or demographic growth in the region. The results described here thus suggest much longer term effects than those which are traditionally considered in transportation impact analysis, and consequently suggest a possible revision in the methods used to perform those analyses.

The second group of findings results from analyses of the effects of regionwide changes in transportation times and/or costs. These could result from, for example, increases in fuel prices or decreases in speed limits. These regionwide changes in transportation time/cost produce pervasive changes in the spatial patterns of activities of the region. In general, the suburbs and rural fringes grow in response to decreases in transportation time/cost, and decline with time/cost increases. Moreover, these changes are not just a matter of sprawl versus centralization. They are better described in terms of clustered versus dispersed activity patterns, with the clustering or dispersion taking place around numerous regional subcenters in addition to the traditional CBD. Nonetheless the overall patterns are not unexpected. Transportation time/cost increases shift population back towards the central areas, with declines at the urban fringes.

Transportation time/cost decreases shift population towards the suburbs at the expense of the urban core.

There are some observable differences in the apparent response of the population in different income groups. First, there are overall increases in trip lengths as transportation costs decrease, and decreases in trip lengths as transportation costs increase. Second, the dispersion of population which results from decreases in transportation cost is accompanied by modest dispersion of employment. Taken together, these results suggest the possibility of reductions of access of the low-income population to employment and residence opportunities. An interesting finding is that UMI households are most sensitive to decreases in transportation cost, quickly taking advantage of the newly accessible suburban residence opportunities. The LI households are much less responsive to changes in transportation costs, perhaps because various socioeconomic factors combine to make such opportunities unavailable.

Finally, it must be noted that the sets of overall conclusions about the consequences of regionwide changes in transportation time/cost should be regarded as only indicative of potential responses. Several potentially important phenomena have necessarily been disregarded in these preliminary analyses. First, there was no provision in these analyses for any mode switching in response to increases in automobile travel time/cost. Second, there is no provision for switching to more fuel-efficient vehicles, which might reduce the actual travel costs to previous levels. Finally, other factors were assumed to be unimportant, for example, time–cost equivalences are ignored. These matters will be the subjects of future research in this area.

The third group of findings is concerned with the results of specific changes in the transportation system of a metropolitan region. The new highway links, changes in bridge tolls, and changes in downtown parking charges.

There are several phenomena which result from construction of new high-speed road links from urbanized areas to the suburban–rural fringe. First, there are large population increases at the suburban–rural end of the new roadway. These population increases will, after a time, lead to employment increases as well. Initially though, the new roadway is used by a decentralizing population to make longer work trips. Some of the suburban sprawl commonly produced by new roadway construction of this type can be prevented by limiting the access to the new facility to only a few points (or perhaps none) along its length.

At the downtown end of the new highway link, there are different effects. First, there appear to be increases in LI households in these zones, perhaps because they offer increased access (via reverse commuting) to suburbanized employment opportunities which were formerly inaccessible to many city-center residents. Second, there are modest increases in nonbasic types of employment, such as services and retail, perhaps due to the increased accessibility of these zones to suburban residents. It is possible though, particularly in light of the next finding, that these employment increases may be short lived. Finally, there are large decreases in UMI households who are apparently able to make use of the new highway facility to commute from new suburban residence locations to their places of work in the CBD.

Both increases and decreases in tolls on bridges leading from populous suburbs to city-center areas had important and widespread consequences. A substantial toll decrease served to make the city-center area more accessible to the suburbs and then encouraged suburbanization of the population with a corresponding decline in the central areas. These toll reductions yielded a decrease in transportation costs to commuters which was most noticeable in the zones that experienced the greatest growth. A substantial toll increase yielded almost exactly opposite results. The increase served to make the suburbs less accessible to the city center (or the city center less accessible to the suburbs), thus producing tendencies towards centralization of population with corresponding declines in the suburbs. Furthermore, in the declining zones the cost of trips could be seen to be substantially increased. It should be noted here that though the exact equivalences are difficult to determine, changes both in travel time and in travel cost operate in similar ways, so the comments here about tolls apply equally well, but perhaps to differing degrees, to road improvements such as bridge improvements, new bridges or tunnels, and so on.

Finally, although the evidence is less clear-cut, it appears that increased downtown (CBD) parking charges may produce results similar to those of bridge-toll decreases. The mitigating factors will be possible travel-time reductions resulting from decreased road congestion in the area, and the extent of mode switching from highway to transit. In any case, these tentative results should prompt further examination of such proposals.

Thus there was clear confirmation, by the results of the simulation analysis, that transportation cost (regionwide) has a direct effect on metropolitan sprawl or compactness. Most existing metropolitan areas offer empirical evidence to the effect that decreases in travel time (or cost, or both) yield increased spread of activities. The dispersed spatial patterns of metropolitan areas which result from highway expansion are likely to prove all but impossible to serve (with transportation facilities) in any energy-efficient fashion. How the problem will be dealt with over the next few decades is currently one of the most critical of public policy problems.

Other simulation analyses showed the effects to be expected from link-specific or subarea-specific changes in travel time or travel cost. Substantial bridge-toll reductions at key points in a metropolitan transportation network can produce suburbanization pressures in several subareas of the region. Parking-charge increases could, if people are unwilling to switch to public transit, result in declines in the population or employment of subareas in which they were imposed. This might not be the intended result.

Analysis of unintended results may be one of the most valuable advantages of the simulation techniques described in this report. For example, a greenbelt open-space policy of land-use control (the results of which were not discussed here), seemed to result in longer work trips. These in turn, led to greater vehicular emissions. Similarly an emission-control policy to improve urban air quality might, by increasing travel time or cost to parts of a metropolitan region, lead to the decline of those areas of the region. A very successful program to increase auto occupancy rates could reduce traffic congestion quite substantially. Yet this reduced congestion might be viewed by other tripmakers as a transportation improvement which they might use to increase their trip lengths. The complexities of these systems are never-ending. As illustrations of additional types of policy simulation which could be done, the next sections of this chapter briefly describes several additional simulation studies done with the integrated package.

9.4 Policy effects on air quality

Additional procedures were developed for connection to the ITLUP package to allow the evaluation of the consequences of transportation, location, and land use policies on air quality (Gross, 1979). In this section, a brief description of the necessary procedures is given, followed by a summary of the effects on air quality of these policies.

Atmospheric air pollution is considered in terms of two general processes, emission and diffusion–dispersion. Emission analyses deal with the quantities of pollutants being generated, the geographic locations at which they are being generated, and the times at which they are generated. Diffusion–dispersion analyses deal with the diffusion of pollutants and their chemical reactions, as well as the physical transport (dispersion) of the pollutants through the atmosphere as a function both of these processes and of weather-related processes. To evaluate the consequences of urban spatial phenomena on air quality, it is necessary to have models of emission and diffusion processes which could be integrated with urban land-use and transportation models.

Consideration of the quantities of pollutants generated along with their geographic locations yields two types of emission sources: stationary sources and mobile sources. Each of these types of source is further classified into two subclasses. Stationary sources are classed either as point sources or as area sources. Although the definitions here are somewhat arbitrary, point sources generally achieve their rather dubious distinction by virtue of emitting in excess of one hundred tonnes per year of any given pollutant. Sources which emit less than this arbitrary amount (whose value or definition may vary somewhat from study to study) are grouped together as area sources. Mobile sources are divided into two subclasses according to a similar notion. Particular high-traffic-volume links of a transport system, such as urban expressways, are considered as specific line sources. General traffic flows on normal streets are considered as area sources.

Regardless of the source, only some of the many types of emission are currently considered as potentially serious (that is, harmful. The principal types of substances that fall into this category at present are: hydrocarbons (HC), carbon monoxide (CO), nitrogen oxides (NO_x), sulfur dioxide (SO_2), and particulate matter (PT). These are known as the primary pollutants. The primary pollutants are generated either by combustion or by non-combustion chemical processes, with the rather modest remainder coming from evaporation of hydrocarbon-based fuels, and so on. The proportional distributions of emission by source vary considerably from one region to another. For example, CO emissions from mobile sources are said to account for over 90% of total CO emissions in six major US metropolitan areas. Thus, to a certain extent, the importance of accurate estimation procedures for various types and sources of emissions will have varying importance depending on the region being analyzed.

9.4.1 Stationary sources

Analysis of stationary sources presents several types of problems. Considering industrial sources first, one is immediately confronted by several difficulties. Perhaps the most succinct description of these difficulties is that it has proven necessary to know (a) numbers of employees per (facility) plant, (b) specific industrial process weights, (c) plant fuel consumption, (d) classification of product type, and (e) age of plant, before reliable estimates of emissions from a plant could be made[(16)]. The required level of detail of input data for estimation of emissions presents difficulties for the collection of base year data. The further problems associated with projecting these data into the future in order to make forecasts of emissions are truly monumental. It appears that the data necessary for reliably estimating emissions from industrial area sources approaches the 'complete inventory' type of data necessary for dealing with industrial point sources.

The development of a general model of residential (nonindustrial) area sources requires similarly difficult levels of detail in the input data; again, this raises questions as to the reliability of forecasts of future emission levels. To avoid these problems, yet at the same time to have some capability for estimation of emissions, a rather pragmatic view of the matter is customarily taken. Data are gathered on emissions and on land use (spatial distributions of activities). With good inventories of these data it becomes possible to prepare correlations to existing levels and compositions of emissions. If the analyst is then willing to assume that these relationships are stable (or has exogenous information about how they may be modified) for the duration of the forecast period, then they

[(16)] See Epstein et al (1974) for a more complete discussion.

may be used to make forecasts of future emissions, based upon forecasts of future land-use and activity patterns. The key notion here is that of emission rates, and the key question is of their long-term reliability.

The accurate estimation of emission rates and their long-term reliability depend both on spatial and on sectoral levels of aggregation. These rates tend to become more accurate, at a given time, as the sectoral detail (that is, numbers of different sectors described) is increased. The same is true for geographic detail. As the subareas or zones become smaller, the emission data become more accurate. However, there are limits to the level of disaggregation beyond which the analysis becomes either prohibitively expensive or statistically unreliable. Furthermore, for long-term reliability, some moderate degree of aggregation is desirable. Thus each study area will require a preliminary analysis to determine the appropriate levels of detail for the collection of emission data for area sources, and for subsequent forecasting. And, despite its obvious heavy-handed assumptions, it is quite clear that this is the type of procedure which will most often be practical for area-source-emission analysis in conjunction with transportation and land-use planning.

Point sources (stationary) will, of course, need to be inventoried to determine their overall importance to the analysis of any given region. If they are important (that is, a major contribution to the emissions in the region), then some consideration will need to be given to their projected future emissions as well. Once these projections are produced, regardless of the method, the emissions may then be used as input to a dispersion model (to be discussed below). After the dispersion calculations, the point-source emissions will have been converted to area concentrations. These will eventually be added to the area-source emissions after the area-source emissions themselves have been dispersed.

9.4.2 Mobile sources

A rather different set of problems is encountered when one attempts to estimate emissions from mobile sources. In theory, these include two distinct estimation problems, as mentioned above. For general traffic in each zone or subarea of the region, the emissions (once they are estimated) may be considered as area-source emissions. This means that the numbers, origins, and destinations of intrazonal trips must be estimated, as well as the average speed and distance travelled. In addition, of course, the vehicle fleet must be described: types of vehicles, ages of vehicles, and so on. All of this information, taken together, will be necessary to estimate area-source emissions from mobile sources.

Estimating the mobile-source emissions along the links of heavily travelled roads presents, in theory at least, a different type of problem. Again, it is necessary to know the average speeds and distances travelled, numbers of starts and stops, as well as a description of the vehicle fleet.

These emissions, once estimated, are then considered to have come from a line (as distinct from a point or area) source. However, this presents rather difficult problems in describing the dispersion of these emissions, as what is required is an integration (in the calculus sense) of a point-source dispersion model over the length of the line source. In current practice it proves much simpler—and not significantly less accurate—simply to add the total line-source emissions to the already calculated area-source emissions from mobile sources.

In summary, then, there are several sources of emissions in urban regions, each of which is estimated in a somewhat different fashion. After the imposition of several simplifying assumptions, one practical way of estimating stationary area-source emissions involves gathering data for a base year statistical analysis of emissions as they relate to existing spatial patterns of activities and land use. It should be noted here that even for this rather simplified emission estimation procedure, the level of detail required for the input data is in excess of what is typically produced as the output of current location and land-use models such as ITLUP. Thus, in some instances, rather important inputs for emission estimation may be outputs from such models at the margins of statistical reliability.

Stationary point sources can be dealt with at any level of detail, as they are exogenous inputs to these types of estimation procedures in any case. It may be possible, for forecasts of future levels of emission outputs from point sources, simply to trend or extrapolate their base year levels, but such a procedure inevitably leads to more difficult questions of the likelihood of location and/or relocation of such sources, and the nature of the particular emission generating process and its behavior over time.

Mobile emission sources may be considered as area sources for general urban traffic, and as a line (integrated point) source for expressways and other heavily travelled highway links. Practical considerations suggest that line sources, although their emissions may be calculated as a function of the numbers of vehicles travelling on specific links, should be treated as contributing to the mobile area-source emissions as well. These calculations also depend upon detailed data being output from a transportation model. Although trip data are not too difficult to obtain from such models, the vehicle-fleet compositions present greater difficulties. Reasonable estimates of fleet compositions for the base year are quite feasible. The problems arise in projecting the composition of the fleet into the future. In all likelihood these projections will depend upon regionalized versions of national projections. When mobile-source emissions are of particular interest in a specific urban area, additional research into the vehicle fleet of that area and its projected variation over time will be necessary.

It therefore appears that estimates of urban (metropolitan) emissions both from stationary and from mobile sources are clearly feasible, but need to be analyzed with considerable care, as they impose strain upon the limits of statistical reliability of the outputs from the transportation, location, and land-use models which would be required to prepare inputs for the emission calculations. Then, even when the emissions are estimated, it remains to deal with the problems of estimating dispersion of these emissions over the region.

9.4.3 Dispersion of emissions

As there are both area-source and point-source emissions to be dealt with, it is unfortunately necessary to consider two types of dispersion calculation. These dispersion and diffusion models serve to link or transform emission data to estimates of the spatial distribution of pollutants which result from the location of emissions and the mechanics of their dispersion. They consequently involve consideration of the meteorological aspects of the region(s) being analyzed.

In theory, the following meteorological phenomena should be considered in an attempt to estimate properly the dispersion of emissions.

1. Horizontal air movements, which are determined by the direction and speed of winds, along with variability either in direction or in speed.
2. Vertical movements or transfers, which result from the combined effects of several phenomena:
   (a) temperature gradient, with temperatures decreasing with increasing distance from the surface of the earth;
   (b) thermal turbulence, which results from atmospheric heating as well as from differences between air temperature and surface temperature;
   (c) mechanical turbulence, which results from irregularities in the surface of the earth due both to natural topography and to man-made structures;
   (d) numerous other minor factors.

Of course, the horizontal and the vertical phenomena interact with each other to obscure the clear understanding of the basic processes further. To this already complex situation must also be added consideration of the rate at which various pollutants are being emitted, and of any chemical reactions taking place in the atmosphere which may be altering the concentrations of pollutants.

The most frequently used formulations for the diffusion–dispersion process derive from what is known as the Gaussian diffusion model (Williamson, 1973). Its general form is as follows:

$$\theta(x, y, z, t) = \frac{q}{2\pi^{1\cdot 5}\sigma_x \sigma_y \sigma_z} \exp\left[-\frac{(x-ut)^2}{2\sigma_x^2} - \frac{y^2}{2\sigma_y^2} - \frac{z^2}{z\sigma_z^2}\right]\ ; \qquad (9.1)$$

here

$\theta(x, y, z, t)$ is the pollutant concentration at point (x, y, z) at time t;
q is the pollutant emission rate;
x is the distance, along the mean wind direction, from the source;
y is the distance, along the horizontal crosswind direction, from the source;
z is the distance, along the vertical crosswind direction, from the source;

u is the wind speed;
t is the travel time of pollutants from the source;
σ_k is the standard deviation of the emission plume dispersion in the kth direction.

This equation is simply a three-dimensional representation of a Gaussian, or normal, distribution. The presence of the variable t, the travel time of the pollutants, allows the calculation to be made for any instant, as a function of time-varying meteorological conditions and of time-varying emissions. In practice it is usually not worth the extra expense of data collection and calculation to maintain this time variation. Consequently, certain steady-state assumptions are usually made, which lead to the following simplified form of the Gaussian model:

$$\theta_{x,y,z} = \frac{q}{2\pi\sigma_y\sigma_z u}\exp\left(-\frac{y^2}{2\sigma_y^2}-\frac{z^2}{2\sigma_z^2}\right) . \tag{9.2}$$

This equation does not, however, take account of the reflection of airborne pollutants from the ground. To do this, a further modification of the equation is required. Additional modifications of equation (9.2) are necessary to yield the ground-level concentrations of pollutants. Yet further problems arise in the case of plume trapping, where atmospheric conditions prevent dispersion of the plume above the 'stack height'. Finally, as this discussion was intended to pertain to area sources, a further modification is necessary to convert area sources to equivalent point sources for use in the dispersion formulation. To simplify all this, these calculations are subsumed into the calculation of a pollution-transfer matrix which yields a set of coefficients that describe the effects of emissions from zone i on air quality in zone j. The elements of this matrix are calculated on the basis of average meteorological values, with certain levels of stability and time averaging of these variables and of the emissions assumed. The best-known model of this form is known as the Hanna–Gifford model (Hanna, 1971).

In addition to this dispersion model for area sources, an additional dispersion model for point sources is necessary. This model is in fact derived from equation (9.2) above, but with fewer assumptions than for the Hanna–Gifford model. However, some simplifications and modifications are also necessary to yield a practical point-source dispersion model. The best-known form of this type of model is the Martin–Tikvart model (Martin and Tikvart, 1968). All the procedures for generating emissions, linked to the procedures for calculating dispersion, together provide the means for estimating the consequences on air quality of the economic, transportation, and land-use policies that were described and discussed in the previous section of this chapter.

9.4.4 Policy impacts[17]

The policy tests of changes in regional growth rates described in section 9.3 clearly illustrated that there are environmental costs associated with growth. Increases in population and/or employment in specific zones are always associated with increases in total vehicle distances travelled and zonal trip starts[18]. In zones where the population increase was above average, the increase in SO_2 concentration was also above average. Similarly, with increases in regional growth there are increases in CO concentration. As CO emissions result in the main from cold trip starts and as these are more evenly distributed over the region, the CO increases are not quite so concentrated.

Similar emission and concentration effects appear as a consequence of regionwide changes in transportation costs. Cost increases resulted in tendencies towards concentration of population and employment and thus result in shorter average trip lengths. Cost decreases result in just the opposite tendencies. With the 10% increase in regionwide transportation cost, there were substantial increases in SO_2 and CO concentrations in zones 1 and 2 in San Francisco and in zones 16 and 17, Oakland and Berkeley. The zonal changes in SO_2 and CO concentrations as a result of this policy are shown in figure 9.9. In the case of the transportation cost

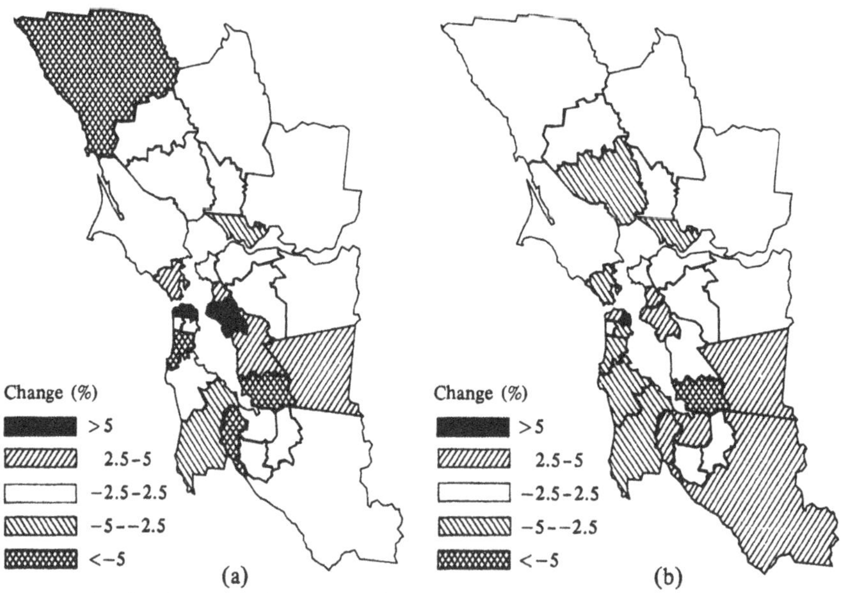

Figure 9.9. San Francisco Bay Area—30 zone system: percentage changes in pollutant concentrations as a result of a transportation cost increase of 10%. (a) SO_2; (b) CO.

[17] These results are summarized from Gross (1979).

[18] An additional set of output measures available from ITLUP. Minor modifications of the program were made to obtain these in addition to the standard outputs.

decrease, the decentralization of activities resulted in reduced SO_2 and CO emissions and concentrations in the most urbanized (central) zones, and increases in suburban zones.

Again, in the parking-charge and bridge-toll policy tests, a cost of increased population and employment in a zone was increased exposure to SO_2 and CO pollution. Similarly, a consolation to zones losing population and/or employment was that the remaining population would breathe cleaner air. In general, even though some of these policies result in small or no changes in average SO_2 concentration, indices of employee or resident exposure may change owing to different populations being at risk in zones where there were changes in concentrations. A 10% increase in SO_2 concentration for a zone with one million population outweighs a 20% decrease in SO_2 concentration for a zone with three hundred thousand population. Thus a full evaluation of the effects on air quality of transportation, location, and land-use policies, although made possible by integration of emission and dispersion subroutines with ITLUP, requires very detailed analysis of outputs. Finally, it should be recalled that detailed climatological and topographical data are input to the diffusion and dispersion calculations. These data will vary considerably from city to city, suggesting that careful consideration should be given to attempts to generalize from these results.

9.5 Further policy analyses with ITLUP

During the several years of development of the integrated package, a great many of the author's graduate students made use of the then current versions of the package or submodels thereof for term projects as well as for master's and doctoral theses. These projects served the dual purpose of training the students in the potential uses of such models and of testing the performance of the models themselves. Some of the earliest projects used only DRAM, as at that time ITLUP was not sufficiently well documented and standardized for convenient student use. Several of these 'DRAM-only' projects are described first, then several DRAM–ITLUP projects. The projects are described here in more or less chronological order.

First, there was a group of three projects using the original version of DRAM, and the Minneapolis–St Paul data set. The first of these attempted to simulate the effects of different greenbelt strategies in the Minneapolis–St Paul area. The substantive results of the project were not clear-cut, were subject to considerable interpretation, and are not of interest here; however, the methodological results pointed the way to important additional research work. In particular, it was discovered that even though the model had fitted the base year data well, there were certain difficulties to be found in the input of policy tests. As a result the specifications of some of the independent variables of DRAM were altered. The most important point learned from this first project, however, was that the testing of policy with urban land-use and transportation models requires an informed user. As an example, the input of certain policies may require changes of several variables in a particular zone, plus consideration of the regionwide effects of those changes and how they may or may not require the change of regional control totals as well. This further extends into the general question of—and the understanding of—the different ingredients of the residence-location decision.

The final project in this early group attempted to combine earlier modelling work on greenbelt policies for the Minneapolis–St Paul area with evaluation of the consequent changes in air pollution and vehicular emission. The above-described pollution and emission subroutines, connected to DRAM and ITLUP, yielded in this case the rather interesting finding that certain greenbelt policies, though preserving urban open space, did, by virtue of producing longer home-to-work trips consequent on the relocation of residents, lead to increased levels of air pollution. These results offered further evidence of the old adage that "one does not get something for nothing". It further reinforced the notion that policy analyses for urban areas require the evaluation of systems so complex that it is inevitable that computer simulation of the type described here will be necessary for their proper execution.

A second set of four projects was done using the current version of DRAM with data sets for the Atlanta and the Seattle (Puget Sound) metropolitan areas. One of the Atlanta simulation projects produced interesting substantive results from policy simulation directed at possible means of revitalizing an older CBD. It was found that even rather substantial changes of thousands of employees in particular zones do not have a very significant effect over an entire large metropolitan region. In the case of Atlanta, 20000 employees, for example, represent only 3·4% of the basic employment (and less than 0·5% of total employment) in the metropolitan area. Thus, although such a substantial change (relocation) of employees will have certain local effects—for example, to rings of immediately adjacent zones—it is not sufficient to have the effect of revitalizing the entire central business district of the metropolitan region. Such a revitalization would require a number of highly coordinated programs, all aimed at stimulating economic growth in the CBD. These would include attempts at relocation of employment into the CBD, but would also require policies such as transportation improvement, provision of parking (which would be difficult to represent in a model such as this in any case), and perhaps several other kinds of economic stimuli.

The other three projects in this group involved attempts to change either several parameters or several variables all at once. This had useful pedagogical results. It taught students that if several parameters or variables are altered all at once, it is difficult to understand the causes of any consequences observed in model outputs. The substantive value of the outputs of these projects was minimal. However, as was mentioned above in the description of the projects with the Minneapolis data, they do highlight the need for informed users of these models. This should not be construed as a criticism of this sort of model: even an office typewriter, for its most efficient use, requires an informed user.

From the substantive standpoint, the brief description of these projects serves to illustrate both the complexity and the stability of urban transportation and land-use systems. The complexity is illustrated by the ineffectual attempts to influence (revitalize) a declining CBD by the relocation (in policy simulation) of employees. Here it was demonstrated that the movement of what appeared to be quite a substantial number of employees, indeed more than one would expect to be able to move by the introduction of policies, was still unable to effect any substantial regionwide change in the location of employment or population when undertaken in the absence of other stimuli.

From a methodological point of view, small projects such as these served to aid in the development and evolution of the models. The various attempts to make use of the models invariably uncovered unthought-of-possibilities, some of them beneficial and some detrimental to the model runs. The projects, even when unsuccessful, thus suggested directions for further research and modifications. Moreover, these efforts clearly illustrated the need for informed users (analysts) of these models so that policies can be properly simulated and analyzed, and model results properly interpreted. They also emphasized the general difficulty involved in clearly describing the policies to be tested. A policy-simulation exercise can often be beneficial in real policy analysis by an agency, as many times the attempt to spell out the composition of policies—the variables which are to be changed, for example—will also shed light on the real intent of the policy and the expectations (justified or unjustified) which the analysts and policymakers may have for the long-term effects of the policies.

At a somewhat later date it became possible for students to undertake projects making use of the full capabilities of ITLUP. One of the first of these was the prototype for the San Francisco bridge-toll analysis discussed earlier in this chapter. Some of the results of this project were, for bridge-toll decreases, as follows (refer to figure 3.7 for county names).

1. When the impedance on bridge links was reduced, Marin and Alameda sustained the greatest population growth; this was attributable to their location contiguous to the bridges. Households of all income groups demonstrated a propensity to move their place of residence outward from San Francisco and thereby increase the distance of home–work–shop trips. Middle income households were most inclined to relocate.

2. The population loss in San Mateo was relatively small; this suggests that, except in the northern urban areas, the changes in bridge tolls were not the prime motivation for relocation.

3. Contra Costa and Sonoma, counties moderately distant from the bridges, experienced growth, but it was considerably less than that of Marin and Alameda. The diminishing growth is the result of increased

distance from the bridges. That is, as travel distance increases, the impact of lowered bridge tolls on residential location diminishes.
4. Napa, Solano, and Santa Clara comprise the 'gray zone': those areas where other factors are more important than lowered bridge tolls in attracting or discouraging residential location.

The results for bridge-toll increases were less clear-cut. The overall trend was towards population increase in the southwestern quadrant of the region, in San Francisco and San Mateo, with decreases in the areas north and east of the San Francisco Bay. The changes were smaller than those found in the case of bridge-toll decreases. This may have been a result of the already heavily developed condition of San Francisco and, to a slightly lesser degree, of San Mateo. These counties were unable to grow as much as the 'suburban' counties in the northern and eastern sections of the region. Moreover, there are several unanswerable questions as to the implicit price elasticities for the travellers using the bridges. It is probable that the responses to toll increases versus toll decreases are asymmetric; thus more information about the value of time is required along with these other results before a full revaluation could be undertaken. Overall, the study produced a number of interesting findings about the possible effects of such toll changes on regional transportation and location patterns.

Another student project making use of ITLUP involved an attempt to examine the effects of 'user charges' to reduce peak hour highway congestion. The economic principle underlying this potential policy is that by charging users for peak hour use of selected links of the transportation system, some users will be persuaded to make use of other segments of the system or of the same segments but during off-peak hours. The anticipated result would be a low-capital-cost policy that reduced peak hour traffic congestion. There are, of course, many practical problems associated with the imposition of user charges. These range from the problem of determining when and where and what size charges should be imposed, to the problem of how to collect the charges from the users. Despite these problems there has been considerable interest in the potential effectiveness of such policies.

The San Francisco data set which has received so much attention in these pages has a highway network described in terms of 5 107 links. A set of 21 one-way links was selected for imposition of user charges. These links were the principal expressway connections of the San Francisco CBD to the rest of the region, and are shown in figure 9.10. The residence-location and work-trip-generation part of ITLUP focuses on the work-to-home trip. Thus the outbound (from the CBD) one-way links were those altered for the policy analysis. This was done by substantially increasing the free flow impedances on these links.

The results of these runs were quite interesting, and suggest the possible effectiveness of such user charges both to relieve congestion, and to shape the metropolitan area. The highway links that were modified for user charges are located in zones 5, 16, and 30, directly adjacent to downtown San Francisco. Highway congestion decreased in all three of these zones, with a significant decrease of 12% in zone 30. Note, this is one of the 'special' zones found in the growth/decline analyses in section 9.3. The congestion changes for all thirty zones are shown in figure 9.11(a).

If a policy such as this was considered by the population of the region to be permanent, then there would eventually be induced population relocation as well. Thus rather interesting shifts in population were observed in the simulation results. The policy resulted in a population shift out of the zones previously containing long-distance commuters to downtown San Francisco. Also, there appeared to be a tendency towards growth in zones which served as dormitories for commuters to employment not located in the San Francisco CBD. These zones also received increased residence location from persons still commuting to San Francisco, but

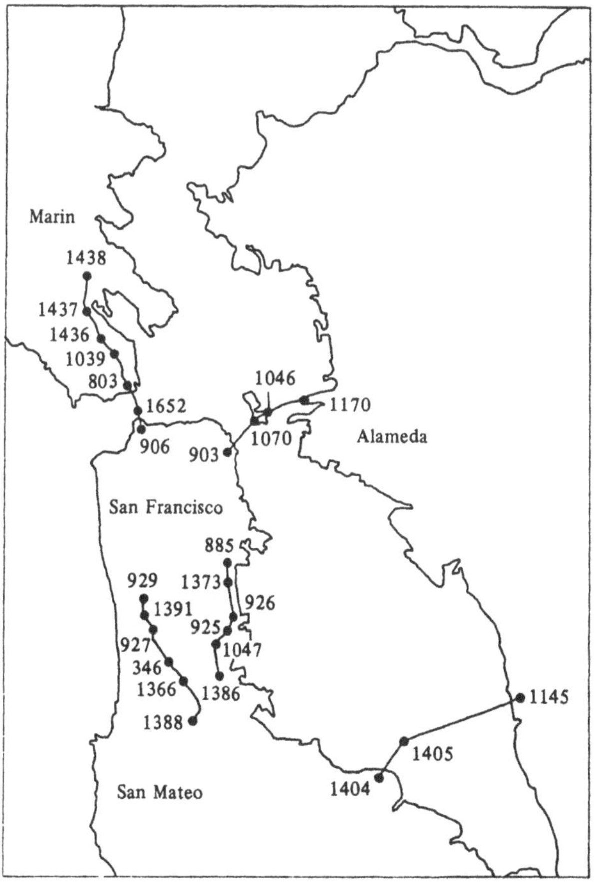

Figure 9.10. San Francisco Bay Area: highway links selected for user charges.

living closer in an effort to deal with the increased transportation costs. The zone-level population changes are shown in figure 9.11(b). Thus one result of the policy was to shift more population, and subsequently more employment, to the east side of the San Francisco Bay. In general, the policy did reduce congestion in the directly affected zones. However, the subsequent relocation of population created increases in congestion elsewhere. Thus, as is always the case in these complex systems, an actual policy suggestion would probably need to be evaluated and revised in concert with model runs such as these in order to achieve its desired ends without their being negated by undesired results elsewhere in the region.

Additional projects such as these have been undertaken to examine regional increases or decreases in transportation costs, inclusionary policies for low-income housing, effects of exclusionary zoning policies, spatial consequences of regionwide employment growth or decline, and others. The principal purpose of most such projects was pedagogical, and thus the results were somewhat mixed. It was not always possible for small groups of students to accomplish as much as they hoped in a single term project. Nonetheless most of the results were informative and reasonable. Regional increases in transportation costs caused regional subnucleation of population. Inclusionary housing policies had only modest and usually zone-specific results and very little regionwide effect. Exclusionary policies 'worked'

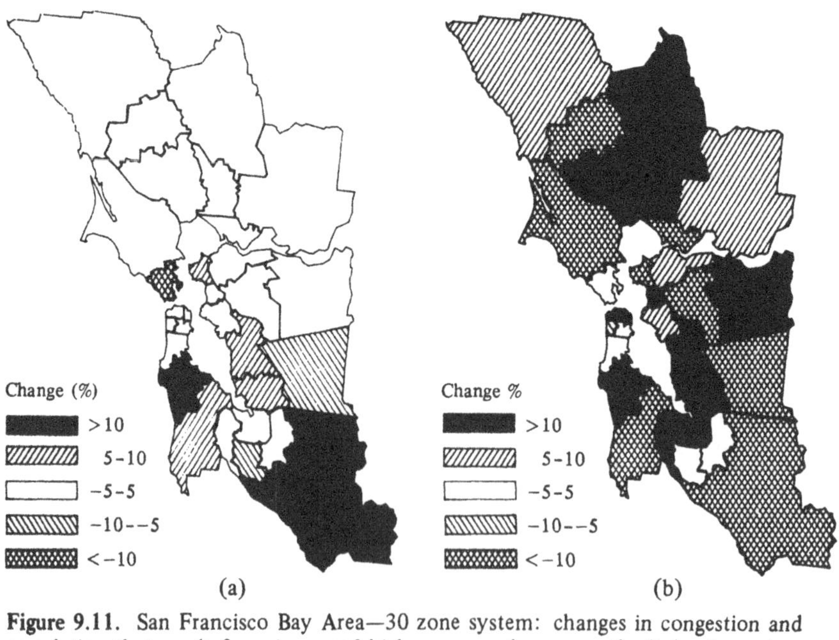

Figure 9.11. San Francisco Bay Area—30 zone system: changes in congestion and population that result from increased highway user charges on the links shown in figure 9.10. (a) Congestion; (b) population.

for a time, but did not prevent changes in household composition in zones where such changes were the consequences of regionwide phenomena. Many students learned both what could and what could not be attempted in policy analyses with large-scale urban simulation packages, as well as gaining new insights into the functioning of the urban system.

There are, of course, many other policies which could be discussed and evaluated. There are also many which cannot be addressed by models of this type. The results discussed in this chapter were presented with the intention of showing, by example, the types of analyses that can and should be done now. In the next chapter some suggestions are given as to future development of components of integrated transportation and land-use analysis procedures.

10

Some future directions for location and land-use research

10.1 Introduction

At several points in the previous chapters of this book allusions were made to the need for further work on one point or another. Although overall there are a great many such areas for further research, this chapter contains discussions of a small set of selected topics which are of particular interest. These topics are roughly grouped into four main categories. The first of these has to do with the general dynamic properties of the integrated model package and its components. The second general heading is that of substantive improvements or additions to the component models of the integrated packages. The third covers further work which should be undertaken to improve the procedures for estimating the parameters of employment-location and residence-location models. The fourth category contains topics relating to the applications of these modelling techniques to actual planning and policy analyses.

10.2 Dynamics of the separate submodels

10.2.1 Discussion

Before the dynamic properties of linked sets of models are discussed, some of the properties of the submodels of such systems should be examined. The development of urban transportation and land-use models has seen by far the greatest part of the effort spent on improving the ability of the models to replicate reality. Reality in these efforts has usually meant a single cross-sectional data set (that is, for fixed time point). In most such efforts, the magnitude of that task alone was sufficiently great to preclude any serious attempt to study the dynamic properties of the models. The few studies that have attempted to examine the dynamic properties of urban systems have done so at the expense of spatial and sectoral detail—so much so as to render current dynamic formulations virtually useless for most practical planning and policy analysis in metropolitan areas.

The present impracticality of dynamic model formulations does not imply that research in this direction should be discontinued. The results of continuing theoretical research on the dynamics of urban systems can provide guidance to the ongoing applied research and application of the kinds of models described in this book. At the time of writing, the preliminary results of the 1980 US Census are just becoming available. In the next few years this trove of new data will provide a second (and for some cities, a third) cross-sectional data set, and this will allow some interesting new investigation into the dynamics of urban transportation and land-use phenomena. The material presented here is indicative of some of the areas which need examining.

First, consider the general form of the residence-location model developed in earlier chapters. The basic hypothesis is that where people live is a function of where they work, the relative attractiveness of all possible residence areas, and the generalized cost of travelling from any possible place of work to any possible place of residence. Thus the usual form of this model is

$$N_i = \sum_j E_j \, \mathrm{f}(W_i, c_{ij}) \,, \tag{10.1}$$

where N_i is the number of employed residents living in zone i, E_j the number of employed residents working in zone j, c_{ij} the generalized travel cost between zone i and zone j, and W_i the relative residential attractiveness of zone i.

Now in an actual application there are a great many elaborations to such models, as described in chapter 7, but for the moment the necessary points may be made by use of this general formulation.

When such a model is used for forecasting and/or policy analysis, one or more of the independent variables is altered, and the resulting change in N_i is studied. Note, however, that equation (10.1) is a static, or 'state' equation whose solution implies an equilibrium. It is obvious to anyone who has given the matter even a few moments thought that urban areas are never in a static equilibrium condition. There are always changing forces impinging on the system. Thus no matter how well an equation structure, with its variables and parameters, can describe a cross-sectional view of an urban area it is bound to have built-in errors arising from its neglect of the dynamics of the system.

It is possible to reformulate equation (10.1) into a difference equation such as

$$N_{i,t} = \sum_j \Delta E_j \, \mathrm{f}(\Delta W_i, \Delta c_{ij}) + N_{i,t-1} \,, \tag{10.2}$$

or even into a differential equation such as

$$N_{i,t} = \sum_j \frac{\mathrm{d}E_j}{\mathrm{d}t} \, \mathrm{f}\left(\frac{\mathrm{d}W_i}{\mathrm{d}t}, \frac{\mathrm{d}c_{ij}}{\mathrm{d}t}\right) + N_{i,t-1} \,. \tag{10.3}$$

However, either of these formulations, which might yield more accurate replications of reality, presents substantial problems not only in terms of analytical solutions, but also in terms of data requirements for parameter estimation. Moreover, it is customary to use several variables to construct the attractiveness term, so, for example, we may have

$$W_i = X_{1,i}^{a_1} X_{2,i}^{a_2} X_{3,i}^{a_3} \ldots \tag{10.4}$$

where $X_{1,i}$, $X_{2,i}$, $X_{3,i}$ are attributes of zone i such as land area, vacant land, percentage of low-income population, and so on, as per the discussion in chapter 7. In addition, the generalized travel cost will usually be used in some functional form such as

$$\mathrm{f}(c_{ij}) = \exp(\beta c_{ij}) \,, \tag{10.5}$$

or

$$f(c_{ij}) = c_{ij}^{\alpha} \exp(\beta c_{ij}) \ . \qquad (10.6)$$

The $a_1, a_2, a_3, \ldots$ from equation (10.4) and the α and/or β from equations (10.5) or (10.6) are empirically derived parameters. It is quite likely that these parameters vary over time too. Consequently they too should be included in any general equation structure in some time-varying way. The difficulties of obtaining the data and performing the analysis for such equation structures puts their practical application some years into the future. Thus the important questions at the present are concerned with the degree to which currently possible structures may provide useful results, and the circumstances in which their results must be regarded with particular caution.

Consider, for example, several typical uses of models with structures like that of equation (10.1). Suppose that it was desired to analyze the effects of a change in employment in a particular zone or set of zones. Thus one of the values of E_j would be changed and would result in a change in one or more of the N_i values. Next, suppose that one of the attribute variables in equation (10.4) was altered. Again, there would be a change in one or more of the N_i values. The same will obviously be true for a change in any c_{ij} value. Note, however, that there are two different ways to interpret such changes, and the manner of interpretation may be of importance.

In the first case, that of a change in employment, consider first attempting to analyze the effects of such a change. One way would be to assume that

$$N_i + \Delta N_i = \sum_j (E_j + \Delta E_j) \mathrm{F}(W_i, c_{ij}) \ . \qquad (10.7)$$

It might then be argued that the ΔN_i which results is a direct result of the ΔE_j. Although this is computationally correct, it is at variance with the static formulation of equation (10.1). The equation structure does not describe the effect on N_i of a change in E_j. The static equation structure describes the spatial pattern of N_i which 'coexists' at a given time with the spatial distributions of the other variables included in the model. These coexisting spatial patterns will have evolved over some considerable length of time. All manner of transient phenomena will have contributed to the distributions which are finally used to estimate the parameters of the static model formulation. Thus, if one wishes to use a static model in the way implied by equation (10.7), it is necessary to assume that the employees represented by the change in employment ΔE_j behave like the average of all the employees included in the original cross-sectional data set.

Another way to interpret a change in employment would be as follows,

$$N_i' = \sum_j E_j' \mathrm{F}(W_i, c_{ij}) \ . \qquad (10.8)$$

Here the notion is that the changed employment E_j' represents an alternative static distribution of employees which is associated with N_i', an alternative static distribution of residences. The end results here are, of course, the same as those from the change assumptions made above. In this case, however, it is clear that the alternative employment distribution will be expected to behave like the original employment distribution, as it is indeed an alternative distribution rather than a specific change.

The issue having been raised in this fashion, it is worth considering just how different the behavior of a change ΔE_j would be from the average behavior of all the E_j included in the original data. The primary considerations in assessing the degree of difference here are the levels of spatial and sectoral detail. If the cross-sectional model is based on, say, total employment, and the change in employment is both small and in an atypical employment sector, then the results will not be well described by the static equation. If, however, there is some degree of sectoral detail in the cross-sectional data set, and if the change in employment is in a more typical employment sector, then the static model may do reasonably well in estimating the consequences of the change.

Consider further the matter of using such static models for forecasting over long periods of time. Strictly speaking, the static model structure of equation (10.1) is time-independent. If, for example, the variables on the right-hand side of the equation, E_j, W_i, c_{ij}, were known (or estimated) for times $t+1$, $t+2$, and $t+3$, it would not matter if the N_i were estimated for $t+3$ before being estimated for $t+1$. In most actual applications there are several practical factors which eliminate this possibility. First, it is often the case that the variables on the right-hand side of the equation are, in part, a function of the variables on the left. For example, given equation (10.1), there may be a companion equation for E_j. This might be of the form

$$E_j = \sum_i N_i \mathrm{F}'(Z_j, c_{ij}) , \tag{10.9}$$

where Z_j is relative employment attractiveness of zone j.

The addition of this complexity would, in itself, still not eliminate the possibility of calculating the $t+3$ forecast before calculating the $t+1$ forecast. However, once these kinds of matching sets of models are developed, it is almost invariably the case that lagged variables are introduced on the right-hand side of the equations. Thus, for example, the structure might be

$$N_{i,t} = \sum_j E_{j,t} \mathrm{F}(W_{i,t}, c_{ij,t}, N_{i,t-1}) , \tag{10.10}$$

and, in a corresponding way,

$$E_{j,t} = \sum_i N_{i,t} \mathrm{F}'(Z_{j,t}, c_{ij,t}, E_{j,t-1}) . \tag{10.11}$$

Now, obviously, it is necessary to calculate the N_i and E_j in the proper temporal sequence. However, by adding a temporal dimension to the basic model structures, an entire new set of questions are raised about their temporal dynamics. In fact, it becomes necessary to decide, for all the variables on the right-hand sides of equation (10.10) and (10.11), whether the time subscript should be $t-1$ or t. This is a question of the speed of response in urban phenomena. How quickly, for example, does population change location in response to changing spatial patterns of employment? How quickly does employment relocate in response to changing spatial patterns of markets (for retail trade), or of labor force? The answers to these questions require additional data, too, as well as further investigation. In this case, however, it has been possible to do some numerical analyses to examine the consequences of various assumptions about the various lagging factors. It has also been possible to examine some of the consequences of the use of a static model formulation in an incremental recursive forecasting procedure. These results are described below.

The original, hypothetical, form of the residence-location model DRAM was dynamic in that it posited the use of change-in-state variables (for example, ΔN, ΔE) on both sides of the equation. Gaps in the data soon made the impracticality of such a formulation quite obvious. Attention next turned to a simultaneous (static) form where the variables on both sides of the equation had the same time subscripts. Subsequent development of the model formulation led to its disaggregation (for most applications) into four household types. This presented no particular difficulties in estimation of parameters, as a cross-sectional data set was still able to serve the purpose. For forecasting purposes it was, however, a different matter. The use of the same time subscripts on both sides of the equation implied the need for a simultaneous solution to a rather large set of nonlinear equations. For this reason an investigation was undertaken to determine the consequences of using lagged ($t-1$) variables on the right-hand side of the equation. It should be noted that this possibility was explored despite the expectation that for most applications the $t-1$ data would not be available for calibrations, and that therefore there would be a mismatch in time subscripts between the parameter-estimation and the forecasting phases of modelling in any case.

The empirically derived parameters in DRAM have been estimated, as described above, for fixed time points in all applications to date. This raises the further question of interpreting the parameters so estimated. Essentially the model formulation implies that at a particular instant (the time point of the data set) the relationships between all the variables yield the particular parameter set. If one asserts that the system was in equilibrium at that instant, then these parameters represent the parameters for the equilibrium situation. It would be difficult to justify the assertion of such an equilibrium; thus there is no alternative left but to assume that the parameters represent the relationships which existed at a particular instant. An unfortunate consequence of this latter assumption is the necessity of accepting the high probability that the parameters estimated for one time point will not be correct for any past or future time point. It then becomes necessary to attempt to thread between the Scylla of an equilibrium assumption and the Charybdis of time-varying parameters.

There is no way to resolve this dilemma at the present time. Although these parameters quantify certain components of behavior, they do so only for a particular point in time and space. This point is most certainly not an equilibrium state, consequently any forecast produced via the use of these fixed parameters is certain to contain some degree of inaccuracy. This problem may eventually be solved by either of two developments, neither of which is likely to be available of policy/planning application for five to ten years, if ever. The first of these, the development of aggregate models based on microbehavioral theories whose parameters (it is hoped) will be essentially time-invariant, may emerge from current research work such as that described in section 10.3. The second possibility is that enough data, and subsequently parameters, for dynamic macrobehavioral models will become available to allow the development of procedures for varying model parameters systematically over simulated forecast periods.

10.2.2 Numerical experiments

The calibrations of DRAM described in chapter 7 were done for a static, simultaneous-solution form of the model. The forecasting form of the model differed in its use of lagged household-composition variables and lagged land-use variables. Simply stated this means that, in forecasting, the $N_i^{n'}$ and N_i^n in equation (7.26) are replaced by $N_{i,t-1}^{n'}$ and $N_{i,t-1}^n$, and the L_i^v, $\tilde{x}_i$, and L_i^r variables are replaced by $L_{i,t-1}^v$, $\tilde{x}_{i,t-1}$, and $L_{i,t-1}^r$, respectively. By so doing, it becomes possible to solve the forecast form of DRAM for one household type at a time, rather than having to deal with the complexities of a simultaneous solution.

Testing the consequences of this formulation involved a process of repeated iterations through the household estimates, with the travel-cost and land-use variables remaining constant. The retail (LOMS) employment portion of IPLUM was included in DRAM as part of this test. Stated in different terms, DRAM was run with 1970 inputs for basic employment and for the transportation network, and 1960 inputs for all other variables. With these inputs a 1970 forecast of households was calculated. This forecast was, in turn, used to replace the 1960 household inputs, and the forecast was then recalculated. The process was then repeated for a predetermined number of times, depending on the particular test being conducted. The results were then examined in order to determine whether the forecasts had reached some equilibrium point, how many iterations were required to reach that point, and whether the household forecasts at that point were better or worse forecasts than those which were produced without iteration (the first 1970 forecast mentioned above).

These numerical experiments were performed with the Minneapolis–St Paul 108 zone data set used in previous testing and development of

DRAM as described in chapter 8. The first of these experiments involved four iterations of the household projections, followed by the nonbasic-employment projection and one additional household iteration. The second experiment again had four iterations of household projection, followed by a nonbasic employment forecast and nine more iterations of household projections. The last experiment of this set had ten iterations of household projection followed again by nonbasic employment and a further nine iterations of household projections. These experimental model runs are referred to as runs 4+1, 4+9, and 10+9, respectively. For the analysis of these runs several representative zones (urban, suburban, rural) were selected and the results are tabulated in tables 10.1, 10.2, and 10.3.

No results from run 4+1 were tabulated, as these merely showed that the household estimates did change from one iteration to the next. In table 10.1 the first ten iterations of household projections are shown for eight different zones. The rural zones, zones 94 and 24, show almost complete convergence for all household types. This is also true for the suburban zones, zones 49 and 33. Zones 57 and 56, urban core zones with a low and high level of population respectively, also converge. Zones 64 and 74, both high-population urban zones, did not quite converge, though the largest changes from the ninth iteration to the tenth iteration (+3·3% for zone 74 LI households and +3·0% for zone 74 LMI households) were actually rather small. From this it was clear that the model did indeed seek, and to all intents reach, an equilibrium solution after ten iterations of household projections.

In table 10.2, the results of nine iterations after the LOMS employment allocation (which was preceded by four household iterations) are shown. Table 10.3 shows the results of nine iterations after the LOMS employment allocation (which was preceded by ten household iterations). The same zones selected in table 10.1 are repeated in tables 10.2 and 10.3. Virtually all household types in all zones reached an equilibrium value in both of these runs. Thus with or without the LOMS employment allocation procedure included in the projection process, the household projections did tend towards equilibrium values and, for the most part, reached that equilibrium at the completion of nine or ten iterations.

A separate but closely related question was whether the starting points of these iterative procedures affected the numerical value reached at equilibrium. Here comparison of the results in tables 10.2 and 10.3 is informative. Tables 10.2 and 10.3 each show nine iterations. In both cases these nine iterations followed an allocation of LOMS employment. The household projections used as input to these LOMS employment allocations were different in run 4+9 from those in run 10+9. The input values for run 4+9 would correspond to the values for iteration 4 in table 10.1. The input values to the LOMS employment allocations in run 10+9 correspond to the values for iteration 10 in table 10.1.

Table 10.1. Household projections for Minneapolis–St Paul: first ten iterations for selected zones, from run 10+9.

Iteration	Household type				Iteration	Household type			
	LI	LMI	UMI	UI		LI	LMI	UMI	UI
Zone 94					*Zone 56*				
1	218	527	530	265	1	8457	4658	3054	2942
2	151	406	548	297	2	9813	5346	3084	2772
3	125	324	545	323	3	10661	5831	3098	2675
4	112	278	537	343	4	11148	6121	3107	2619
5	105	253	529	357	5	11407	6280	3114	2587
6	101	239	523	366	6	11535	6362	3119	2569
7	99	232	518	372	7	11594	6404	3123	2558
8	98	228	516	375	8	11620	6424	3125	2552
9	97	226	514	377	9	11630	6433	3127	2549
10	97	225	513	378	10	11633	6438	3128	2547
Zone 24					*Zone 74*				
1	241	707	614	346	1	2839	3030	4406	6647
2	129	527	706	375	2	3765	3612	4187	6425
3	97	389	723	411	3	4504	4237	4196	6054
4	84	315	714	442	4	5158	4821	4271	5699
5	77	279	701	465	5	5745	5346	4350	5408
6	74	262	689	479	6	6263	5804	4415	5185
7	72	253	682	488	7	6712	6194	4463	5016
8	71	249	677	493	8	7091	6516	4498	4890
9	71	247	674	495	9	7404	6778	4523	4797
10	71	246	673	496	10	7657	6987	4540	4727
Zone 49					*Zone 57*				
1	1108	2509	3995	3883	1	966	244	85	49
2	1072	2374	3897	4061	2	690	212	86	53
3	1037	2287	3830	4174	3	584	199	88	54
4	1012	2230	3789	4242	4	530	192	89	55
5	995	2192	3767	4282	5	500	187	89	56
6	984	2167	3754	4304	6	483	184	90	56
7	977	2151	3747	4317	7	472	182	90	56
8	973	2141	3744	4324	8	466	181	90	56
9	969	2135	3742	4328	9	462	180	90	56
10	967	2131	3741	4330	10	460	180	90	56
Zone 33					*Zone 64*				
1	825	1767	2236	1939	1	16558	10822	6403	5299
2	668	1593	2278	2027	2	15823	10944	6269	5416
3	591	1439	2266	2116	3	15017	10851	6259	5474
4	547	1334	2240	2189	4	14314	10653	6273	5519
5	521	1267	2215	2242	5	13745	10433	6287	5559
6	505	1225	2196	2278	6	13300	10230	6297	5595
7	496	1200	2183	2302	7	12958	10057	6304	5627
8	490	1184	2175	2317	8	12698	9917	6307	5654
9	486	1175	2169	2326	9	12502	9806	6309	5677
10	484	1169	2166	2332	10	12354	9719	6309	5695

Table 10.2. Household projections for Minneapolis-St Paul: second nine iterations for selected zones, from run 4+9.

Iteration	Household type				Iteration	Household type			
	LI	LMI	UMI	UI		LI	LMI	UMI	UI
Zone 94					*Zone 56*				
5	91	252	531	356	5	11434	6275	3109	2586
6	85	234	527	367	6	11547	6366	3112	2565
7	83	223	522	375	7	11585	6409	3116	2553
8	81	218	519	379	8	11593	6427	3119	2546
9	80	215	516	382	9	11589	6434	3121	2543
10	80	214	515	383	10	11582	6436	3123	2541
11	80	213	515	384	11	11575	6436	3124	2539
12	80	212	514	384	12	11569	6435	3124	2539
13	79	212	514	384	13	11564	6435	3125	2538
Zone 24					*Zone 74*				
5	67	279	702	463	5	5843	5333	4330	5393
6	63	258	693	479	6	6426	5829	4389	5153
7	61	248	685	488	7	6925	6258	4438	4970
8	60	243	680	493	8	7346	6613	4476	4834
9	60	240	677	496	9	7692	6900	4502	4733
10	60	239	675	497	10	7972	7127	4520	4660
11	59	239	675	498	11	8194	7304	4533	4605
12	59	239	674	498	12	8368	7441	4541	4566
13	59	239	674	498	13	8502	7545	4547	4536
Zone 49					*Zone 57*				
5	965	2195	3777	4292	5	494	188	90	56
6	944	2159	3765	4321	6	470	184	90	56
7	933	2135	3757	4338	7	456	182	90	56
8	926	2119	3752	4349	8	447	180	91	57
9	921	2110	3749	4355	9	441	179	91	57
10	918	2104	3748	4358	10	438	178	91	57
11	917	2100	3747	4360	11	436	178	91	57
12	915	2098	3746	4361	12	435	178	91	57
13	914	2097	3746	4362	13	435	177	91	57
Zone 33					*Zone 64*				
5	506	1266	2220	2243	5	13943	10448	6273	5570
6	486	1219	2202	2281	6	13552	10291	6273	5595
7	475	1190	2189	2307	7	13231	10152	6279	5616
8	469	1172	2180	2323	8	12979	10032	6285	5636
9	465	1162	2174	2333	9	12786	9933	6289	5654
10	462	1156	2170	2338	10	12639	9852	6291	5669
11	461	1152	2168	2342	11	12528	9789	6292	5682
12	460	1150	2167	2344	12	12444	9740	6292	5692
13	459	1148	2166	2345	13	12380	9702	6292	5700

Table 10.3. Household projections for Minneapolis-St Paul: second nine iterations for selected zones, from run 10+9.

Iteration	Household type				Iteration	Household type			
	LI	LMI	UMI	UI		LI	LMI	UMI	UI
Zone 94					*Zone 56*				
11	83	223	515	377	11	11655	6436	3123	2545
12	81	218	516	380	12	11637	6444	3121	2541
13	80	215	516	382	13	11615	6446	3122	2539
14	80	214	515	383	14	11597	6444	3123	2538
15	79	213	515	383	15	11583	6442	3124	2538
16	79	212	514	384	16	11573	6439	3125	2538
17	79	212	514	384	17	11566	6437	3125	2538
18	79	212	514	384	18	11560	6435	3125	2538
19	79	212	514	384	19	11557	6434	3126	2538
Zone 24					*Zone 74*				
11	61	245	674	494	11	7992	7135	4530	4661
12	60	242	675	496	12	8229	7308	4531	4608
13	60	240	675	497	13	8405	7449	4538	4566
14	59	239	675	498	14	8538	7558	4544	4534
15	59	239	674	498	15	8640	7639	4550	4511
16	59	239	674	498	16	8716	7699	4553	4494
17	59	239	674	498	17	8774	7744	4556	4482
18	59	239	674	498	18	8817	7777	4558	4473
19	59	239	674	498	19	8849	7802	4559	4466
Zone 49					*Zone 57*				
11	934	2131	3751	4340	11	453	180	91	57
12	924	2118	3751	4347	12	446	179	91	57
13	919	2109	3750	4352	13	441	179	91	57
14	916	2103	3749	4356	14	438	178	91	57
15	914	2099	3748	4358	15	437	178	91	57
16	913	2097	3747	4359	16	435	178	91	57
17	913	2095	3747	4360	17	435	177	91	57
18	912	2095	3746	4361	18	434	177	91	57
19	912	2094	3746	4361	19	434	177	91	57
Zone 33					*Zone 64*				
11	466	1165	2169	2336	11	12415	9667	6295	5721
12	462	1158	2168	2340	12	12387	9660	6284	5720
13	460	1153	2167	2343	13	12352	9656	6284	5718
14	459	1150	2166	2345	14	12320	9648	6286	5717
15	458	1148	2166	2346	15	12293	9638	6287	5717
16	458	1147	2166	2346	16	12271	9629	6289	5718
17	457	1146	2166	2346	17	12253	9620	6289	5720
18	457	1146	2165	2346	18	12239	9613	6290	5721
19	457	1146	2165	2347	19	12228	9607	6290	5722

There are, especially in the urban zones, some substantial differences in table 10.1 between the values for iteration 4 and those for iteration 10. After nine more iterations, most household types in most zones have reached nearly equal values, regardless of the starting values. This suggests that—at least for this data set and these parameters—the starting points of the household-projection iterations, given the distribution of basic employment, are not significant determinants of the equilibrium values. These results also suggest that if one were to use the LOMS employment model from IPLUM, then it would not matter where in the iterative process the LOMS employment allocations were made. (Note that this stems from the specific structure of the LOMS employment submodel of IPLUM and would not necessarily be true for other LOMS employment model structures.) To summarize, this first set of tests suggests that on the basis of numerical experiments, there is a unique equilibrium solution to the DRAM equations for residence-location forecasting.

The next question to be examined was whether this equilibrium solution was a better or worse estimate of the 1970 household distributions than the estimate from the first iteration or any subsequent estimate during the iterations toward equilibrium. The same data set (input data and parameter values), was used for this investigation as for the above. Recall that the parameters used throughout these tests were estimated on the cross-sectional data set for 1970. In this experimental run, the distributions of both basic and LOMS employment were actual 1970 data. rather than model estimates. This was done in order to eliminate any confounding of the results which might stem from errors in employment estimates.

The household projections produced at each iteration were correlated with the known (actual) 1970 household distributions. The results are shown in table 10.4. The results are immediately obvious, as the household projections move toward equilibrium their correlation with the known 1970 household distribution declines.

In an attempt to understand these results better, it will be helpful to establish a kind of baseline against which all these experiments may be judged. The simplest forecast procedure is the straight line extrapolation. The effectiveness of this procedure may be gauged by the correlation between households by type and zone in 1960, and households by type and zone in 1970 (again, from the Minneapolis–St Paul data set). This gives values of the correlation coefficient r, as follows:

LI 0·9282; LMI 0·9474; UMI 0·8746; UI 0·8307.

It is clear from these results that whichever model formulation is used, it will have to perform rather well to yield better fits to the base data than the straight line extrapolation. Of course, if one is interested in policy evaluation, the use of straight line extrapolation will be rather constraining insofar as it permits no manipulation of variables for policy tests.

Table 10.4. Values of the correlation coefficient, *r*, between household projections and 1970 census—run A.

Iteration	Household type			
	LI	LMI	UMI	UI
1	0·8939	0·9324	0·9239	0·8602
2	0·8849	0·9277	0·9163	0·8498
3	0·8708	0·9235	0·9112	0·8412
4	0·8539	0·9189	0·9082	0·8341
5	0·8347	0·9137	0·9064	0·8280
6	0·8139	0·9018	0·9053	0·8225
7	0·7921	0·9022	0·9046	0·8174
8	0·7700	0·8962	0·9040	0·8126
9	0·7479	0·8901	0·9035	0·8080

Table 10.5. Values of the correlation coefficient, *r*, between household projections and 1970 census—run B.

Iteration	Household type			
	LI	LMI	UMI	UI
1	0·9500	0·9296	0·9229	0·9297
2	0·9424	0·9233	0·9171	0·9189
3	0·9314	0·9199	0·9127	0·9073
4	0·9173	0·9171	0·9097	0·8964
5	0·9004	0·9141	0·9077	0·8865
6	0·8810	0·9105	0·9063	0·8774
7	0·8596	0·9066	0·9052	0·8688
8	0·8368	0·9023	0·9044	0·8608
9	0·8133	0·8978	0·9038	0·8533

Table 10.6. Values of the correlation coefficient, *r*, between household projections and 1970 census—run C.

Iteration	Household type			
	LI	LMI	UMI	UI
1	0·9185	0·9394	0·9389	0·8959
2	0·9117	0·9301	0·9363	0·8909
3	0·8986	0·9197	0·9301	0·8851
4	0·8823	0·9035	0·9239	0·8795
5	0·8648	0·8909	0·9184	0·8744
6	0·8477	0·8800	0·9136	0·8699
7	0·8326	0·8707	0·9095	0·8659
8	0·8190	0·8631	0·9060	0·8624
9	0·8080	0·8570	0·9030	0·8599

It was also possible to examine whether the decline in correlation as the model solution moves towards equilibrium, shown in table 10.4, was due to the use in the projection calculations of 1960 household data when the calibrations had been performed with 1970 data throughout. Again, with both basic and LOMS employment data taken from the 1970 census, but with land-use data for 1960, ten iterations of DRAM were run. The results are shown in table 10.5. Here again the values of r decrease as the projections reach equilibrium.

The next step taken was to recalibrate DRAM with lagged (1960) independent variables. These new calibration results showed some changes in parameters and goodness of fit, but no striking differences appeared. DRAM was then rerun to produce projections as in the above tests. In the final run the inputs were the 1970 data for all employment and transportation variables and 1960 for the land and household variables. The results of nine iterations in this final run are compared to the 1970 census data in table 10.6. Here too the results show a steady decline in correlation as the model moves towards an equilibrium solution.

The three experiments described here may first be compared in terms of their inputs. For each run the time periods of the input variables are summarized below.

	run A	run B	run C
Calibration variables			
employment	1970	1970	1970
transportation network	1970	1970	1970
land use	1970	1970	1970
households	1970	1970	1960
Projection variables			
employment	1970	1970	1970
transportation network	1970	1970	1970
land use	1960	1960	1960
households	1960	1970	1960

In all three of these numerical experiments, the first-iteration projection of households was the best estimate of the actual 1970 household distribution. With successive iterations of the household projections an equilibrium solution was achieved, but that equilibrium solution was not as good an estimate of the actual 1970 household distribution as the first-iteration estimates. The various experiments done with variables for different time periods in calibration more or less eliminate the time periods of variables as a cause of this phenomenon.

These results arise from two interrelated effects. The first is common to any mathematical model with parameters estimated by 'best-fit' statistical techniques. In the estimation of these parameters, a certain percentage of the variation in the dependent variable remains unexplained. In most cases, this unexplained variation will be small but perhaps still significant.

When the best-fit parameters are substituted into the forecasting version of the model, this unexplained variation is no longer accounted for. The first iteration of model estimates retains some of the unexplained variation that was present in the input data. In successive iterations (or recursions in a stepwise recursive forecasting process) this initial unexplained variation is gradually eliminated from the model forecasts, as the forecast values eventually outweigh the influence of the input data. It is likely that a portion of the divergence of the model forecasts from the actual 1970 household data was, in part, due to this phenomenon.

The second effect results from the problem, described earlier, that the model parameters reflect an instant of time rather than the time-varying process itself. Thus the use of the parameters which are estimated from static (cross-sectional) data sets is an implicit assertion of a spatial, sectoral, and temporal equilibrium. Yet this equilibrium does not exist. In each successive iteration of the model the 'out-of-equilibrium' aspects of the data are slowly overridden by the implicit equilibrium assumption of the parameter values. This phenomenon, probably more important than the unexplained variation effect, is the second reason for the results obtained above. Although there is nothing to be done at present about these effects, this preliminary analysis points towards some future work on the dynamics of these models. Another research direction, concerning the underlying model formulations, is taken up in the next section.

10.3 New approaches to the residence-location model

The past twenty years have seen the development of computer models for analysis of urban spatial phenomena pass through several different stages. The optimism and naiveté of the early 1960s was overcome by the disappointment and disillusion of the late 1960s. The early to middle 1970s saw the emergence of new techniques along with considerable interest from planning agencies in potential model applications. Indeed, by the middle 1970s many planning agencies appeared to consider such models of potential use, and a substantial portion of agencies were actually using them in one form or another (Pack, 1978).

The early development of these models, like the initial phase of any developing scientific discipline, was characterized by widely divergent approaches and rather few commonly agreed underlying hypotheses for urban spatial phenomena. By the early 1970s many of the researchers working on these problems found themselves split into two groups: one actively involved in the development and application of macrodescriptive (spatial interaction) models such as those described in previous chapters of this book, and another which was involved in attempts to operationalize specific economic theories of urban form (Senior, 1973; 1974). By far the greatest number of actual model applications were of the former group, though some development success was achieved with the latter.

Beginning in 1973, a series of research efforts were able to demonstrate that the spatial interaction models were structurally equivalent to some forms of the economic models (Evans, 1973; Senior and Wilson, 1974; Coelho and Wilson, 1977).

Also during the middle 1970s a separate line of research was developing on the topic of behavioral choice theories as applied to the problem of transport mode selection. This work, building upon the use of n-dimensional logit models as the formulation for behavioral choice, is slowly but surely altering the traditional approaches to this problem (Domencich and McFadden, 1974; McFadden and Reid, 1975; Koppelman, 1976). As a by-product of the development of these behavioral choice models for mode split, recent research efforts have proposed that the same class of models could be used for residence-location estimates as well (Lerman, 1976; McFadden, 1978). Taken all together, these several strands of research investigation imply a steady movement toward a unified theory of urban spatial location which would incorporate aspects of the various current theories.

The most extensively examined group of macrodescriptive models are those which derive from 'most-probable-distribution' concepts of statistical mechanics. The models subsequently derived by entropy-maximizing techniques represent a least biased solution of the trip-distribution problem and, by suitable rearrangements of the equation structure, they can be used as residence-location models. This, of course, was the basis for the development of DRAM as described in chapter 7. To review, the simplest form of such a model may be derived as follows. First for a region divided into many small zones, let: T_{ij} be the number of workers living in zone i and working in zone j, O_i the number of workers residing in i, D_j the number of workers employed in j, and c_{ij} the cost of travel between i and j. Then we define the following constraints: first, trip-end (origin and destination) constraints,

$$\sum_j T_{ij} = O_i \ , \qquad (10.12)$$

$$\sum_i T_{ij} = D_j \ ; \qquad (10.13)$$

then, a total-travel-cost constraint,

$$\sum_i \sum_j T_{ij} c_{ij} = C \ . \qquad (10.14)$$

The most probable distribution of T_{ij}, given these constraints, is [see Wilson (1970a, pages 1–14) for the derivation].

$$T_{ij} = A_i B_j O_i D_j \exp(-\beta c_{ij}) \ , \qquad (10.15)$$

where

$$A_i = \left[\sum_j B_j D_j \exp(-\beta c_{ij})\right]^{-1}, \qquad (10.16)$$

$$B_j = \left[\sum_i A_i O_i \exp(-\beta c_{ij})\right]^{-1}. \qquad (10.17)$$

Now, in using such a model as a residence-location model, it is the O_i which are being forecasted. Thus the O_i and A_i in equation (10.15) are replaced by a measure of attractiveness W_i (to be discussed further below), to yield:

$$T_{ij} = B_j W_i D_j \exp(-\beta c_{ij}), \qquad (10.18)$$

$$B_j = \left[\sum_i W_i \exp(-\beta c_{ij})\right]^{-1}. \qquad (10.19)$$

Finally, we let N_i equal the forecast number of residents of zone i, the result is

$$N_i = \sum_j T_{ij} = \sum_j B_j W_i D_j \exp(-\beta c_{ij}). \qquad (10.20)$$

Numerous applications of this rather simple one-parameter model were made in the late 1960s and early 1970s; these are reviewed in Batty (1972).

It is possible, in theory, to extend such model formulations in directions of sectoral disaggregation as well as greater economic content. A prime example of such an extension is shown below (Wilson, 1970b). First, a housing stock variable, h, is added; second, employees are disaggregated by wage, w, and house type, k. Thus two new constraints are added:

$$\sum_i \sum_k T_{ij}^{kw} = E_j^w, \qquad (10.21)$$

and

$$\sum_i \sum_w T_{ij}^{kw} = h_i^k. \qquad (10.22)$$

The cost constraint then becomes

$$\sum_i \sum_j \sum_k T_{ij}^{kw} c_{ij} = C^w. \qquad (10.23)$$

Then, an economic budget constraint is added such that for each individual,

$$q^w(w - c'_{ij}) \geqslant p_i^k, \qquad (10.24)$$

where p_i^k is the cost of type k housing in zone i, and q^w is average amount which a household of income w spends on housing after journey-to-work costs c'_{ij} are deducted. If it is further assumed that individual variations in housing preference produce a normal distribution of housing expenditures, then equation (10.24) leads to the constraint:

$$\sum_i \sum_j \sum_k T_{ij}^{kw}[p_i^k - q^w(w - c'_{ij})]^2 = \sigma_w^2 , \tag{10.25}$$

where σ_w^2 is the variance of the housing expenditure distribution for income group w. These new constraints lead to the following model

$$\begin{aligned} T_{ij}^{kw} &= A_i^k B_j^w h_i^k E_j^w \exp(-\beta^w c_{ij}) \exp\{-\mu^w[p_i^k - q^w(w - c'_{ij})]^2\} \\ &= A_i^k B_j^w h_i^k E_j^w \exp\{-\beta^w c_{ij} - \mu^w[p_i^k - q^w(w - c'_{ij})]^2\} , \end{aligned} \tag{10.26}$$

with

$$A_i^k = \frac{1}{\sum_j \sum_w B_j^w E_j^w \exp\{-\beta^w c_{ij} - \mu^w[p_i^k - q^w(w - c'_{ij})]^2\}} , \tag{10.27}$$

$$B_j^w = \frac{1}{\sum_i \sum_k A_i^k h_i^k \exp\{-\beta^w c_{ij} - \mu^w[p_i^k - q^w(w - c'_{ij})]^2\}} . \tag{10.28}$$

It is worth noting that if the housing stock constraint, equation (10.22), is replaced by an exogenously provided measure of available housing stock, $\hat{h}_i^k$ (this allows a possible mismatch between housing supply and housing demand), then equation (10.26) becomes

$$T_{ij}^{kw} = B_j^w \hat{h}_i^k E_j^w \exp\{-\beta^w c_{ij} - \mu^w[p_i^k - q^w(w - c'_{ij})]^2\} , \tag{10.29}$$

with

$$B_j^w = \frac{1}{\sum_i \sum_k \hat{h}_i^k \exp\{-\beta^w c_{ij} - \mu^w[p_i^k - q^w(w - c'_{ij})]^2\}} . \tag{10.30}$$

If we sum to get the residence-location distribution, and combine equation (10.29) with equation (10.30), we obtain

$$N_i^{kw} = \sum_j T_{ij}^{kw} = \sum_j E_j^w \left[\frac{\hat{h}_i^k \exp\{-\beta^w c_{ij} - \mu^w[p_i^k - q^w(w - c'_{ij})]^2\}}{\sum_i \sum_k \hat{h}_i^k \exp\{-\beta^w c_{ij} - \mu^w[p_i^k - q^w(w - c'_{ij})]^2\}} \right] . \tag{10.31}$$

Although the development of equation structures of this sort are rather straightforward, it should be noted here, however, that attempts to operationalize models of this complexity have been largely unsuccessful, partly because of the extreme difficulty of obtaining the necessary data[(19)].

The development of the DRAM model began in chapter 7 with a simple form:

$$N_i = \sum_j E_j p_{ij} , \tag{10.32}$$

where N_i and E_j are as before, and where p_{ij} is usually described as the probability of a worker in zone j having a place of residence in zone i.

[(19)] One attempt to verify empirically a development of this type of model may be found in Anas (1975).

Note that if p_{ij} is a function of the residential attractiveness of i and a function of the cost of travel between i and j, then it is possible to define it as follows

$$p_{ij} = \frac{W_i \, \mathrm{f}(c_{ij})}{\sum_j W_i \, \mathrm{f}(c_{ij})} \, . \tag{10.33}$$

This, if $\mathrm{f}(c_{ij}) = \exp(-\beta c_{ij})$, is identical to what would be obtained by substituting equation (10.19) into equation (10.20). In DRAM the attractiveness term and the trip function are expanded to give

$$W_i = (L_i^{\mathrm{v}})^q (\tilde{x}_i)^r (L_i^{\mathrm{r}})^s (\tilde{N}_i^1)^{b^1} (\tilde{N}_i^2)^{b^2} (\tilde{N}_i^3)^{b^3} (\tilde{N}_i^4)^{b^4} \, , \tag{10.34}$$

where L_i^{v} is the vacant buildable land in i, $\tilde{x}_i$ is (one plus) the proportion of buildable land in i which is already built upon, L_i^{r} is the residential land in i, $\tilde{N}_i^n$ is (one plus) the proportion of zone i residents in the nth income group, and q, r, s, b^1, ..., b^4 are empirically derived parameters which are understood also to have a subscript n (omitted for notational convenience); and,

$$\mathrm{f}(c_{ij}) = c_{ij}^{\alpha} \exp(-\beta c_{ij}) \, , \tag{10.35}$$

where α, β are also empirically derived parameters also understood to have a subscript n. Thus the final version of the DRAM model had the form:

$$N_i^n = \sum_j T_{ij}^n = \sum_j E_j^n \left[\frac{(L_i^{\mathrm{v}})^q (\tilde{x}_i)^r (L_i^{\mathrm{r}})^s \prod_{n'} (\tilde{N}_i^{n'})^{b^{n'}} c_{ij}^{\alpha} \exp(-\beta c_{ij})}{\sum_i (L_i^{\mathrm{v}})^q (\tilde{x}_i)^r (L_i^{\mathrm{r}})^s \prod_{n'} (\tilde{N}_i^{n'})^{b^{n'}} c_{ij}^{\alpha} \exp(-\beta c_{ij})} \right] . \tag{10.36}$$

Now, if a substitution is made, so that

$$C_{ij} = c_{ij} + y^n \ln c_{ij} + z_1^n \ln L_i^{\mathrm{v}} + z_2^n \ln \tilde{x}_i + z_3^n \ln L_i^{\mathrm{r}} + \dots \, , \tag{10.37}$$

which simply results from taking the natural logarithm of the numerator of the fraction in equation (10.36), then we can make the transformation

$$\begin{aligned} \exp(-\beta^n C_{ij}) &= \exp[-\beta^n (c_{ij} + y^n \ln c_{ij} + z_1^n \ln L_i^{\mathrm{v}} + \dots)] \\ &= \exp[-\beta^n c_{ij} - \beta^n (y^n \ln c_{ij} + z_1^n \ln L_i^{\mathrm{v}} + \dots)] \, . \end{aligned} \tag{10.38}$$

Now, equation (10.36) becomes

$$N_i^n = \sum_j T_{ij}^n = \sum_j E_j^n \left\{ \frac{\exp[-\beta^n c_{ij} - \beta^n (y^n \ln c_{ij} + z_1^n \ln L_i^{\mathrm{v}} + \dots)]}{\sum_i \exp[-\beta^n c_{ij} - \beta^n (y^n \ln c_{ij} + z_1^n \ln L_i^{\mathrm{v}} + \dots)]} \right\} . \tag{10.39}$$

If this model were disaggregated by house type k, or the model given by equation (10.31) were to be aggregated over all house types, the results would show that the two models have remarkably similar structures.

The derivation of behavioral choice models assumes the existence of the classic economic consumer who maximizes utility in making decisions. To develop such models some form of utility measure must be defined.

The utility of a particular choice c to a particular individual k will be given by

$$U_{kc} = V_{kc} + \epsilon_{kc} , \tag{10.40}$$

where U_{kc} is the utility of choice c to individual k, V_{kc} the average utility of choice c to a homogeneous group of individuals, and ϵ_{kc} a random (perceived) utility of choice c to individual k. In general, V_{kc} is taken to be a function of measurable (observable) attributes of choice c, known as the 'strict utility' of kc and expected to be common to all members of the population. The term ϵ_{kc} is the random utility component of kc and is taken to be a function of unmeasurable (unobservable) attributes of choice c which will vary from individual to individual. Given such a utility function, various choice models may result, depending upon assumptions to be made about the random utility term. The customary procedure is to assume that these terms are independently and identically distributed according to the Weibull distribution. This results in the choice probabilities being given by a multinomial logit model:

$$p_{kc} = \frac{\exp V_{kc}}{\sum_k \sum_c \exp V_{kc}} , \tag{10.41}$$

where p_{kc} is the probability of individual k making choice c.

With some types of choice behavior, it is possible to decompose the utility function into several strict utility components plus a random utility component. Suppose for example that a simultaneous mode-split and residence-location decision is being made. We can then write

$$U_{jkm} = (V_j + \epsilon_j) + (V_k + \epsilon_k) + (V_{jkm} + \epsilon_{jkm}) ; \tag{10.42}$$

where

U_{jkm} is the utility of residence zone j, workplace zone k, and commutation by mode m;

V_j is a measurable function of attributes of residence zone j;

V_k is a measurable function of attributes of workplace zone k;

V_{jkm} is a measurable function of attributes which vary with j, k, and m;

ϵ_j, ϵ_k, and ϵ_{jkm} are random terms measuring unobservable attributes corresponding to the above definitions.

Equation (10.42) may be written as

$$U_{jkm} = V_j + V_k + V_{jkm} + (\epsilon_j + \epsilon_k + \epsilon_{jkm}) ; \tag{10.43}$$

and if we make the same assumptions that were made above about the random utility terms, we obtain a multinomial logit model of the form

$$p_{jkm} = \frac{\exp(V_j + V_k + V_{jkm})}{\sum_i \sum_k \sum_m \exp(V_i + V_k + V_{ikm})} , \tag{10.44}$$

where p_{jkm} is the probability of choosing residence in zone j, workplace in zone k, and commutation by mode m.

Now, p_{jkm} may be written as a sequential choice:

$$p_{jkm} = p_{km} p_{j|km} , \tag{10.45}$$

where p_{km} is the probability (or expected frequency) of choosing workplace k and mode m, and $p_{j|km}$ is the conditional probability of choosing residence zone j, given workplace k and mode m; then

$$p_{j|km} = \frac{\exp(V_j + V_k + V_{jkm})}{\sum\limits_i \exp(V_i + V_k + V_{ikm})} \tag{10.46}$$

and

$$p_{km} = \frac{\exp(V_j + V_k + V_{jkm})}{\sum\limits_i \sum\limits_k \sum\limits_m \exp(V_i + V_k + V_{ikm})} . \tag{10.47}$$

Note that the V_k divide out of equation (10.46), thus leaving $p_{j|km}$ as a function of measurable attributes which vary according to residence zone j, and measurable attributes which depend on j, k, and m together. This is very like the form of the travel probability p_{ij} in equation (10.33), and is as far as it is necessary to go in this discussion. In any actual model application, numerous other questions will need to be resolved about the exact determination of utility functions, the bias which may be introduced by aggregating the choice space, and so on[20].

The point of all this discussion is first that the p_{ij} of equation (10.33) becomes the lengthy fraction in brackets in equation (10.39). Thus the p_{ij} portion of a multivariate-attractiveness spatial interaction model is a multinomial logit formulation, even though its derivation is not at all like the standard derivation of a multinomial logit model. Second, from the behavioral choice derivation, it is found that the conditional probability that a worker in workplace k and using transportation mode m (if those workers were randomly sampled) would live in zone j is a multinomial logit model of the same form [equation (10.46)]. In itself, this is a nice result, as it lends confirmation to the reasonableness of the form of equation (10.39). The purpose of the discussion above is to suggest that empirical verification of this convergence of model derivations and results should be an important focus of future research on residential location. The next section of this chapter discusses areas of research on the supply or availability of land.

10.4 Land consumption and constraints

All of the location models discussed above have in common that they are 'demand-oriented' models. Their constructs assume that if a zone is attractive to residence or employment location, then location will take place. What is omitted is explicit consideration of the 'supply' side.

(20) See Anas (1979a), Litinas and Ben-Akiva (1979), McFadden and Reid (1975), and McFadden (1978) for further discussion.

How much land may be available—and/or how much housing stock may be available—is not taken into consideration in determining residence location in these models.

In a similar vein, proper consideration of land consumption by locating activities, and of the action of land constraints on locators, is absent from virtually all of the work on urban spatial theory and models done to date. Land consumption is taken here to mean the amount of land occupied by locating activities (that is, residence or employment). Land constraint is taken here to mean the regulation (by policy and/or legislation) of the use of land in terms either of type of use or of density of use. The most frequently used procedure for estimating land consumption has been the assumption that future land-consumption rates (expressed, for example, as land per household, or as land per employee) will equal past rates. The most frequently used land-constraint procedure has been simply to remove 'excess' locators from the constrained area and reallocate them to other areas of the region being analyzed. Thus it is obvious that there is a need to develop a general procedure for estimating land consumption, along with a consistent procedure for responding to land constraints. The few residence-location models that do attempt to represent the supply side of the location problem do so in terms of housing stock. Although several classes of housing stock may be represented in such models, the land-consumption procedures are usually cast in terms of previous average consumption rates per housing stock type. By way of elaboration, the land-consumption procedures from several urban models are described below.

Some of the earlier macrodescriptive residence-location models actually had no formal land-consumption procedure at all (Putman, 1979). The TOMM model, for example, first calculated the available residential land per zone (or subarea) as a residual after other types of land requirements (exogenously specified) had been satisfied. This residual residential land was used as a factor in forecasting new residence location in the zone. Once the new number of residents was calculated, the residential land consumption was calculated as the product of this number with the existing ratio of residents per thousand square feet. This procedure was, of course, virtually identical to that used in the Lowry model, from which TOMM was derived (Crecine, 1964; Lowry, 1964).

The PLUM model, another descendent of the Lowry model, used a somewhat more complex land-consumption procedure, as described in chapter 3. First, the past levels of residential density were used to calculate (when multiplied by the available residential land) a residential holding capacity for each subarea. This holding capacity was used as one factor in the calculation of attractiveness of the zone (subarea) to potential new residential locators. These new locators consumed land at the past average rate for the zone. Only if the available residential land in the zone was less than the estimated new land consumption did the rate of land consumption change (Goldner et al, 1972). This change was related to the operation of an arbitrarily specified land-constraint procedure.

One of the first residence-location models to deal both with the demand and with the supply side of the problem was the BASS model. In this model an attempt was made to forecast both residential demand (that is, new residence-location potential) and residential supply in terms of housing stock. The supply and demand forecasts were then balanced to yield the final residence-location forecasts. Even so, the rate of land consumption of the new locators was determined from exogenously determined 'land absorbtion coefficients' for each housing type (CREUE, 1968). The lack of progress in this general area is continued even by the NBER Urban Simulation Model, developed in an attempt to overcome the shortcomings of the various macrodescriptive models which preceded it, which follows an almost identical land-consumption procedure. Despite the NBER model having a rather more complex underlying structure than many other urban simulation models, its actual land-consumption calculation involved only the use of past land-consumption rates by new residential locators (Kain et al, 1977). The new behavioral or disaggregated models have not progressed to a sufficiently operational condition even to begin to give any serious consideration at all to the problem of land consumption.

The situation for describing and understanding nonresidential land consumption is, if anything, even worse. One reasonably comprehensive review of location models for retail employment made no mention at all of the land-consumption problem (Cordey-Hayes, 1968). Another review of intraurban employment-location models found virtually no discussion in any of the then current or forthcoming models of any aspect of land consumption (Putman, 1972). The only occasional mention of the problem appeared in discussion of updating of the land-use accounts in these models. In these cases some form of past average land-consumption rates in the zone were used, or in one case the average of land-consumption rates in the several adjacent zones were considered as well. Nonetheless, the general treatment of the problem was at best perfunctory.

The more theoretical work in urban economics unfortunately offers very little of practical value towards the solution of these problems. Neither, for that matter, is the empirical work on urban densities of any particular assistance. The principal result of the work on urban density was the establishment of the fact that, in very general terms, residential densities decline exponentially with distance from concentrations of employment (Clark, 1967). The urban economic theorists attempt to explain this and other phenomena, but are able to offer very little empirical verification. Amongst the earliest of contemporary workers in this field, Wingo (1961) argued that the consumption of residential land was a function of its cost, and the cost was a function of the difference between the transportation cost (from the center of the urban area) to the land and that to the urban fringe. Various differences in theoretical structure appeared in subsequent work on the problem (Alonso, 1964; Muth, 1969). However, from the point of view of the determination of a procedure for estimating land-consumption rates in a spatially disaggregated metropolitan area, no definitive results have appeared in the decade or more following these works.

Several land-consumption models were developed as part of a larger urban growth model package for the Philadelphia area in the 1960s (Seidman, 1970). The residential land-consumption model, after some unsuccessful attempts at a more complex structure, eventually took the form of a multiple regression where the dependent variable was the natural logarithm of residential density and the independent variables were various zone-specific socioeconomic and access measures. Though not particularly elegant, this model appeared to achieve a good explanation of variance, and this suggested some simple exploration of the data. Thus, in connection with the work on DRAM described above, a preliminary examination was undertaken to explore the possibilities of this rather simple and direct first approach to determining whether there were any observable regularities in land consumption at relatively fine levels of spatial disaggregation.

A rather comprehensive data file was available for the Philadelphia region, with nearly two hundred variables (for 1970) for each of 173 subareas. The independent variables in this preliminary analysis were limited to variables which could reasonably be expected to be endogenous to a sophisticated employment/residence-location model. This requirement was imposed so that the land-consumption relationships, if successfully developed, could be integrated into a combined location and land-use model package such as ITLUP. Given this constraint, numerous combinations of variables were hypothesized and tested. Several quite promising regularities emerged from these examinations, with independent variables such as income distribution of zonal population, percentages of available and developable land, and percentages of other land uses in the zone (industrial and/or commercial, for example). One equation tested was:

$$\ln(L_i^{\mathrm{r}}/N_i^{*}) = a_0 + a_1(L_i^{\mathrm{a}}/L_i^{\mathrm{u}}) - a_2 N_i^{1} - a_3(L_i^{\mathrm{ind}}/L_i^{\mathrm{u}}) - a_4(L_i^{\mathrm{c}}/L_i^{\mathrm{u}}) \; ; \qquad (10.48)$$

here, L^{r} denotes residential land, L^{a} available, developable land, L^{ind} industrial land, L_i^{c} commercial land, and L_i^{u} total useable land in zone i; N_i^{1} is the number of households with income less than \$10000 in zone i, and N_i^{*} the total number of households in zone i.

Although a regression result such as this is certainly a long way from being a theoretically satisfactory model of residential land consumption, the signs of the coefficients $a_0 \ldots a_4$ in this examination did make good sense. Residential land per household was greater in zones where there were higher percentages of vacant land, and was less in zones with larger numbers of low-income households and higher percentages of industrial and/or commercial land use. All the coefficients were statistically significant at 5% or better, and the value of R^2 was just over 0·9 for the whole regression. Several other regression models yielded promising results of a similar nature, and further results of interest were obtained from a subset of the zones where selected variables for 1960 were also available, thus allowing some limited examination of the effects of change variables. It will be noted here that no accessibility variables were used in these equations. This was largely because the accessibility variables were poorly documented in the data files, and thus their examination was postponed to a future effort. The mention of accessibility variables leads to the discussion of a possible second approach.

Starting in the late 1960s, a series of urban model projects were conducted which attempted to represent both the demand and the supply side of the location decision with a matched pair of spatial interaction models (Echenique et al, 1969; Mackett, 1979; 1980). In these studies a normalized accessibility function was used, along with available land and spatial distribution of various activities. As the revised spatial patterns of activities also affected the availability of floorspace and/or housing stock, the linked models had to be iterated to a solution. Although the supply models used were relatively simple one-or-two-parameter models, as were the location models with which they were linked, the results did, however, clearly indicate a further avenue of research. The extensive work with DRAM as described in chapter 7 established the descriptive validity of multiparameter models of this form for residence-location analysis. It is reasonable to expect that such multiparameter formulations would be of use in land-consumption analysis as well.

Finally, it should be noted that at the same time that the examination of residential land consumption in the Philadelphia data set by multiple regression was done, some preliminary analyses of regularities in industrial and commercial land consumption were done. Analysis of these few very preliminary results indicated the possibility that a similar approach to that used for the analyses of residential land consumption would also be useful for other classes of land consumption as well. It will be recalled that equation (10.48) for residential land consumption included industrial and commercial land as independent variables. These facts, taken together, suggest the existence of a simultaneous or at least recursive equation structure which would jointly estimate several classes of land consumption at once. Furthermore, the spatial interaction floorspace model referred to above was also used for estimating commercial floorspace. This approach may also yield useful results then, for both residential and nonresidential floorspace (land consumption) as well. All in all, the purpose of these comments has been to indicate another area which needs further research, and to make some general suggestions for the preliminary direction of that work. In the next section, a similar discussion of the problems of the calibration of location models is presented.

10.5 Calibration and policy sensitivity

In this last section of the chapter, two potential research topics are grouped together that would ordinarily be thought not to be very closely related; yet it was in the process of doing the algebra for one of these that the possibility of the other became apparent.

To date, the calibration work both with DRAM and with EMPAL, as described in chapters 6 and 7, has involved the use of a gradient-search technique. This approach assumes that there is a criterion measure which is a function of several input variables (which are actually constant during a model calibration) and several associated parameters (which vary during the calibration process). In principle, then, the criterion measure is a function of a large number of constants (the input data) and a few variables (the parameters).

Thus the equations for the value of the criterion function for a given set of parameter values may be written

$$\mathrm{C}^k = \mathrm{f}(\alpha, \beta, \ldots, x)\ . \tag{10.49}$$

Then the gradient of C^k is a vector of partial derivatives:

$$\nabla \mathrm{C}^k = \begin{bmatrix} \dfrac{\partial \mathrm{C}^k}{\partial \alpha} \\ \dfrac{\partial \mathrm{C}^k}{\partial \beta} \\ \vdots \\ \dfrac{\partial \mathrm{C}^k}{\partial x} \end{bmatrix} . \tag{10.50}$$

The gradient-search technique moves from one value of the criterion to the next by proceeding in the direction of steepest ascent (for function maximization). Thus the step yields

$$\mathrm{C}^{k+1} = \mathrm{C}^k + \lambda^k \nabla \mathrm{C}^k\ , \tag{10.51}$$

where λ^k is a multiplier that may itself be some form of complex function which can be overlooked in this discussion.

As discussed in chapters 7 and 8, certain difficulties were encountered in the application of the gradient search; in particular the functions tend to have rather flat optima which are difficult to locate precisely in n-dimensional space. In an attempt to remedy this problem, an investigation was undertaken into the practicality of applying Newton's method for function optimization. This method uses both the gradient and the Hessian matrix to find the function maximum. The Hessian matrix is the square matrix of the second-order partial derivatives, as follows

$$\mathrm{HC}^k = \nabla^2\mathrm{C}^k = \begin{bmatrix} \dfrac{\partial^2\mathrm{C}^k}{\partial\alpha^2} & \dfrac{\partial^2\mathrm{C}^k}{\partial\alpha\partial\beta} & \cdots & \dfrac{\partial^2\mathrm{C}^k}{\partial\alpha\partial x} \\ \dfrac{\partial^2\mathrm{C}^k}{\partial\beta\partial\alpha} & & & \vdots \\ \vdots & & & \\ \dfrac{\partial^2\mathrm{C}^k}{\partial x\partial\alpha} & \dfrac{\partial^2\mathrm{C}^k}{\partial x\partial\beta} & \cdots & \dfrac{\partial^2\mathrm{C}^k}{\partial x^2} \end{bmatrix} . \tag{10.52}$$

By Newton's method, the criterion step is found from the gradient and the inverse of the Hessian matrix:

$$\mathrm{C}^{k+1} = \mathrm{C}^k + \lambda^k[\mathrm{HC}^k]^{-1}\nabla\mathrm{C}^k . \tag{10.53}$$

The partial derivatives for EMPAL and DRAM were given in chapters 6 and 7. The second partial derivatives can be formulated as well(21). The general form for the second-order partial differentials in the Hessian matrix is

$$\frac{\partial^2\mathrm{L}}{\partial\alpha^2} = \left[\sum_i\left(\frac{N_i}{\hat{N}_i^2}\right)\right]\left(\frac{\partial\hat{N}_i}{\partial\alpha}\right)^2 - \left[\sum_i\left(\frac{N_i}{N_i^2}\right)\right]\left(\frac{\partial^2\hat{N}_i}{\partial\alpha^2}\right) , \tag{10.54}$$

and if (as was done for the first derivatives) the general form of the model is taken to be

$$\hat{N}_i = \sum_j E_j\left[\frac{W_i^\delta c_{ji}^\alpha \exp(\beta c_{ji})}{\sum_i W_i^\delta c_{ji}^\alpha \exp(\beta c_{ji})}\right] , \tag{10.55}$$

let

$$A_{ji} = W_i^\delta c_{ji}^\alpha \exp(\beta c_{ji}) , \tag{10.56}$$

and

$$B_j = \sum_i W_i^\delta c_{ji}^\alpha \exp(\beta c_{ji}) , \tag{10.57}$$

then

$$\frac{\partial^2\hat{N}_i}{\partial\alpha^2} = \sum_j E_j A_{ji} B_j^{-1}\left\{(\ln c_{ji})^2 + B_j^{-1}\left[2B_j^{-1}\left(\sum_i A_{ji}\ln c_{ji}\right)^2 - 2(\ln c_{ji})\left(\sum_i A_{ji}\ln c_{ji}\right) - \sum_i A_{ji}(\ln c_{ji})^2\right]\right\} , \tag{10.58}$$

(21) This rather tedious but nonetheless wholesome task has been carried out by Gary Silbert, PhD student in the Department of City and Regional Planning, University of Pennsylvania.

with

$$\frac{\partial^2 \hat{N}_i}{\partial \delta^2} = \sum_j E_j A_{ji} B_j^{-1} \left\{ (\ln W_i)^2 + B_j^{-1} \left[2B_j^{-1} \left(\sum_i A_{ji} \ln W_i \right)^2 \right.\right.$$

$$\left.\left. - 2(\ln W_i) \left(\sum_i A_{ji} \ln W_k \right) - \sum_i A_{ji} (\ln W_i)^2 \right] \right\} , \tag{10.59}$$

and

$$\frac{\partial^2 \hat{N}_i}{\partial \beta^2} = \sum_j E_j A_{ji} B_j^{-1} \left\{ c_{ji}^2 + B_j^{-1} \left[2B_j^{-1} \left(\sum_i A_{ji} c_{ji} \right)^2 \right.\right.$$

$$\left.\left. - 2c_{ji} \left(\sum_i A_{ji} c_{ji} \right) - \left(\sum_i A_{ji} c_{ji} \right)^2 \right] \right\} . \tag{10.60}$$

These second-order partial derivatives may then be substituted into equation (10.54) along with the first-order derivatives from chapter 7 to get the second-order partial derivatives of the criterion with respect to each parameter.

The general form for the cross partial derivatives (the off-diagonal elements of the Hessian matrix) is

$$\frac{\partial^2 \mathrm{L}}{\partial \delta \partial \alpha} = \left[\sum_i \left(\frac{N_i}{\hat{N}_i^2} \right) \right] \left(\frac{\partial \hat{N}_i}{\partial \delta} \frac{\hat{N}_i}{\partial \alpha} \right) = \left[\sum_i \left(\frac{N_i}{N_i^2} \right) \right] \left(\frac{\partial^2 \hat{N}_i}{\partial \delta \partial \alpha} \right) ; \tag{10.61}$$

then

$$\frac{\partial^2 \hat{N}_i}{\partial \alpha \partial \beta} = \sum_j E_j A_{ji} B_j^{-1} \left\{ c_{ji} \ln c_{ji} + B_j^{-1} \left[2B_j^{-1} \left(\sum_i A_{ji} \ln c_{ji} \right) \left(\sum_i A_{ji} c_{ji} \right) \right.\right.$$

$$\left.\left. - (\ln c_{ji}) \left(\sum_i A_{ji} c_{ji} \right) - c_{ji} \left(\sum_i A_{ji} \ln c_{ji} \right) - \left(\sum_i A_{ji} c_{ji} \ln c_{ji} \right) \right] \right\} ; \tag{10.62}$$

$$\frac{\partial^2 \hat{N}_i}{\partial \alpha \partial \delta} = \sum_j E_j A_{ji} B_j^{-1} \left\{ (\ln c_{ji}) \ln W_i + B_j^{-1} \left[2B_j^{-1} \left(\sum_i A_{ji} \ln W_i \right) \left(\sum_i A_{ji} \ln c_{ji} \right) \right.\right.$$

$$- (\ln W_i) \left(\sum_i A_{ji} \ln c_{ji} \right) - (\ln c_{ji}) \left(\sum_i A_{ji} \ln W_i \right)$$

$$\left.\left. - \left(\sum_i A_{ji} (\ln c_{ji}) \ln W_i \right) \right] \right\} ; \tag{10.63}$$

$$\frac{\partial^2 \hat{N}_i}{\partial \beta \partial \alpha} = \frac{\partial^2 \hat{N}_i}{\partial \alpha \partial \beta} ; \tag{10.64}$$

$$\frac{\partial^2 \hat{N}_i}{\partial \beta \partial \delta} = \sum_j E_j A_{ji} B_j^{-1} \left\{ c_{ji} \ln W_i + B_j^{-1} \left[2B_j^{-1} \left(\sum_i A_{ji} c_{ji} \right) \left(\sum_i A_{ji} \ln W_i \right) \right.\right.$$

$$\left.\left. - c_{ji} \left(\sum_i A_{ji} \ln W_i \right) - (\ln W_i) \left(\sum_i A_{ji} c_{ji} \right) - \left(\sum_i A_{ji} c_{ji} \ln W_i \right) \right] \right\} ; \tag{10.65}$$

and

$$\frac{\partial^2 \hat{N}_i}{\partial\delta\partial\alpha} = \frac{\partial^2 \hat{N}_i}{\partial\alpha\partial\delta} \; ; \tag{10.66}$$

$$\frac{\partial^2 \hat{N}_i}{\partial\delta\partial\beta} = \frac{\partial^2 \hat{N}_i}{\partial\beta\partial\delta} \; . \tag{10.67}$$

These cross partial derivatives may then be substituted into equation (10.61) to obtain the cross partial derivatives of the criterion with respect to the pairs of parameters. With these expressions for the second-order partial derivatives and the cross partial derivatives, the Hessian matrix may be evaluated and Newton's method attempted for calibration.

As a side note, on performing the algebra necessary to formulate the second partial derivatives, it becomes possible to calculate asymptotic t-tests of the parameters. This may eliminate an important shortcoming of the current calibration process, by allowing tests of statistical significance of the parameters of the models.

The final item to be discussed here is the concept of sensitivity testing via numerical evaluation of partial derivatives. During the development of the second partial derivatives necessary for performing Newton's method, a general discussion of the model equations and their derivatives took place. In this discussion it became clear that the numerical evaluation of the appropriate partial derivatives would yield information on the sensitivity of variables one to the other.

If, for example, another simple version of a residence-location model is taken to be

$$N_i = \sum_j E_j B_j W_i^{\alpha} \exp(\beta c_{ij}) \; , \tag{10.68}$$

with

$$B_j = \left[\sum_i W_i \exp(\beta c_{ij})\right]^{-1} \; , \tag{10.69}$$

then it is possible to calculate an $n \times n$ matrix of partial derivatives[22] of N_i with respect to E_j (for $i, j = 1, ..., n$). This matrix yields the change in employed residents in zone i which would result from a change in employment in zone j; thus

$$\frac{\partial N_i}{\partial E_j} = B_j W_i^{\alpha} \exp(\beta c_{ij}) \; . \tag{10.71}$$

Obviously, the partial derivatives of N_i with respect to W_i and of N_i with respect to c_{ij} can also be evaluated. Given these, it becomes possible to deal with policy questions such as: What zones will be most sensitive to changes in travel cost between two particular zones i and j? Or, given that

[22] These partial derivatives were worked out by Hugh Miller, PhD student in the Department of City and Regional Planning, University of Pennsylvania.

one might wish to maximize the location of residents in a particular zone i, in what zone or zones j should employment location be encouraged? Taken together with the traditional ways of using these types of model for policy evaluation, these partial derivative procedures offer a whole new area of investigation.

In summary, this chapter has presented a selected set of topics on which further research efforts are not only desirable, but logical extensions of existing lines of enquiry. However, these topics are but a small subset of that which needs to be done. In essence much of the work to date on urban location and transportation models has served only to define the questions more precisely. Although some of this work has yielded practical results, much more remains to be learned. In the next and final chapter of this book an attempt will be made to draw together what has been learned in such a way as to provide some guidance both to researchers and to policymakers.

11

Conclusions

11.1 Introduction

The intent of this book was to bring together the results of a decade of research investigation into the theory and application of models of the interrelationships of transportation and land use. Bringing the material together in this way serves the purpose of informing students, researchers, and practitioners, of progress to date, potential applications, and possible avenues of further research. This final chapter summarizes the material presented from three different perspectives. The first discussion is centered on the substantive findings in previous chapters and will, in particular, summarize present knowledge of the relationships between transportation and land use. The second discussion will focus on methodological findings and issues with particular emphasis both on the progress with and on the problems of modelling these systems. The final discussion, taking what might be considered to be a more philosophical approach, offers some comments on the development and use of these techniques in the context of a more general view of planning, science, and society.

11.2 Substantive results: transportation–land-use interrelationships

In the past three of four decades there have been hundreds of transportation-impact studies (especially of highways) performed. Some of these have been 'if–then' studies of the form: if a particular transportation facility is built or modified in a particular way, then the following impacts will be observed. Other studies have been 'before-and-after' studies of the form: when a particular facility was built or modified, the following impacts were (or were not) observed. Taken all together, the results of these studies have been equivocal[(23)]. Yet, it is perfectly obvious that, more often than would be accounted for by chance alone, land-use changes have accompanied transportation changes.

One more or less typical study involved a survey of fourteen highway interchanges (Bertolina and Carbine, 1974). Traffic histories were compiled and analyzed, subsequent traffic levels and projections were then compared with the historic data. Development patterns from before and after the construction of the interchanges were also analyzed, along with existing local controls. From these analyses the study concluded, among other things, that:

> "1. An interchange creates a highly favorable climate for business ventures and attracts intense commercial development in its influence area.

[(23)] See, for example, National Council for Urban Economic Development (1980) and USDoT (1974).

2. Control of land use and development in interchange influence areas is essential to protect the area from intense commercial development that could cause overloading and congestion from development-generated traffic".

Another study, federally sponsored and somewhat more general in its coverage, examined the impacts of beltways on metropolitan areas (Payne-Maxie and Blayney-Dyett, 1980). This study began with statistical analyses of fifty-four metropolitan areas, half with and half without beltways. Then, each of the twenty-seven areas with beltways was examined in greater detail. Finally, eight of these twenty-seven areas were examined in considerable detail. Not all of the conclusions of the study are of direct relevance here, but some findings (pages 11-14) are worth noting.

"A beltway can increase development opportunities in its corridor, reinforce prevailing urbanization patterns, and facilitate compact development. However, it is not a sufficient inducement to counteract the effect of an area's poor image or to create a market for land, housing, or commercial and industrial space where none historically existed ..."

"Industrial and office park developers are willing to pay a premium for corridor and interchange area sites with accessibility to and visibility from the beltway. However, the presence of a beltway is less important than the availability of developable land and the accessibility to a skilled labor force. In most communities, radial highway sites with rail access were preferred over beltway sites for industrial park development ..."

"A beltway may have a locational effect on multi-family housing; single family residential development patterns rarely are affected over the long run. Differences in housing patterns between beltway and non-beltway cities are not statistically significant ..."

"Historically, central cities surrounded by suburban beltways posted lower gains in retail sales and employment than those with beltways within their jurisdiction or those where no beltway was built. Changes in the central city population and work force were the most important determinants of central city retail sales ..."

"Beltways can have a 'one time' effect on the distribution of new office space in a metropolitan area that may weaken the downtown office space market, drawing some employment out to suburban locations that might have stayed downtown had the beltway not been built. This effect is short-run in nature and need not harm central city revitalization efforts ..."

"Differences in 1967-77 sales trends in central business districts are not significantly different in beltway and non-beltway areas studied, and the existence or location of a beltway is not a statistically significant factor after accounting for other influences, such as primarily changes in residential population and manufacturing employment ..."

"Beltways showed a small, negative statistically significant effect on employment in the wholesaling and service sectors of central city economies between 1972 and 1977 ..."

"Changes in industrial employment in central cities show a small, but statistically significant negative relationship with the presence of a suburban beltway after accounting for other factors, suggesting that some fraction of the shift of employment within metropolitan areas between 1967–72 could be attributed to beltways."

From these findings a number of policy implications were drawn. Of particular relevance were the following.

"The most significant policy implication of the findings on beltway impacts on development is that a coherent and effective land use policy is required to control beltway-induced development in growth areas and to enforce conformance with local land use plans and objectives.

Though planners have not historically integrated beltways or their effects into their plans, the greatest opportunities for controlling the influence of a beltway lies in the realm of land use planning. Potential adverse impacts of beltway construction can be largely mitigated, even eliminated, through careful application of land use controls and incentives. Opportunities for compact, nodal development are provided by beltway interchanges with supportive planning policies. Though such policies have rarely been implemented in conjunction with beltway construction, their potential for effecting an energy-efficient pattern of land use is substantial and has been demonstrated elsewhere."

The development of integrated transportation and land-use models may be judged, at least in part, by the extent to which it provides techniques which can first replicate and later anticipate the phenomena observed in these empirical studies. The policy-simulation results described in this book should be viewed in this context. The development work on the model has taken as a starting point the hypothesis that transportation can affect land use, but does not always do so. Thus the interesting question is, "By what means and under what circumstances can transportation affect land use?" More generally, what is needed is a reasonably complete explanation of the relationships between transportation and land use. The set of relationships embodied in the DRAM–ITLUP package of models is an important step towards a more complete understanding of the nature of the relationships between transportation and land use.

It has sometimes been observed that changes in transportation are not accompanied by any observable changes in land use. Moreover, it has been argued that in at least some of the cases where land-use changes have been observed together with transportation changes, those land-use changes would have occurred even without the transportation changes. The equation structure of the DRAM–ITLUP integrated model package is capable of yielding these responses as well as others, some seemingly contradictory.

The work done in developing this analytical description has yielded considerable progress towards the desired understanding of transportation–land-use interrelationships. In addition, the development and application of these techniques yielded numerous substantive findings which are summarized and discussed below.

One of the most important substantive findings was that of the general interrelatedness not only of transportation and land use, but of subarea to subarea within a large metropolitan region. Examples of this interrelatedness were found in the earliest simulation experiments with the models. Stringent land-use controls in one subarea of a region were found to divert locating activities to other parts of the region. Similarly, changes in a specific link of a regional transportation system could produce effects, albeit in some cases rather minor, in numerous subareas throughout the region.

There were several important findings that apply to the effects of transportation changes. First, in a rather general way, regionwide increases in transportation cost/time result in tendencies towards centralization of activities (both employment and population) around existing concentrations of those activities. Regionwide decreases in transportation cost/time result, not surprisingly, in tendencies towards dispersion and decentralization of activities. The extent to which these tendencies towards centralization or decentralization are apparent varies from one metropolitan area to another. Conceivably, then, there may be circumstances or regions in which these tendencies might not appear at all. This possibility should not be construed as illustrating that transportation does not affect land use. Rather, the inference to be drawn here is that the effects of transportation on land use (activity location) act through a complex process involving direct effects, indirect effects, and feedbacks.

The general structure of the DRAM residence-location model clearly embodies such a process. There are two central hypotheses of this structure. The first is that the place of residence of an employed resident is directly related to that individual's place of work, and to the probability of commuting between that place of work and any possible place of residence. The second hypothesis is that the probability of commuting is a function of the 'cost' (travel time and/or cost) of commuting, and the attractiveness of the possible residence zone to the particular type of residential locator. Depending upon the particular form of the model construct being used, and also upon the data available for the region in which the application is to be undertaken, a number of different variables may be included in the attractiveness measure. Typically these include land-use variables such as percentage of developable land developed, and amounts of vacant developable land and/or residential land. They also include, when possible, the income distribution of the residents previously or currently residing in the particular zone of interest. Often this income distribution will be classified by four income groups roughly corresponding to quartiles. There are empirically derived parameters for the travel-cost measure and for each of the variables in the attractiveness measure. These parameters, obtained in a statistical fitting of the model structure to the data, are the means by which it is possible to know the sensitivity or insensitivity of a particular region to possible changes in its transportation system.

By adopting this structure it is possible, without any inconsistency, to produce superficially contradictory results. Suppose that in a particular urban area, changes in the transportation system appeared to have little or no effect on residential location patterns. This simply implies that the transportation cost variable with its parameter(s) is relatively insignificant in comparison with the land-use and income-distribution variables with their parameters. The same model structure, utilizing a data set for a different urban area, might show considerable sensitivity to the transportation cost variable and rather little sensitivity to say, one or more of the land-use variables. This possibility of different responses of the model structure under different circumstances is a proper reflection of reality, and as such goes a long way towards explaining the variety of results obtained from the many highway and mass transportation impact studies.

The central notion here is that transportation per se does not cause changes in activity location and land use; rather, it permits such changes by its presence or prohibits such changes by its absence. Viewed in this way it can readily be seen that whereas, for example, a new transportation facility in a particular part of a metropolitan region may lead to growth in activities in that part of the region, this would not take place unless there was some other impetus for this growth. The provision of the new transportation facility would not in itself produce such growth. Similarly, the absence of such a facility would not itself prevent such growth, but might well inhibit it.

In the case of declining regions or regional subareas, a similar situation prevails. The development of a new transportation facility in a declining area will not, by itself, suffice to stem or reverse the decline. However, there may be other factors working against the decline which, when aided by new transportation development, will have the effect of stemming or reversing it. As many as possible of these various factors must be taken into account when analyzing the potential effects of transportation changes, lest the results of such changes be both unanticipated and unpleasant.

In situations of regionwide growth or decline, although many of the subareas of a region followed the general trend, a significant number of subareas diverged considerably from those trends. In particular, the zones which grew least in response to regionwide growth and declined most in response to regionwide decline were those with already congested (inadequate) transportation facilities. In contradistinction, the zones which grew most in response to regionwide growth and which declined least in response to regionwide decline were those where there was considerable capacity in the transportation system. In these zones there was virtually no increase in congestion even as the activity levels in the zones grew. They might, in fact, have been considered to have excess transportation capacity. At the same time, it should be noted that not all congested zones declined rapidly in the face of regionwide decline, and not all uncongested zones grew rapidly in the face of regionwide growth.

In the same sense that transportation changes, by themselves, may produce no observable effects on activity location, policy-directed activity location by itself may produce no effects either. Simulation experiments involving location of new employment in the urban core resulted in very little response in terms of relocation of residents. In other experiments, attempts to foster development of the urban core simply by improving the transportation system often resulted in exacerbation of existing tendencies towards decline through migration of population to the urban fringe. Thus although in some circumstances transportation changes alone or policy-directed activity location alone may produce very little change in anything else, in other circumstances either or both may produce results contrary to what was intended. Given a reasonably complete model structure the evidence of these different types of behavior is a blessing rather than a curse. It provides the possibility that potential contrary and unanticipated results of policies will be discovered in prior analysis rather than regretted after such policies have been implemented.

Another set of findings about the relationships between transportation and land use emerges in the evidence of different effects on different population types. First, for all population types, there are tendencies towards longer trips as transportation costs decline, and towards shorter trips as transportation costs increase. This suggests that with lower transportation costs residential locators are more able to 'consume' what is apparently more desirable land further from their places of work, and with higher transport costs they are less able to do this. Within the different household types it appears that upper-middle income households are most responsive to transportation cost decreases. These households seem most willing and able to take advantage of such transportation improvements to suburbanize, despite the consequent increase in their average trip lengths. Low-income households seem to be the least responsive income group, though analysis of the parameters obtained from a number of urban data sets suggests not that low-income households are insensitive to transport costs, but rather that other circumstances inhibit (or prohibit) their potential relocation response.

To summarize these findings on the effects of transportation on location and land use, the most important notion is that transportation can affect land use. The mechanism by which these effects are realized is, however, rather complex. This results in sometimes (seemingly) contradictory results which can only be made understandable by taking a comprehensive, systematic view of transportation–land-use interrelationships. From the many simulation experiments conducted the following are a sample of the effects of transportation on land use.

1. Regional variations in transportation cost, due perhaps to such phenomena as gasoline price changes or highway speed restrictions, have these effects:
   (a) price increases or speed decreases promote clustering of population around existing regional concentrations;
   (b) price decreases or speed increases promote dispersion of population and employment throughout the region;
   (c) UMI households show the greatest sensitivity, rapidly suburbanizing with price decreases or speed increases;
   (d) LI households show least sensitivity to price and speed changes—other factors may affect their residential mobility.
2. Effects of improvements in radial highways include:
   (a) initial use by decentralizing population for longer work trips to the urban core;
   (b) increases in population at the suburban end of highway;
   (c) subsequent increases in employment at the suburban end of the highway;
   (d) increases in LI households at the urban end of highway;
   (e) decreases in UMI households at the urban end of highway.
3. Changes in bridge tolls have these effects:
   (a) toll decreases yield increased suburbanization, and decline of the urban core;
   (b) toll increases yield modest urban recentralization.
4. Increases in downtown parking charges result in:
   (a) modest decline in the urban core;
   (b) modest growth adjacent to the urban core.

The complement to transportation affecting land use is, of course, land use affecting transportation. Though this notion has been given lip service for many years, not very much has been done towards developing a formal representation of the phenomena involved. The initial development of the integrated model package ITLUP was specifically intended to be a first step towards resolving this problem. The conceptual basis of this integration results from the following chain of interactions. First, the location of activities (both employment and population) depends on the prior spatial distribution of those activities, the attributes of the possible places of location, and the cost of overcoming their spatial separation. The cost of overcoming spatial separation is obviously dependent on the existing transportation facilities and their attributes. The new and existing location of activities, in conjunction with the characteristics of the transportation system, gives rise to a set of trips of various types: home-to-work, home-to-shop, and so on. The presence of these trips on the transportation network results in some amount of congestion, and thus alters the characteristics of the network. This is the feedback link from location and land use to transportation. The revised (congested) characteristics of the transportation network produce, in turn, a new pattern of activity locations, and the system continues to develop in this cyclical way.

In effect, some of the phenomena observed in such systems are rather straightforward extrapolations of what an experienced 'student' of urban systems might expect. However, other phenomena, sometimes of even greater import, result from the nonlinearities of the interacting components in the system. In activity location there are nonlinear responses to changes in the density of land uses (an implicit type of land-congestion measure), as well as to changes in the characteristics of the transportation network. In the transportation network there are nonlinear changes in characteristics (link travel times or costs) which result from the congestion that is produced by increasing traffic volumes on links. The interactions of these nonlinear phenomena often produce interesting and sometimes unanticipated responses.

In the many simulation experiments conducted with the integrated model, some of the activity-location/land-use phenomena observed resulted from the land-use–transportation feedback, and some from the direct action of the location of activities one on the other. The nonlinear feedback mechanism showed up most clearly in the response to regional growth and decline. Without such a feedback mechanism, activity location in all zones would rather closely follow the regional trends. With the feedback mechanism, in response to changes in regional growth or decline:

- many areas follow regional trends;
- some zones grow at more than the regional rate when the region grows;
- other zones decline at more than the regional rate when the region declines;
- the zones showing high growth have little transportation congestion;
- the zones showing rapid decline have considerable transportation congestion.

Other phenomena appear to have rather little effect on transportation per se, and result from the interaction between locating activities. Thus, in response to 'policy'-induced' population changes, it was found that:

- exogenously induced displacement of LI population from urban core zones is followed by increases in UI and UMI population in those zones;
- the displaced LI population relocates in zones with existing concentrations of LI population;
- exogenously induced location of UI population in urban core zone can result in subsequent growth of UMI and UI population in that zone, along with declines in the LI and LMI population of the zone.
- exogenously induced location of employment in zones usually results in some displacement of population from those zones, and very widespread rather small changes in population elsewhere in the region.

Some of these findings can themselves be altered if the policy manipulations are done in conjunction with transportation changes. Taken all together, these findings show that:

(a) transportation changes can affect activity location and land use;
(b) activity location and land use can affect transportation;
(c) activity location and land use can affect activity location and land use.

The key phrase is 'can affect', and the equation system embodied in the models used to produce the results described in this book is a significant step towards a realistic, comprehensive, and systematic understanding of these relationships. Some of the methodological implications of this work will now be summarized and discussed.

11.3 Methodological results

The method-related conclusions from the work described here may be considered from two perspectives: that of the developer of methods, and that of the user of methods.

From the perspective of the developer of methods, one of the principal findings concerns partial versus holistic approaches. Before the work (begun in 1971) on integrated transportation and land-use models described in chapter 3, there had been considerable work both on land-use models and on transportation analysis. Although in neither case was anything like the definitive model developed, reasonably good progress had been made. Yet nothing whatever was known of the nature of possible interactions between transportation and land-use models, nor of the extent to which linking such models might yield useful information about the interactions of the actual phenomena being simulated. Thus, that first research investigation was undertaken in the knowledge that the components of the model system were imperfect, but equally well in the awareness that it was necessary to begin investigating the interactions of the system.

The strategy followed was quite straightforward. To begin with, existing component models were assembled and their operation was examined and tested. Then these models were linked together, with adaptations and heroic assumptions made where necessary. Even in its first prototype version, the model package so constructed yielded rather interesting findings (as described in chapter 3). What is more, several years after the report on this initial work was published, a number of other researchers began to examine the same sets of problems (Berechman, 1976; Boyce, 1978; Echenique, 1977; Los, 1979; Mackett, 1979; Peskin and Schofer, 1977). Even as this is being written, new projects pursuing this same strategy are being reported (Choukroun and Harris, 1981).

The second major finding from the perspective of model developers is the obvious importance of the interplay between the development and the testing and verification of theory. Each step in the continuing development of the integrated sets of models described here has involved development of new procedures to solve an existing problem or deficiency, testing of those procedures, and then their addition to (or substitution for) previous procedures in the model system. This allows systematic improvement to take place in an orderly fashion, in the context of an existing set of models. In this manner new information from application or testing stimulates new theoretical development which can, in turn, build on an existing operational base. Similarly the need for additional data and application or testing can be indicated by new theoretical insights which require verification.

A number of new research topics have been suggested by the work done to date. and bear mention here. First, the dynamics of these systems, both the models and the reality, need much more examination. Fortunately now, almost a full decade after some of these issues were raised in the initial research report on the integrated model package, some investigation of the system dynamics is under way. There continues to be a need for improvement in calibration techniques, some discussion of which has been given in chapters 8 and 10. One question which has not been dealt with at all is that of the calibration of interacting multiparametric models. The problem of forecasting land consumption is almost entirely untouched. In virtually all currently operational location and land-use models, newly locating activities consume land at the previous average density. This is clearly an inadequate procedure whose resolution may well involve further work in supply-demand interactions. This leads on to the apparent convergence of a number of modelling approaches. As discussed in chapter 10, spatial interaction models with multivariate attractiveness measures (such as DRAM and EMPAL) can be shown to have much the same structure as logit formulations of behavioral choice models. In fact, these same multivariate attractiveness functions can, with only moderate algebraic manipulation, also be related to Cobb-Douglas production (utility) functions as well.

From the perspective of the potential user of methods, the substantive findings and the methodological findings combine to yield the major point. The substantive findings show the considerable complexity of transportation-land-use interactions; they also show that, owing to these interactions, changes in one activity sector or one subarea, or on a particular portion of the transportation system, can result in unexpected changes in other portions of the system. Sometimes these resultant changes, far removed sectorally or spatially from their sources, can be of considerable importance. This clearly implies the need for coordinated transportation and land-use planning, and most probably, for the use of techniques such as those described in this book. In this respect, these techniques are no longer as impractical as they were, say, ten years ago. Their current application is now possible (and reasonably practical) as well as being essential for coordinated planning and forecasting.

11.4 Models in planning

There is much more to be said about the use of models in planning than can be said here[24]. However, having written ten and one-half chapters on the development and application of such models, I feel that a few comments of a more general nature should be made. Perhaps the most important of these is that the development of urban land-use and transportation models is a scientific endeavor that has met its greatest failures through premature attempts at application. Despite these failures, development and application of models have continued (Batty, 1979; Pack, 1978). This is a result both of a 'quest-for-knowledge' incentive and of the demand for practical operational techniques for urban forecasting and policy analysis.

As this book was being completed, EMPAL and DRAM, both singly and in combination, were being examined for forecast accuracy for several metropolitan areas. The regions being examined were spatially disaggregated into 75–175 zones. The models were calibrated to sets of base year data by the techniques described in chapter 7. The sectoral disaggregation was to four employment types and four household income groups. Forecasts were then computed, in two steps of five years each. Though data were not available to verify all eight activity types in all cities, more than half could be verified. The forecast accuracy, measured in terms of the goodness of fit, R^2, ranged from just under 0·8 to over 0·9, with most of the verifiable forecasts being just at or above 0·9. A reasonable, systematic procedure capable of forecasting the spatial distribution of population and employment in a metropolitan area with this degree of reliability is certainly a useful technique for planning and policy analysis. Very few other techniques used in planning are capable of accounting, over a ten-year projection, for 90% of the variation in the variables of interest.

It has often been contended that urban spatial models require considerable, indeed excessive, input data. For some of the earlier model work this was a legitimate concern. The applications of the EMPIRIC model, for example, made up for a lack of explicit model structure by the inclusion of all obtainable variables in their parameter-estimation stages. In some of these applications there were well over 100 variables for each subarea in the region. The situation was exacerbated by the requirement for an extensive series of data-processing steps before the actual calibrations could be undertaken. In contrast, simplified forms of DRAM (see chapter 8) may be calibrated with less than a half-dozen variables per zone. Such simplified forms of DRAM, or of EMPAL, can provide quick and inexpensive assistance in analysis of spatial patterns of regions and can be especially useful in situations where full data sets are difficult or impossible to obtain. Even the fully articulated forms of EMPAL and DRAM require less than two dozen variables per zone. The level of data required does

(24) *An excellent* but not well-known discussion is given in Openshaw (1978).

not differ significantly from that which a planner would require to analyze similar problems in traditional ways. The difference is in the requirements of the models for specific numbers. That some planners have found models frustrating to deal with is perhaps more a consequence of the hunger of the models for 'specificity' than their supposed hunger for data.

Another of the difficulties with earlier models was the very considerable expense involved in their attempted application. Some early model applications were not completed even after expenditures totalling close to \$1 000 000. Many early modelling projects cost several hundreds of thousands of dollars. The problem was that these were not model-application projects. There were no fully developed models to be applied then. These were research and development projects. With hindsight, it is obvious that these projects could never have been successful. What is perhaps the most unfortunate aspect of these early operational failures is the pall they cast over the whole field of endeavor. Only recently are more objective critiques of the work beginning to be published (Webber, 1981). The considerable expense of these earlier models is no longer concomitant with their application. Some of the recent EMPAL-DRAM applications have been accomplished by planning agencies for costs, largely in terms of staff time, of tens of thousands of dollars or less, and in elapsed times of months rather than years. Moreover, the ease of operation of programs has been brought to the point where programs are operating on new computing equipment in days or weeks, rather than months, and individual calibration or forecast runs are completed in seconds of computer time and in hours of elapsed time. At these levels, the arguments stressing the time and monetary expense involved in use of the models begin to seem somewhat ludicrous.

This new class of models has, in a sense, paid heed to an important lesson from the earlier work: for operational models, simpler is better. The structure of this class of models is macrobehavioral rather than microbehavioral. As such their hypotheses are based on group behavior rather than individual choice behavior; and this results in a rather clear structural form. There are certainly statistical error terms associated with such models which are only recently beginning to yield to analysis (Openshaw, 1979). Yet the microbehavioral models also contain statistical error. No model of human behavior, macro or micro in concept, is ever likely to be error free—a notion which, by the way, many planners find to be an anathema. More importantly, the structure of these macrobehavioral models may well prove to be an aggregate, reduced form of the microbehavioral models. As discussed in chapter 10, current research work in the field is demonstrating a convergence of these structures with the individual choice structures currently preferred by economists working on similar problems.

Another consequence of the continuing work on articulation of a proper structure for these models is the subsequent development of calibration procedures. Even the development of the first entropy-maximizing derivations of these models left major problems of calibration. One consequence of this was the preoccupation, in British practice, with single-parameter location models. Now it is clear that an attempt to replicate behavior as complex as residence or employment location with only one or two parameters was somewhat naive. The use of multivariate, multiparameter attractiveness terms in these model structures leads, as described in chapter 8, to models which are readily adjustable to virtually any metropolitan area. Furthermore, current research suggests that some of the most serious criticisms of these models (Openshaw, 1976) are blunted with these modified multiparametric structures. What is more, these calibrations can be easily performed and do not require the trip-matrix data which caused such problems for the earlier one-parameter models.

That models such as these continue to play an important role in the planning process has been well documented (Pack, 1978). Yet it is clear that they cannot and should not be applied willy-nilly to all policy analyses. Webber (1981) cites their role as "a method of producing development plans (or forecasts of the physical structure of metropolitan areas) which reflect a consistent, integrated view of the functioning of a large urban area, in order to guide public investments in urban infrastructure and to safeguard the private development process". They are, in their current form, demonstrably sensitive to changes in transportation systems, land-use controls, exogenous relocations of employment or population by type, and regional growth or decline. Policies which can be described in these terms include many of the most important questions of planning policy. To the extent that these models provide reliable forecasts and can properly respond to these types of policy input, they can and should be used by planning agencies to help in their decisionmaking procedures. The fact that there is still much to be learned about urban spatial systems cannot be sensibly used as a justification for denying or ignoring the considerable knowledge already available.

11.5 A final note

One of the unique attributes of the late nineteenth and early twentieth centuries was an almost unshakeable faith in the continuing benefits to be had from scientific endeavor. Scientists were at least allowed, and usually encouraged, to pursue their efforts in whatever vein they chose. This attitude was partly a result of a world view which assumed that eventually, by continuing incremental efforts, *all* knowledge was achievable. Moreover, it was assumed that whatever was unknown or seemingly unfathomable at a particular moment would inevitably yield to the progress of science and intellectual achievement. This attitude held both for 'hard' and for 'soft' science. The populace flocked eagerly to public lectures by greater and lesser scientific personages. This is not to say that there were no difficulties.

A glance at the biography of any scientist of the era will give ample evidence of the difficulty of successfully communicating new ideas. More than one innovator came to great frustration at the hands of intransigent colleagues. Yet the attitude, overall, seems to have been encouraging and permissive.

In the second quarter of the twentieth century, science—perhaps most especially physics—yielded results whose popularization and application have helped to change that naive world view. First, it was argued that there were some things which could not be known; it is not, for example, possible to know all the attributes, simultaneously and exactly, of atomic particles. Second, mechanized and, especially, atomic warfare ushered in an era of terror and doubt about the continued existence of humankind. Third, the development of electronic computers lent credence to the fears of many about the inferiority of humans to their own science and technology. The simultaneous advances in medicine and agriculture helped produce enormous increases in population which resulted in the consequent need for technological assistance for society to continue to function. Increased knowledge yielded increasingly complex views of the world, and past success in dealing with some problems led to increasing attempts to deal with others. Finally, as the cost of the pursuit of knowledge escalated, governments began taking an increasingly large role in the sponsoring of such research endeavors, to support attempts to deal with societal problems.

Although this is all greatly simplified, this description of events sets a useful context for some final comments on methods and models in planning and policy analysis. In particular, it suggests reasons for several underlying attitudes which color the attitudes of planners and their view of planning in the last decades of the twentieth century. Urban and regional planners today find themselves increasingly under attack as, time after time, the unanticipated effects of their plans and policies lead to further problems rather than the hoped-for solutions. They find themselves at variance with political decisionmakers who wish quick solutions to problems whose causes are not even known. They often recognize the need for technological assistance but have, perhaps often, been disappointed at its use in the past. They loathe the possibility that some questions may have no answers. In what light, then, are they to consider the research and findings presented here?

It must be realized that virtually all of the work in the field has been accomplished in the past two decades. The work has not been very expensive in comparison with many other scientific endeavors. The complexities of the problems involved are staggering. In few fields of enquiry are the systems more complex. The work must be done with nonexperimental techniques, as every urban area is in some way different from every other, and none are available for the researcher to manipulate at will. The available data range in quality from moderately good to unuseably bad. Researchers are constantly being put in the position of having to attempt to do that which they know cannot be done in order to have further research funded. It is an era when the major portion of publicly funded research can only be justified in terms of immediate applicability. Yet all too often, immediate applicability is almost out of the question for lack of the requisite basic knowledge, and funds are not available merely for the pursuit of knowledge.

Nonetheless, very considerable progress has been made. Many false starts were superceded. Models of urban systems ranging from rather general theoretical notions to rather vague agglomerations of data have been tested and often discarded. Increasingly well structured views of such systems have been developed and, to a lesser extent, tested. There appears to be a convergence of various current theoretical approaches. New research is directed by known gaps in earlier system structures or by specific errors in the performance and dynamics of existing systems. Progress overall is much more orderly. All these are indicators of a maturing discipline.

The work described in this book makes no pretence of being the definitive description of the urban system. The author, better than anyone else, is aware of many shortcomings. Yet, here are a set of methods which have produced interesting and useful substantive findings. These methods can be used by planners now to analyze policy issues in ways that were not possible before. These methods cannot answer all questions. No methods will be able to answer all questions. We may, however, expect that continued progress in this field will permit greater numbers of increasingly complex questions to be analyzed in more comprehensive and sensible ways. This, after all, should be the intent of all such endeavors.

References

Alonso W, 1964 *Location and Land Use* (Harvard University Press, Cambridge, MA)

Anas A, 1975 "The empirical calibration and testing of a simulation model of residential location" *Environment and Planning A* **7** 899-920

Anas A, 1979a "The estimation of multinomial logit models of joint location, housing and travel model choice from small area aggregated data" Department of Civil Engineering, Northwestern University, Evanston, IL

Anas A, 1979b "The impact of transit investment on housing values: a simulation experiment" *Environment and Planning A* **11** 239-255

Atherton T J, Ben-Akiva M E, 1976 "Transferability and updating of disaggregate travel demand models" *Transportation Research Record* (610) 12-18

Barras R, Broadbent T A, Cordey-Hayes M, Massey D B, Robinson K, Willis J, 1971 "An operational urban development model of Cheshire" *Environment and Planning* **3** 115-234

BATSC, 1969 *Bay Area Transportation Report* (Bay Area Transportation Study Commission, Berkeley, CA)

Batty M, 1972 "Recent developments in land use modelling: a review of British research" *Urban Studies* **9** 151-177

Batty M, 1976 *Urban Modelling* (Cambridge University Press, Cambridge)

Batty M, 1979 "Progress, success, and failure in urban modelling" *Environment and Planning A* **11** 863-878

Batty M, Mackie S, 1972 "The calibration of gravity, entropy, and related models of spatial interaction" *Environment and Planning* **4** 205-233

Baxter R, Williams I N, 1975 "An automatically calibrated urban model" *Environment and Planning A* **7** 3-20

Berechman J, 1976 "Interfacing the land-use activity system and the transportation system" *Journal of Regional Science* **16** 183-194

Bertolina C J, Carbine M W, 1974 *Utah Land Use at Interchange Study—Program Summary* (Office of Planning and Programming, Systems Planning Division, Utah State Department of Highways, Salt Lake City, UT)

Black J, 1981 *Urban Transport Planning* (Johns Hopkins University Press, Baltimore, MD)

Boyce D E, 1978 "Equilibrium solutions to combined urban residential location, modal choice, and trip assignment models" in *Competition Among Small Regions* Eds W Buhr, P Friedrich (Nomos, Baden-Baden) pp 246-264

Boyce D E, Southworth F, 1978 "Quasi-dynamic urban location models with endogenously determined travel costs" Department of Civil Engineering, University of Illinois at Urbana-Champaign, Urbana, IL

Branston D, 1976 "Link capacity functions: a review" *Transportation Research* **10** 223-236

Bureau of Public Roads, 1964 *Traffic Assignment Manual* (US Department of Commerce, Washington DC)

Chan Y, 1976 "A method to simplify network representation in transportation planning" *Transportation Research* **10** 179-191

Choukroun J M, Harris B, 1981 "Modeling complex urban locational systems" in *Essays in Societal System Dynamics and Transportation* Ed. D Kahn (US Department of Transportation, Research and Special Programs Administration, Washington DC)

Clark C, 1967 *Population Growth and Land Use* (Macmillan, London)

Coelho J D, Wilson A G, 1977 "An equivalence theorem to integrate entropy maximizing submodels within overall mathematical programming frameworks" *Geographical Analysis* **9** 160-173

Cordey-Hayes M, 1968 "Retail location models" CES WP-16, Centre for Environmental Studies, London

Crecine J P, 1964 "A time oriented metropolitan model for spatial location" technical bulletin 6, Community Renewal Program, Department of City Planning, Pittsburgh, PA

CREUE, 1968 *Jobs, People and Land (BASS)* Center for Real Estate and Urban Economics, University of California, Berkeley, CA
Cripps E L, Foot D H S, 1969 "The empirical development of an elementary residential location model for use in sub-regional planning" *Environment and Planning* **1** 81-90
Dial R B, 1971 "A probabilistic multipath traffic assignment model which obviates path enumeration" *Transportation Research* **5** 83-111
Domencich T, McFadden D, 1974 *Urban Travel Demand—A Behavioral Analysis* (North-Holland, Amsterdam)
Ducca F W Jr, 1978 "Report on the inclusion of planned shopping centers as part of a retail employment allocation function" Appendix 2 in Putman (1978)
Ducca F W Jr, 1979 *Retail and Service Employment Location: Simulation, Analysis and Projection* PhD dissertation, Department of City and Regional Planning, University of Pennsylvania, Philadelphia, PA
Ducca F W, Wilson R H, 1976 "A model of shopping center location" *Environment and Planning A* **8** 613-623
Eash R W, Janson B N, Boyce D E, 1979 "Equilibrium trip assignment: advantages and implications for practice" *Transportation Research Record* (728) 1-8
Echenique M, 1977 "An integrated land use and transport model" *Transactions of the Martin Centre* (University of Cambridge, Cambridge) volume 2, pp 195-230
Echenique M, Crowther D, Lindsay W, 1969 "A spatial model of urban stock and activity" *Regional Studies* **3** 281-312
Epstein A H, Leary C A, McCandless S T, 1974 *A Guide for Considering Air Quality in Urban Planning* EPA-450/3-74-020 (US Environmental Protection Agency, Office of Air and Water Programs, Research Triangle Park, NC)
Evans S P, 1973 "A relationship between the gravity model for trip distribution and the transportation problem in linear programming" *Transportation Research* **7** 39-61
Evans S P, 1976 "Derivation and analysis of some models for combining trip distribution and assignment" *Transportation Research* **10** 37-57
Ferland J, Florian M, Achim C, 1974 *On Incremental Methods for Traffic Assignment* publication 9, Centre de Recherche sur les Transports, Université de Montréal, Montréal, Québec, Canada
Fertal M, Weiner E, Balek A, Sevin A, 1970 *Modal Split* (US Department of Trans Transportation, Federal Highway Administration, Office of Highway Planning, Washington DC)
Florian M (Ed.), 1976 *Traffic Equilibrium Methods* (Springer, Berlin)
Florian M, 1977 "A traffic equilibrium model of travel by car and public transit modes" *Transportation Science* **11** 166-179
Florian M, Nguyen S, 1974 "A new look at some old problems in transportation planning" paper presented at the PTRC Summer Annual Meeting, University of Warwick, England; Planning and Transport Research and Computation Ltd, London
Florian M, Nguyen S, 1975 "An application and validation of equilibrium trip assignment methods" *Transportation Science* **10** 374-389
Florian M, Nguyen S, 1978 "A combined trip distribution, modal split and trip assignment model" *Transportation Research* **12** 241-246
Florian M, Nguyen S, Ferland J, 1975 "On the combined distribution and assignment of traffic" *Transportation Science* **9** 43-53
Frank C, 1978 "The addition of modal choice to the integrated land use and transportation package" Appendix 5 in Putman (1978)
Gleave G R, 1973 *Some Aspects of Loading Traffic Networks* MSc thesis, Department of Operations Research, University of Pennsylvania, Philadelphia, PA
Goldner W, Rosenthal S R, Meredith J R, 1972 "Projective land use model—PLUM: theory and application" Institute of Transportation and Traffic Engineering, University of California, Berkeley, CA

Gross M, 1979 *The Impact of Transportation and Land Use Policies on Urban Air Quality* PhD dissertation, Department of City and Regional Planning, University of Pennsylvania, Philadelphia, PA
Hanna S R, 1971 "A simple method of calculating dispersion from urban area sources" *Journal of the Air Pollution Control Association* **21** 774-777
Harris B, 1964 "A model of locational equilibrium for retail trade" mimeo, Institute for Environmental Studies, University of Pennsylvania, Philadelphia, PA
Homans G C, 1967 *The Nature of Social Science* (Harbinger, New York)
Huber M J, Boutwell H B, Witheford D K, 1968 "Comparative analysis of traffic assignment techniques with actual highway use" report 58, National Cooperative Highway Research Program; Transportation Research Board, Washington DC
Jewel W S, 1967 "Models for traffic assignment" *Transportation Research* **1** 31-46
Kain J F, Apgar W C, Ginn J R, 1977 "Simulation of the market effects of housing allowances; vol. I: description of the N.B.E.R. urban simulation model" City and Regional Planning Research Report R77-2; Harvard University, Cambridge, MA
Koppelman F S, 1976 "Guidelines for aggregate travel prediction using disaggregate choice models" *Transportation Research Record* (610) 19-24
Kuner P, 1973 "Investigation of traffic assignment algorithms" in *The Interrelationships of Transportation Development and Land Development* Ed. S H Putman (US Department of Transportation, Federal Highways Administration, Office of Highway Planning, Washington DC)
LeBlanc L J, Morlok E K, Pierskalla W P, 1975 "An efficient approach to solving the road network equilibrium traffic assignment problem" *Transportation Research* **9** 309-318
Leontief W, 1971 "Theoretical assumptions and nonobserved facts" *American Economic Review* **61** 1-7
Lerman S R, 1976 "Location, housing, automobile ownership and mode to work: a joint choice model" *Transportation Research Record* (610) 6-11
Litinas N, Ben-Akiva M, 1979 "Behavioral modelling of continuous spatial distributions of trips, residential locations, and workplaces" Center for Transportation Studies, Massachusetts Institute of Technology, Cambridge, MA
Los M, 1979 "Combined residential-location and transportation models" *Environment and Planning A* **11** 1241-1265
Lowry I S, 1964 "A model of metropolis" report RM 4125-RC, The Rand Corporation, Santa Monica, CA 90406
McFadden D, 1978 "Modelling the choice of residential location" in *Spatial Interaction Theory and Planning Models* Eds A Karlquist, L Lundqvist, F Snickars, J W Weibull (North-Holland, Amsterdam) pp 75-96
McFadden D, Reid F, 1975 "Aggregate travel demand forecasting from disaggregated behavioral models" *Transportation Research Record* (534) 24-37
Mackett R L, 1979 "A model of the relationships between transport and land use" WP-122, Institute for Transport Studies, University of Leeds, Leeds, England
Mackett R L, 1980 "The mathematical representation of the relationships between transport and land use" WP-123, Institute for Transport Studies, University of Leeds, Leeds, England
McLaughlin W A, 1966 *Multi-path System Traffic Assignment* report RB 108 (Ministry of Transportation and Communications, Downsview, Ontario, Canada)
March L, 1971 "Urban systems: a generalized distribution function" in *London Papers in Regional Science 2. Urban and Regional Planning* Ed. A G Wilson (Pion, London) pp 157-170
Martin B, Manheim M, 1965 "A research program for comparison of traffic assignment techniques" *Highway Research Record* (88) 69-84

Martin D O, Tikvart J A, 1968 *A General Atmospheric Diffusion Model for Estimating the Effects of One or More Sources of Air Quality* publication 68-148 (US Department of Health, Education and Welfare, National Air Pollution Control Administration, Washington DC)

Massey D B, 1973 "The basic:service categorization in planning" *Regional Studies* 7 1-15

Murchland J D, 1968 "Method of diverted assignment over adjacent routes used in London Underground programs" report LBS-TNT-102.1, Transport Executive, 55 Broadway, London SW1, England

Muth R, 1969 *Cities and Housing* (University of Chicago Press, Chicago, IL)

Nathanson J, 1970 "Basic employment model: a model for intra-county location of basic employment and land" BATSC technical report 222 (preliminary), Bay Area Transportation Study Commission, Berkeley, CA

National Council for Urban Economic Development, 1980 "Synthesis of literature on transportation/economic development" [prepared for US Department of Transportation], NCUED, 1730 K Street, NW Washington DC

Nguyen S, 1974 "An algorithm for the traffic assignment problem" *Transportation Research* 4 391-394

Openshaw S, 1976 "An empirical study of some spatial interaction models" *Environment and Planning A* 8 23-41

Openshaw S, 1978 *Using Models in Planning: A Practical Guide* (Retail Planning Associates, Corbridge, Northd, England)

Openshaw S, 1979 "A methodology for using models for planning purposes" *Environment and Planning A* **11** 879-896

Pack J R, 1978 *Urban Models: Diffusion and Policy Application* (Regional Science Research Institute, Philadelphia, PA)

Payne-Maxie and Blayney-Dyett, 1980 "The land use and urban development impacts of beltways" DOT P-30-80-38 [prepared for US Department of Housing and Urban Development, and US Department of Transportation, Office of the Assistant Secretary for Policy and International Affairs, Washington DC]

Peat, Marwick, Mitchell and Co., 1971 "Calibration and application of an EMPIRIC activities allocation model for the Twin Cities Metropolitan Area" [prepared for the Metropolitan Council, St Paul, MN] Peat, Marwick, Mitchell and Co., 1990 K Street NW, Washington DC

Peat, Marwick, Mitchell and Co., 1972 "Implementation of the *N*-dimensional logit model" [prepared for US Department of Transportation] Peat, Marwick, Mitchell and Co., 1990 K Street NW, Washington DC

Peskin R L, Schofer J, 1977 "The impacts of urban transportation and land use policies on transportation energy consumption" DOT-OS-50118, Office of the Secretary, Office of University Research, US Department of Transportation, Washington DC

PIDC-PCPC, 1976 *Survey of Industry 1975* Philadelphia Industrial Development Corporation with Philadelphia City Planning Commission (Philadelphia Industrial Development Corporation, Philadelphia, PA)

Putman S H, 1972 "Intraurban employment forecasting models: a review and a suggested new model construct" *Journal of the American Institute of Planners* **38** 216-230

Putman S H, 1976 "Laboratory testing of predictive land-use models" DOT-P-5010 (US Department of Transportation, Office of Transportation Systems Analysis and Information, Washington DC)

Putman S H, 1977a "Calibrating a disaggregated residential allocation model—DRAM" in *London Papers in Regional Science 7. Alternative Frameworks for Analysis* Eds D B Massey, P W J Batey (Pion, London) pp 108-124

Putman S H, 1977b "Calibrating a residential location model for nineteenth-century Philadelphia" *Environment and Planning A* **9** 449-460

Putman S H, 1978 "Development of an improved integrated transportation and land use model package" report of results of NSF grant APR 73-07840-A02; Urban Simulation Laboratory, Department of City and Regional Planning, University of Pennsylvania, Philadelphia, PA

Putman S H, 1979 *Urban Residential Location Models* (Martinus Nijhoff, Boston, MA)

Putman S H, 1980a "Calibrating urban residential location models 3: empirical results for non-US cities" *Environment and Planning A* **12** 813-827

Putman S H, 1980b "Integrated analysis of metropolitan transportation and location" DOT-P-30-80-32, US Department of Transportation, Office of Transportation Economic Analysis, Washington DC

Putman S H, Ducca F W, 1978a "Calibrating urban residential models 1: procedures and strategies" *Environment and Planning A* **10** 633-650

Putman S H, Ducca F W, 1978b "Calibrating urban residential models 2: empirical results" *Environment and Planning A* **10** 1001-1014

Robillard P, 1974 "Multipath traffic assignment with dynamic input flows" *Transportation Research* **8** 567-573

Seidman D, 1970 "The construction of an urban growth model" DVRPC plan report 1, technical supplement, Delaware Valley Regional Planning Commission, Philadelphia, PA

Senior M L, 1973 "Approaches to residential location modelling 1: urban ecological and spatial interaction models (*A review*)" *Environment and Planning* **5** 165-197

Senior M L, 1974 "Approaches to residential location modelling 2: urban economic and some recent developments (*A review*) *Environment and Planning A* **6** 369-409

Senior M L, Wilson A G, 1974 "Exploration and synthesis of linear programming and spatial interaction models of residential location" *Geographical Analysis* **7** 209-238

Smith T, 1977 "A general efficiency principle of spatial interaction" WP-1, Regional Science Department, University of Pennsylvania, Philadelphia, PA

Soberanes J, 1978 "Investigation of traffic assignment and some tests on a capacity-restraint procedure using Dial's multipath algorithm" Appendix 6 in Putman (1978)

Southworth F, 1975 "A highly disaggregated modal-split model" WP-58, Institute for Transport Studies, University of Leeds, Leeds, England

USDoT, 1971 "Request for proposal on inter-relationships of transportation development and land development" GS-10, RFP-26; US Department of Transportation, Federal Highway Administration, Office of Highway Planning, Urban Planning Division, Washington DC

USDoT, 1972 "Program documentation: urban transportation planning" US Department of Transportation, Federal Highway Administration, Office of Highway Planning, Urban Planning Division, Washington DC

USDoT, 1973 "Traffic assignment" US Department of Transportation, Federal Highway Administration Office of Highway Planning, Urban Planning Division, Washington DC

USDoT, 1974 "Summary and evaluation of economic consequences of highways" US Department of Transportation, Federal Highway Administration, Office of Highway Planning, Urban Planning Division, Washington DC

USDoT, 1977 "Computer programs for urban transportation planning: PLANPAC/BACKPAC" US Department of Transportation, Federal Highway Administration, Office of Highway Planning, Urban Planning Division, Washington DC

Van Vliet D, 1976a "Road assignment II: the GLTS model" *Transportation Research* **10** 145-150

Van Vliet D, 1976b "Road assignment III: comparative tests of stochastic methods" *Transportation Research* **10** 151-157

Van Vliet D, 1976c "The choice of assignment techniques for large networks" in *Traffic Equilibrium Methods* Ed. M A Florian (Springer, New York) pp 396-452
Wardrop J G, 1952 "Some theoretical aspects of road traffic research" *Proceedings of the Institution of Civil Engineers, Part II* **1** 325-378
Webber M J, 1981 "Operational models in urban planning" *Environment and Planning A* **13** 763-779
Williams I N, 1976 "A comparison of some calibration techniques for doubly constrained models with an exponential cost function" *Transportation Research* **10** 91-104
Williamson S J, 1973 *Fundamentals of Air Pollution* (Addison-Wesley, Reading, MA)
Wilson A G, 1970a *Entropy in Urban and Regional Modelling* (Pion, London)
Wilson A G, 1970b "Disaggregating elementary residential location models" *Papers and Proceedings of the Regional Science Association* **14** 103-125
Wilson A G, 1974 *Urban and Regional Models in Geography and Planning* (John Wiley, Chichester, Sussex)
Wilson A G, Rees R H, Leigh C M (Eds), 1977 *Models of Cities and Regions* (John Wiley, Chichester, Sussex)
Wingo L, 1961 *Transportation and Urban Land* (Johns Hopkins University Press, Baltimore, MD)

Author index

Subject index

For Product Safety Concerns and Information please contact our EU representative GPSR@taylorandfrancis.com Taylor & Francis Verlag GmbH, Kaufingerstraße 24, 80331 München, Germany